Bernardino Telesio and the Natural Sciences in the Renaissance

Medieval and Early Modern Philosophy and Science

VOLUME 29

The titles published in this series are listed at *brill.com/memps*

Bernardino Telesio and the Natural Sciences in the Renaissance

Edited by

Pietro Daniel Omodeo

BRILL

LEIDEN | BOSTON

Cover illustration: Bernardino Telesio, *De iride* (Venice 1590). Biblioteca Nazionale di Cosenza, Fondo Greco 476.

The chapter "Telesio, Aristotle, and Hippocrates on Cosmic Heat" by Hiro Hirai is a is an updated English translation of the previously published article "Il calore cosmico in Telesio fra il De generatione animalium di Aristotele e il De carnibus di Ippocrate," in *Bernardino Telesio tra filosofia naturale e scienza moderna*, Giuliana Mocchi, Sandra Plastina and Emilio Sergio (eds.) (Rome: Fabrizio Serra Editore, 2012), 71–83.

The chapter "A Conversation by Telesio: Sensualism, Criticism of Aristotle, and the Theory of Light in the Late Renaissance" by Martin Mulsow is an English translation of the previously published article of "Ein unbekanntes Gespräch Telesios: Aristoteleskritik, Sensualismus und Theorie des Lichts in der Spätrenaissance" In: *Sol et Homo. Mensch und Natur in der Renaissance. Festschrift für Eckhard Keßler*. Sabrina Ebbersmeyer, Helga Pirner-Pareschi und Thomas Ricklin (Hrsg.), 511–536. München: Wilhelm Fink, 2008.

This volume is an outcome of a project that has received funding from the European Union's Horizon 2020 Research and Innovation Programme (GA n. 725883 *Early Modern Cosmology*).

The Library of Congress Cataloging-in-Publication Data is available online at http://catalog.loc.gov
LC record available at http://lccn.loc.gov/2019024186

Typeface for the Latin, Greek, and Cyrillic scripts: "Brill". See and download: brill.com/brill-typeface.

ISSN 2468-6808
ISBN 978-90-04-35263-6 (hardback)
ISBN 978-90-04-35264-3 (e-book)

This book is printed on acid-free paper and produced in a sustainable manner.

Contents

Foreword

This volume on Bernardino Telesio is a major contribution to the study of the Cosentine philosopher in the context of Renaissance science. It is more than merely a testimony to the scientific contribution that the conference *Bernardino Telesio, the Natural Sciences and Medicine in the Renaissance*, organized in 2015 by Pietro Daniel Omodeo in Berlin, offers to research on Telesio and, more generally, to scholars of the European Renaissance. It also documents the rich and productive collaboration between the Centro Internazionale di Studi Telesiani Bruniani e Campanelliani (CISTBeC) and the Max Planck Institute for the History of Science (MPIWG). The foundation of the CISTBeC owes a debt of gratitude to the Max Planck Institute, as well as to the Warburg Institute, the Centre d'études supérieures de la Renaissance of Tours, the Istituto Italiano per gli Studi Filosofici of Naples, and the University of Calabria. The principal goal of the CISTBeC is to create a digital library that assembles all manuscript and printed works by Telesio, Giordano Bruno, and Tommaso Campanella and an exhaustive secondary bibliography (essays, books, monographs, and translations) devoted to these authors.

The work of these three outstanding European philosophers is essential to an understanding of the material and intellectual origins of modern science. Historians of science now agree that a truly accurate understanding of the age of Copernicus, Galileo, and Kepler is only possible when it is grounded in an attentive inquiry that recognizes cultural as well as social factors (in the broadest definition of the word). Today we have moved beyond the outmoded search for the "universal key" that leads to the "logic of scientific inquiry." Utilizing the teachings of historians who are mindful of certain economic and political contexts (we are referring, for instance, to experts in the field like Leonardo Olschki, Edgar Zilsel, and Ludovico Geymonat), we have learned to be mistrustful of abstract constructs theorized in some areas of the philosophy of science. For these reasons, we are confident that nothing could be more instructive for researchers who seek to fortify the intimate connection between nature and science, which has existed since the dawn of modern science, than a close reading of the work of Bernardino Telesio, Giordano Bruno, and Tommaso Campanella. Despite certain undeniable differences, their thought is at the same time both empirical and speculative, as well as practical yet visionary.

Furthermore—as the research guidelines of the MPIWS confirm—it is increasingly necessary to address the material, theoretical, practical, and cultural aspects of Renaissance science, considered in their historical complexity. In fact, a reconstruction of the historical and cultural milieu in which modern science underwent its difficult period of gestation is unavoidable. The most significant aspects would include the birth of capitalism, the emergence of the modern political state, the headlong transformation of technology and practical knowledge, and the colonial expansion of European powers on a global scale. The intellectual currents that have profoundly marked modern science stand out against this economic and political background: the culture of artists and engineers directed toward codifying knowledge related to the arts and technologies, the humanistic culture devoted to the restoration of classical texts, and the intensive work by university professors to resystematize, transmit, and critique the philosophical heritage of the Middle Ages. Modern science is derived from the very concreteness of the practical knowledge of miners, mechanics, architects, navigators, and instrument builders and also extrapolates images of wholeness from Renaissance naturalism that provided a way of rethinking nature and man's relationship with his environment and knowledge. For instance, the critical spirit that enlivens the experimental science of a doctor, mathematician, and philosopher like Girolamo Cardano or the Copernican "philosopher of magnetism" William Gilbert shares the intellectual and moral temperament of philosophers such as Telesio, Campanella, Bruno, and Francesco Patrizi. In fact, Renaissance naturalism could be considered the "big brother" of modern experimental science because they share a common origin in the same cultural and social renewal.

The library of the Centro Internazionale di Studi Telesiani Bruniani e Campanelliani offers scholars valuable material for a broader understanding of the connections between the different branches of knowledge that characterize the European Renaissance: literature and philosophy, science and religion, cosmology and astrology, as well as art and politics. They constitute a unique terrain where theory and practice, and reality and utopia, share fertile ground. Furthermore, the creation of the CISTBeC has been instrumental in forming a network of Renaissance scholars that ranges far beyond the European institutions and research centers that first founded it. From the group of most active scholars, we would like to recognize Paolo Galluzzi (Museo Galileo in Florence), Pietro Daniel Omodeo (Ca' Foscari University of Venice), Isabelle Pantin (École Normale Supérieure, Paris) and Maria José Vega (Universidad autónoma, Barcelona), who in past years have actively contributed to animating the meetings which discussed projects and targeted research on the European Renaissance.

Returning for a moment to the inspiration for this new volume, we would like to mention, among the scholarly contributions of international importance published by the CISTBeC, the recent study entitled *Le edizioni antiche di Bernardino Telesio: censimento e storia* [*Early Editions of Bernardino Telesio: a census and history*], which was edited by Giliola Barbero and Adriana Paolini (Paris: Les Belles Lettres, 2017). This volume is the result of exhaustive research and offers, for the first time, a census of the early editions of Telesio's work, and also reconstructs their circulation among the libraries of the world. Made possible by valuable funding by the Carical Foundation directed by Mario Bozzo, the research identifies 718 copies, of which 543 were analyzed first-hand by Barbero and Paolini. An index card has been created for each copy that includes a detailed description of the book and a reconstruction of its trajectory from the printing press to readers, resulting in a goldmine of data, information, technical details, and the names of illustrious (and not so illustrious) owners. There is also an important bibliography that, as Angela Nuovo authoritatively suggests, recalls (through the number of copies it analyzes) Owen Gingerich's outstanding work on Copernicus' *De revolutionibus*.

Now the history of these copies/editions of Telesio's work will finally be available to interested scholars, including notes on ownership and censorship, the presence of *marginalia* and annotations, glosses, markers, corrections, comments, critical remarks, and transfers from one library to another. Scholars can even reconstruct the geography of the institutions (convents, academies, universities, and colleges) and obtain information about their distribution (appearances at fairs and the sale price of a single volume). The index cards also describe the binding (the quality of the material used for the cover and plates often reflects the importance that the owner gave to the text) and indicate with which other works (on medicine, natural philosophy, or optics) Telesio's texts were occasionally bound (knowing which treatises by other authors the first Telesian editions were associated with helps us understand the literary reception of the Cosentine philosopher's thought).

One highly important fact among many concerns the number of copies that have been found: 136 copies of the third edition (1586) of *De rerum natura*—versus 43 of the first edition (1565) and 73 copies of the second (1570)—tell us (as Barbero and Paolini correctly point out) that the broadest distribution of Telesio's thought is owed to the *editio maior* in nine books. In fact, between the first and last editions, the number of surviving copies (43, 73, and 136) more or less progressively doubled, which informs us of Telesio and his printers' increasing investment in the copy run. The inquiry into the circulation of these copies throughout various European libraries and the identity of their owners speaks volumes about the fortune of the Cosentine philosopher's opus. The census shows an ample foreign presence of the third edition of *De rerum natura* (and the Geneva in-folio copy that reproduces it) in different public and private libraries (of the 136 copies, less than half, 68, appear in Italy), and the copies found in England, Hungary, France, and Germany attest to the interest in Telesian thought beyond the borders of Italy. This interest was reinforced by specialized readers, like Henri Savile (an Oxford mathematician and astronomer who introduced Francis Bacon to Telesio's work), Markus Welser (a German humanist and editor), Andreas Dudith (an Italo-Hungarian clergyman who was close to the Reformation milieu and a student of astronomy) and the great French bibliographers Pierre and Jacques Dupuy (savants and librarians to the King of France), Jacques Auguste de Thou (an historian and the owner of an extraordinary library that merged with the Bibliothèque de l'Arsenal), and Gabriel Naudé (doctor, librarian, and the author of the famous treatise *Advice on creating a library*.)

On the basis of these results, we are convinced that a similar census of the works of Tommaso Campanella, conducted according to the same scholarly criteria, would produce equally extraordinary results. With the economic support of the President of the Region of Calabria, the Centro Internazionale di Studi Telesiani Bruniani e Campanelliani has already formed a team of specialists to begin research. We hope that within a reasonable period of time students of Campanella will finally have access to invaluable information about the presence of his works in international libraries, and above all be able to access a blueprint of their circulation and ownership.

We welcome the new contributions and new directions for further research provided by this volume of Renaissance and Telesian studies. Most of the essays collected here were first presented at the aforementioned conference in Berlin in 2015. Its success and the first phase of editorial work was gratefully supported by the *Collaborative Research Centre* 980 "Episteme in Motion" (Freie Universität of Berlin, funded by the German Research Foundation, DFG). The conclusion of the project was made possible through the generous support of the European Research Council that financed the consolidator research project "Institutions and Metaphysics of Cosmology in the Epistemic Networks of Seventeenth-Century Europe" on scientific culture in the early modern era, which is led by Omodeo (European Union's Horizon 2020 Research and Innovation Programme, GA n. 725883 *Early Modern Cosmology*).

In this difficult time for the future of the social sciences and primary academic research, we feel that it is even more important to form alliances in order to carry out projects that require multifaceted abilities and diversified intellectual energy. Shared ideas and passions are not only useful to advance knowledge; they also assist in creating a network of scholars whose collegiality and collective efforts are an invaluable instrument for the transformation of knowledge into a way of life.

Nuccio Ordine
University of Calabria, President of the Centro Internazionale di Studi Telesiani Bruniani e Campanelliani

Jürgen Renn
Director, the Max Planck Institute for the History of Science in Berlin, Vice-President of the Centro Internazionale di Studi Telesiani Bruniani e Campanelliani

Note on Contributors

Roberto Bondì
is Professor of the History of Scientific Thought and the History of Ideas at the University of Calabria, and is Director of Centro Internazionale di Studi Telesiani, Bruniani e Campanelliani. His main research areas are the relationships between philosophy, science, and theology in the early modern period, the scientific and philosophical debate about the Gaia hypothesis, and the current relationships between science and religion. His recent publications include *Il primo dei moderni. Filosofia e scienza in Bernardino Telesio* (Edizioni di Storia e Letteratura, 2018).

Arianna Borrelli
is a historian of natural philosophy and modern science currently working at the Institude for Advances Studies on Media Cultures of Computer Simulation (DFG KFOR 1927), Leuphana Universität Lüneburg. The focus of her research is the relationship between scientific knowledge and the strategies employed to communicate it. She has worked on medieval mathematical cosmology, early modern meteorology and optics, and quantum theories from their early days up to the present. She published *Aspects of the Astrolabe: 'architectonica ratio' in tenth- and eleventh-century Europe* (2008), and edited, with G. Hon and Y. Zik, *The optics of Giovanni Battista Della Porta: A Reassessment* (2017).

Rodolfo Garau
is currently guest lecturer at Bard College Berlin, and a member of the research group *Experimentalsysteme* at Humboldt Universität zu Berlin. His research focuses on the history of early modern philosophy and science. He has authored articles on several early modern inquirers, and in particular on the figures of Descartes, Hobbes, and Gassendi. He is currently working on the monograph *Conatus. History of an Early Modern Concept*, taken from his doctoral dissertation. Along with Pietro D. Omodeo, he has also edited *Contigency and Necessity in Early Modern Science*, forthcoming in Springers' *Boston Studies in the Philosophy and History of Science*. Along with Justin E. H. Smith, he is preparing the first modern translation of Gassendi's *Syntagma philosophicum*, under contract with Oxford University Press.

Giulia Giannini
is assistant Professor at the Università degli Studi di Bergamo. She received her Ph.D. from the Università degli Studi di Urbino and the EHESS (Paris). She

worked at the Centre Alexandre-Koyré in Paris and was Humboldt Fellow at the Max Planck Institute for the History of Science in Berlin. Her current research focuses on the emergence of scientific academies in Europe.

Miguel Ángel Granada
is Professor for the History of Philosophy at the University of Barcelona, where he teaches the history of Renaissance philosophy. He has researched and published on Renaissance philosophy and science, mainly on Giordano Bruno and the cosmological revolution from Copernicus to Kepler. He has translated several works of Giordano Bruno into Spanish and contributed to the critical edition of Bruno's *Oeuvres complètes* published by Les Belles Lettres. He recently published a critical edition on Telesio, *Sobre los cometas y la Vía Láctea/ De cometis et lacteo circulo*, with an introduction, translation, and commentary in Spanish (Madrid 2012). He is Vice-President of the "Centro Internazionale di Studi Telesiani Bruniani Campanelliani Alain Segonds e Giovanni Aquilecchia".

Hiro Hirai
Ph.D. (1999) in history of science and philosophy, University of Lille 3 (France), is a Research Fellow at Radboud University (Netherlands). He has published widely on early modern natural philosophy, medicine, and alchemy, including the book *Medical Humanism and Natural Philosophy* (2011). He also edited *Justus Lipsius and Natural Philosophy* (2011) and *Jacques Gaffarel Between Magic and Science* (2014).

Martin Mulsow
is Professor of Intellectual History at the University of Erfurt and director of the Gotha Research Center. From 2005–2008 he was Professor at Rutgers University, USA. His research topics are Renaissance philosophy, libertinism, enlightenment, history of scholarship, oriental studies, alchemy, and numismatics. He was a member of the Institute for Advanced Study in Princeton and the Wissenschaftskolleg in Berlin. Among his publications are "Frühneuzeitliche Selbsterhaltung. Telesio und die Naturphilosophie der Renaissance" (Tübingen, 1998), and "Prekäres Wissen. Eine andere Ideengeschichte der Frühen Neuzeit" (Berlin, 2012).

Elio Nenci
Dr. phil. Milano 1987, Dr. History of Science Bari 1996, is Professor of the History of Science at Università degli Studi in Milan. The focus of his research is on the recovery of science and ancient technique in the Renaissance and on the relationship between the knowledge of the technicians and that

of philosopher-scientists in the sixteenth and seventeenth centuries. He published the critical edition of Bernardino Baldi's *Vite de' matematici* (Milan, 1998), Girolamo Cardano's *De subtilitate* (Milan, 2004) and Baldi's *In mechanica Aristotelis problemata exercitationes* (Milan, 2010).

Pietro Daniel Omodeo
is a historian and philosopher of science and a Professor of the Philosophy of Science at the Ca' Foscari University of Venice (Italy). He is the principal investigator of the ERC Consolidator research endeavour "Institutions and Metaphysics of Cosmology in the Epistemic Networks of Seventeenth-Century Europe" (2017–2022). His research focuses on science, philosophy, and literature in early modernity, as well as on historical epistemology. Among recent publications and ongoing projects, he is co-author (with Jürgen Renn) of *Science in Court Society: Giovanni Battista Benedetti's* Diversarum speculationum mathematicarum et physicarum liber (Turin, 1585) (Berlin, 2019), he is the co-editor, with Rodolfo Garau, of *Contingency* (2019) and is currently completing the monograph *Political Epistemology: The Hegemony Struggle in Science Studies* (in press).

Nuccio Ordine
(born in 1958 in Diamante, Italy) is a Professor of Literature at the University of Calabria. He is the author of *The Usefulness of the Useless* (2013), translated into 21 languages and published in 31 countries. He has dedicated three works to Giordano Bruno, which have been translated into nine languages (including Chinese, Japanese, and Russian): *Giordano Bruno and the phylosophy of de Ass* (1987; 2017 new edition), *La soglia dell'ombra. Letteratura, filosofia e pittura in Giordano Bruno* (2003) and *Giordano Bruno, Ronsard et la religion* (2004). He has also published *Teoria della novella e teoria del riso nel Cinquecento* (2009), *Le rendez-vous des savoir* (2009), *Trois couronnes pour un roi: La devise d'Henri III et ses mystères* (2011), *Les portraits de Gabriel García Márquez* (2012) and *Classici per la vita* (2015). He was a Fellow of the Harvard Center for Italian Renaissance Studies and of the Alexander von Humboldt Stiftung. He has also been Visiting Professor to various universities in the USA (Yale, NYU) and in Europe (EHESS, ENS, Paris-IV Sorbonne, CESR Tours, IEA Paris, Warburg Institute, Max Planck Institute for the History of Science, Berlin). He is a member of honor of the Institute of Philosophy of the Russian Academy of Sciences (2010) and has received three honorary doctorates in Brazil. In France he received the *Palmes académiques* (with the title of Commander in 2014) and the Legion

of Honor, conferred by the French President of the Republic (2012). In Italy, the President conferred upon him the Order of Merit of the Italian Republic (2010). In Paris, with Yves Hersant, he published three collections of classics (in the edition *Les Belles Lettres*) and, in Italy, the collection "Classics of European Literature" (Bompiani editions). Finally, he collaborates with *Corriere della Sera*.

Alessandro Ottaviani
teaches History of Science at the University of Cagliari. His main area of research concerns the history of biology and the history of the natural sciences. Among his publications are the critical editions of B. Telesio, *De natura iuxta propria principia* (Roma, 1565), (Turin, 2006), B. Telesio, *De rerum natura iuxta propria principia* (Napoli, 1570), (Turin, 2010), and *Stanze sul tempo. Sei variazioni fra rovine, fossili e vulcani* (Rome, 2017).

Jürgen Renn
is Director at the Max Planck Institute for the History of Science in Berlin and currently serves as Chair of the Humanities Section of the Max Planck Society. His research projects focus on long-term developments of knowledge while taking into account processes of globalization. His research projects have dealt with the historical development of mechanics from antiquity until the 20th century. In this context he also investigates the origins of mechanics in China, the transformation of ancient knowledge, and the exchange of knowledge between Europe and China in the early modern period. A main focus of his research is the history of modern physics, investigating the origin and development of the general theory of relativity, and of quantum theory in particular. Recently, he has taken on the challenges of the Anthropocene in investigating the history of knowledge and science.

Riccarda Suitner
is currently lecturer at the Max Weber Center for Advanced Cultural and Social Studies of the University of Erfurt. Her research and publication interests lie above all in the fields of the early and radical Enlightenment in Europe, and in the cultural and intellectual history of the sixteenth century. Her book *Die philosophischen Totengespräche der Frühaufklärung* (Hamburg, 2016) reconstructs the clandestine dissemination of philosophical "dialogues of the dead" in the early Germany Enlightenment.

Oreste Trabucco
presently works at the University of Cagliari. His main area of research concerns the interrelations between philosophy and science in early modern Italy. His publications include a critical edition of Della Porta's *Pneumaticorum libri tres* (2008) and a book entitled "L'opere stupende dell'arti più ingegnose". *La recezione degli Pneumatiká di Erone Alessandrino nella cultura italiana del Cinquecento* (2010). He is currently working on a critical edition of Telesio's *Varii de naturalibus rebus libelli.*

Introduction

Pietro Daniel Omodeo

Bernardino Telesio of Cosenza is one of the Renaissance thinkers who most strenuously defended the ideal of inductive science. He envisaged an inquiry of nature which was firmly anchored in empirical observation at a time in which this was far from common. A priori approaches, resting on standard *corpora* and a set of acknowledged authorities, prevailed in higher education and scholarly debates, despite the rise of a new practical culture in broad sectors of society. Telesio, who has been aptly called *uomo di un sol libro* (the man of one book),[1] continuously revised his major work *De rerum natura iuxta propria principia* throughout his life. This 'work of a life' underwent various changes, substantial amendments, and extensions, and appeared in three editions (Rome 1565, Naples 1570 and 1586). But its fundamental message was clear from the outset: Telesio urged natural philosophers to embrace an empirical, sensible investigation of the world. This method, contrasting with the 'bookish culture of the universities', would provide the basis for a renewal of a philosophy ideally addressing *nature itself* instead of abstract metaphysics. His contemporaries celebrated his attitude as a restorer of pre-Socratic naturalism, as is documented by the triangular exchanges between himself, his pupil the physician Antonio Persio of Matera, and the neo-Platonic philosopher Francesco Patrizi.[2] The 1570 edition began with a programmatic chapter (later adopted as the *prooemium* to the last edition) asserting that "the structure of the world and the nature of the bodies it entails should not be investigated through reason, as the ancients did, but they should be perceived by means of the senses and derived from the things themselves."[3] Telesio's refusal of abstract rationalism coupled with an anti-Aristotelian and anti-Scholastic drive resulted in opposing reactions: the indignation of university professors and ecclesiastical authorities on the one hand and the admiration of generations of *novatores* on the other. Actually, as many historians of science acknowledge today, Scholasticism was less nefarious to the advance of natural knowledge than Telesio's battle might suggest; therefore, a confrontation between the

1 Garin, "Postilla telesiana," 444.

2 See Garin, "Nota telesiana" and Puliafito, "Introduzione."

3 Telesio, *De rerum natura* (1570), f. 2*r*: "Cap. 1. Mundi constructionem corporumque in eo contentorum naturam non ratione, quod Antiquioribus factum est, inquirendam, sed sensu percipiendam et ab ipsis habendam esse rebus." Cf. Roberto Bondì, "Introduzione" to Telesio, *La natura* (2009), XVI.

 | DOI:10.1163/9789004352643_002

influences on scientific thought stemming from the various Aristotelianisms and from his own views would be valuable to gain a more nuanced understanding of the intricacies of modern science. As for his specific contribution to the inquiry of natural phenomena, Telesio composed a series of *opuscula* dealing with aspects of the world, ranging from meteorology to the doctrine of the soul, as integral components of his all-encompassing natural philosophy.[4] Those *opuscula* were partly published as lone-standing essays (as was the case with *Ad Felicem Moimonam iris*, 'on the rainbow', in 1566), partly attached to Telesio's main work (*De mari, De his quae in aere fiunt*, and *De colorum generatione* accompanying the 1570 edition), and partly printed in a posthumous collection of *meteorologica* and *parva naturalia* edited by Antonio Persio under the title *Varii de naturalibus rebus libelli* (Various Booklets on Natural Subjects) (Venice, 1590).[5]

Materialist historians of science have suggested that the social roots of Telesio's sensualism are the same as those which determined the emergence of modern empirical science. In the context of a practically-oriented society, that of early European capitalism, the traditional divides between practitioners and learned élites were challenged and revised; in this context, a new class of scholars emerged, that of proto-scientists who combined the empirical knowledge of artist-engineers and the systematic reasoning of university scholars and learned humanists. The Renaissance produced many instances of this new type of intellectual: from Leonardo da Vinci to Girolamo Cardano, Niccolò Tartaglia, Guidobaldo Del Monte, Giovanni Battista Benedetti, Simon Stevin, and Galileo Galilei.[6] The Marxist historian Edgar Zilsel designated the 'magnetic philosopher' William Gilbert as the champion of modern experimentalism and stressed that he shared the critical attitude and radical anti-Aristotelianism of Italian philosophers such as Telesio, Tommaso Campanella, Giordano Bruno, and Patrizi. However, he remarked that their naturalism should be seen as the "older brother, not the father" of experimental science, since Gilbert's science *directly* stemmed from the practical knowledge of miners, foundrymen, navigators, and instrument-makers, and not from the philosophical discourses as in Telesio.[7] Interestingly enough, Zilsel describes Francis Bacon, together with Gilbert and Galileo, as another exemplar of the modern scientist, in spite of the fact that his contribution to science remained

4 Granada, "Introduzione" to Telesio, *Varii de naturalibus rebus libelli* [1590], XII.

5 Cf. Telesio, *Ad Felicem Moimonam iris* [1566] and *Varii de naturalibus rebus libelli* [1590].

6 Zilsel, "Sociological Roots of Science" [1942]. Among the many scholarly works drawing on Zilselian premises, see Renn, *Galileo in Context*. For a discussion of Zilsel's intellectual milieu cf. Long, *Artisan/Practitioners*, Chap. 1.

7 Zilsel"Origins of William Gilbert's Scientific Method," 24.

limited to its philosophical legitimation. Bacon's science-oriented philosophy rested on inductive logic, rejected metaphysics, and aimed at the advancement of knowledge for the benefit of mankind.[8] Regrettably, Zilsel failed to take into account Bacon's keen interest in Telesio's ideas, despite the ideal link between the latter's sensualism and the former's empiricism. Moreover, although the technological utopia of the *New Atlantis* (1627) is foreign to Telesio, his philosophical work played a comparable role in promoting scientific culture—if not that of the Royal Society, then undeniably that of the *Accademia dei Lincei* surrounding Galileian science.

In contrast to Bacon, and in addition to his advocacy of empiricism and his rejection of transcendence as a source of natural explanation, Telesio provided another significant contribution to the scientific culture of his age. Indeed, his work stands out as the first modern attempt at a new foundation for, and systematic elaboration of, natural philosophy. His most daring idea was that the entire architecture of natural philosophy could be erected on foundations different from those inherited from the past and that the principles of nature should be established anew, independently of academic traditions and scholarly authority. Such an uneasy but ambitious path was followed by his admirers, *in primis* Patrizi, Bruno, and Campanella. The philosophical assessment of the first principles of nature and of its 'laws' soon became an integral part of scientific debate, as can be seen in the philosophical systems of René Descartes, Pierre Gassendi, and other scholars of their century. In the age of mechanical philosophy, problems of physics, astronomy, and physiology were embedded within heated controversies over competing systems of nature. Rodolfo Garau, in the concluding chapter of this volume, argues that Telesio's idea of self-preservation represents the passage from a teleological to an autotelic understanding of natural processes, and also hints at some possible bearing of Telesio's doctrine of self-preservation on early modern inertial (or proto-inertial) natural philosophy, in particular on authors such as Descartes and Spinoza.

The Oxford erudite librarian Robert Burton, who witnessed the European reception of Telesio, mentioned him in a curious passage on "air rectified" in his multifaceted Renaissance encyclopaedia, *The Anatomy of Melancholy* (1621):

> Or, to omit all smaller controversies, as matters of less moment, and examine that main paradox of the Earth's motion, now so much in question: Aristarchus Samius, Pythagoras, maintained it of old, Democritus, and many of their scholars. Didacus Astunica, Anthony Foscarinus,

8 Zisel, "Sociological Roots of Science", 943–945.

> a Carmelite, and some other commentators, will have Job to insinuate as much [...]. Howsoever, it is revived since by Copernicus, not as a truth, but a supposition, as he himself confesseth in the Preface to Pope Nicholas, but now maintained in good earnest by Calcagninus, Telesius, Kepler, Rothman, Gilbert, Digges, Galileo, Campanella, and especially by Lansbergius [...], by Origanus, and some others of his followers.[9]

Burton includes Telesio among the realist Copernicans and defenders of terrestrial motion. In the passage above he discusses heliocentric theory, referring to more or less legendary forerunners of Copernicus in antiquity in accordance with a widespread cliché. Burton then lists the supporters of the reconcilability of terrestrial motion and biblical exegesis, who are followed by the natural philosophers and astronomers. Burton brings together Celio Calcagnini, the humanistic author of a skeptical exercise defending terrestrial motion, *Quod coelum stet, Terra moveatur* (written around 1518–1519 and first printed in 1544), with reputable mathematicians and astronomers, who either upheld heliocentrism or merely the axial rotation of the Earth, e.g. the German ephemerist David Origanus. Telesio's follower, Campanella, earned a place next to Galileo Galilei and Johannes Kepler owing to his *Apologia pro Galilaeo* (written around 1616 and first printed in 1622), a defence of Galileo which was actually meant to defend the philosophical freedom to discuss and argue in favor of Copernicus in the year of the Roman censure of the main theses of *De revolutionibus orbium coelestium*.[10] The reference to Telesio strikes the modern historian of Renaissance science in this context, because he is seldom mentioned among the protagonists of the 'Scientific Revolution' and never among those of the 'Copernican Revolution'.

9 Burton, *Anatomy of Melancholy*, pt. 2, sec. 2, 52. On the British reception of Telesio, see Sandra Plastina, "Bernardino Telesio nell'Inghilterra del Seicento.".

10 The mention of Democritus among the ancients is puzzling. In a cosmological discussion it would have been more apt to mention him as a supporter of the plurality of worlds and, perhaps, of space infinity, but not of terrestrial motion and heliocentrism. Actually, it was Bruno's intention to back Copernican astronomy with an atomistic physics reminiscent of ancient doctrines; the English reviver of Epicurean philosophy, Nicholas Hill, followed him in this aim. Since both names are remarkably absent, one is tempted to see Burton's Democritus as an allusion to these controversial atomists, who were suspected of impiety and persecuted by religious authorities. For a comprehensive discussion of the Renaissance cosmological context, see my *Copernicus in the Cultural Debates of the Renaissance*. As to Telesio's conception of space, it has been argued that his rejection of the Aristotelian theory of places and the idea that it is independent of and prior to bodies contributed to the affirmation of modern views of space as homogeneous and absolute. See Schuhmann, "Le concept de l'espace chez Telesio," 141–167.

Telesio did not adhere to the doctrine of terrestrial motion and did not take Copernicus into account in his monumental work. Astronomy remained peripheral to his intellectual endeavor. His most important study in this field is the booklet *De cometis et lacteo circulo* (*On Comets and the Milky Way*), written around 1580 and published posthumously by Persio (1590). Revising his earlier opinions, he rejected the sublunary location of all comets and the explanation of their light as inflammations, and embraced an 'optical theory' according to which they reflect solar rays. Although he did not cast doubt on the existence of material spheres in the heavens or the origin of comets from terrestrial exhalations, he derived evidence, on the basis of the observation of the comets of 1577 and 1572 (actually a supernova), that cometary bodies can trespass the boundaries of the sublunary world, and therefore that the heavenly spheres are permeable. As to the Milky Way, discussed in the same treatise, he regarded it as a heavenly phenomenon, a condensation of celestial matter in the sphere of the fixed stars.[11] Thus, in spite of the ambitious program of a universal reform of natural philosophy along an anti-Aristotelian line of thought, Telesio was not receptive to the most disruptive cosmology of his time. His astronomical views even presented a glaring contradiction such as the maintenance of Alpetragian celestial spheres despite their penetration by comets.[12] How then could Burton include Telesio among those scholars who read Copernicus in a *realist manner*, those who neglected the epistemological admonishment of the anonymous *Letter to the reader* of *De revolutionibus* to consider the heliocentric theory as a 'mere' hypothesis for the mathematical sake of simplicity?[13] Is this really a sign of confusion on Burton's part? Or is it rather an insinuation, perhaps a hint of the ongoing debates on the status of geometry and physics in the explanation of the heavens? In this context, the reference to Telesio might be a retrospective interpretation caused by the later discussion of his natural views within the astronomical controversies on the status of hypotheses.

The Copernican philosopher Bruno, for one, acknowledged Telesio in *De la causa principio e uno* (1584) as "giudiciosissimo Telesio consentino" (the very discerning Telesio of Cosenza)[14] and began his 'physical' defense of heliocentric astronomy, *La cena de le ceneri* (1584), with a reference to the Telesian

11 Cf. Granada, *Bernardino Telesio: Sobre los cometas*, especially the introduction. For a summary, see also my review.

12 Lerner "La physique céleste de Telesio."

13 Burton's synthesis of the astronomical debates of the time seems to confuse the *Letter to the reader* with Copernicus's dedication to the Pope, thereby ignoring Johannes Kepler's disclosure of the identity of the author of the epistle as the theologian Andreas Osiander in *Astronomia nova* (1609).

14 Bruno, *Opere italiane*, *De la causa* III, 677.

bipolarity of nature, the opposition between solar warmth and terrestrial coldness: "Two [are] the contrary and active first principles, heat and cold. Two, the first parents of the things in nature, the Sun and the Earth."[15] Doctrines that fell even closer to Telesio's were propagated by Campanella—the forceful defender of Galileo's *libertas philosophandi* in the years of the *Affaire*, the revolutionary who drafted his utopic views in *Città del Sole* and the survivor of the jails of the Neapolitan and Roman Inquisitions who hazardously fled to France and attended the literary circles of Richelieu's Paris. Campanella dedicated the Paris edition of *De sensu rerum et magia* (1637) to the powerful cardinal. In this work he stressed the ontological basis of Telesian sensualism, i.e. a vitalistic conception of nature based on the assumption of universal sensitivity: "That which is in the effects has to be in the causes, therefore the elements and the whole have sensation."[16] Following Telesio's philosophy, Campanella posited two opposed principles of the world, *sole* (Sun) and *terra* (Earth), both endowed with sensitivity which they communicate to the plants and animals they generate.

One should mention medical studies as another area influenced by Telesio. This should include not only the well-known instance of Persio but also Telesio's possible influence within broader European circles, including those of radical Italian émigrés such as Agostino Doni of Cosenza. In chapter 10, Riccarda Suitner presents Doni's conceptions on the basis of his only extant work, *De natura hominis* (1581). She stresses both the divergences and the similarities between the doctrines of Doni and those of Telesio that have hitherto not been remarked upon by interpreters, based in particular on the 1565 and 1570 editions of *De rerum natura* which predate the publication of Doni's work.

These examples show that the meaning of Telesio's work for the scientific debates of the Renaissance goes beyond the limits of mathematical astronomy and each science in particular. For his early readers, Telesio's main achievement was his daring project to establish a new basis for the sciences by composing a philosophical system capable of providing a viable alternative to Aristotelianism. He regarded nature as a process ruled by the eternal struggle between the two active principles of heavenly warmth and terrestrial coldness ruling over matter.[17] Telesio claimed that individual natural beings universally follow a principle of self-preservation, emphasizing a concept of *conatus* which in the seventeenth century bridged views on physical inertia

15 Ibid., *Cena* I, 443: "Doi [sono] contrari et attivi principii: il caldo et il freddo. Doi primi parenti de le cose naturali: il Sole e la Terra." Engl. transl. by Gosselin and Lerner (1995), 82.

16 Campanella, *Del senso delle cose e della magia*, I 1: "Ciò ch'è negli effetti esser nelle cause, e però gli elementi e il mondo sentire." For a general introduction to Campanella's philosophy, see Ernst, *Tommaso Campanella*.

17 The most accurate introduction is Bondì, *Introduzione a Telesio*.

and animal life.[18] Self-preservation also allowed him to connect the explanation of natural tendencies with human behavior. In the last edition of *De rerum natura*, Telesio expanded on the domain of ethics and politics. In his view, the virtues first emerge as a support and regulation of primary vital functions and then evolve into complex moral systems aimed at satisfying the needs of society. The further assumption that individual drives are unwittingly co-opted in the universal realization of the common good earned Telesio a place among modern theological-political thinkers.[19]

Telesio was not only concerned with the most general questions concerning the philosophy of nature; he was also devoted to the solution of particular scientific problems, especially in the *opuscula*. The nine booklets gathered by Persio in 1590 comprised four brief treatises on issues that pertained to meteorology according to the Aristotelian tradition. Specifically, they dealt with comets, the Milky Way, winds, earthquakes, the rainbow, and the sea, including the problem of the sea tides. The remaining five booklets dealt with problems of natural history that used to be labelled *parva naturalia*. One dealt with the unity of the soul against Galen, the others with the function of respiration, the nature of colors, taste, and sleep. These writings meant to offer naturalistic explanations of these natural phenomena in line with the doctrine expounded in *De rerum natura*, while at the same time testing the soundness of that doctrine. They should be understood in the context of Telesio's reception and criticism of the scholarly traditions represented by Aristotle and Galen. At the same time, they bear witness to Telesio's contribution to the scientific debates of his time.

In the opening chapter of this volume, Roberto Bondì offers an overview of the historical impact of Telesio's work, views, and methodology seen through the lenses of his immediate successors, particularly Bacon and Galileo. This overview is followed by several studies dedicated to Telesio's treatment of specific natural issues and their cultural context. Hiro Hirai (Chapter 3) explores issues linked to the generation of life in connection with Aristotelian and Hippocratic debates. Arianna Borrelli (Chapter 4) and Oreste Trabucco (Chapter 5) deal with meteorological questions, in particular with the doctrines of heat, moving spirits, and winds. Borrelli shows how Telesio's booklet *De iis quae in aere fiunt* (1570) not only revived Aristotelian explanations but also introduced the idea that spirits and vapours were capable of moving out of their own will, prompted by solar heat and by a desire to escape

18 See *infra*, Garau, Chap. 12. For a comparative treatment of self-preservation in Telesio and early-modern philosophy, see Mulsow, *Frühneuzeitliche Selbsterhaltung*, 193–200.

19 Giglioni, "Introduzione" to Telesio, *De rerum natura* [1586].

compression or dilatation. Borrelli also places Telesio's meteorology in its larger historical context, specifically in relation to another very original treatise on weather and climate published a few decades later, Giovan Battista Della Porta's *De aeris transmutationibus* (1610). Trabucco's contribution testifies to the wide reception of Telesio's thought in sixteenth century natural philosophy by focusing on the *De causa ventorum peripatetica disceptatio* by Federico Bonaventura, which was published in 1592 in Urbino. This timely contribution to Renaissance studies redresses Bonaventura's so-far neglected criticism of some theses put forth by Telesio in his work *De iis quae in aere fiunt*.

My own case study (Chapter 6) is an investigation of Telesio's tidal theory in the context of the heated debates on this topic during the Renaissance. Telesio's theory is interesting because it rejects the possibility of an account of the sea-tides based on the moon, a view which the author connects to a long-standing epistemological commitment that favored causal explanations of natural phenomena over occult virtues. Further connections are made to prominent scholars of the time, such as Pico della Mirandola, Cesalpino, Sfondrati, Patrizi, and Galileo. Elio Nenci (Chapter 7) connects Telesio's views on the rainbow to the sixteenth-century reception of the Greek commentaries from Alexander of Aphrodisias and Olympiodorus on Aristotle's *Meteorology*. Nenci analyzes the representative exponents of these views within Renaissance Aristotelianism in order to identify the elements of continuity and discontinuity between the works of authors such as Alessandro Piccolomini, Ludovico Boccadiferro, Francesco Vimercati, and Telesio's *De iride*.

Telesio's immanent perspective on nature and its implications for the doctrine of the soul worried religious authorities even more than his anti-Aristotelianism—which, in the climate of counter-reformist Italy, could be seen as a threat against Thomistic orthodoxy. His treatment of the soul as a *spiritus* of entirely natural origin in the first editions of *De rerum natura* aroused the attention of ecclesiastical censors. On April 28, 1570, Telesio wrote to the main ecclesiastical authority in Cosenza, Cardinal Flavio Orsini, to defend himself against rumors concerning the impiety of his conceptions. "In truth—he wrote—these two books deal with nothing but the first bodies and the principles, that is, warmth/coldness, humidity, and dryness. Very few things are asserted about the soul and only those that pertain to the matter of the principles and the sensitive and motive soul [...]."[20] The difficulty must

20 De Miranda, "Una lettera inedita di Telesio," 374: "Et veramente—he wrote—in questi doi libri non si tratta d'altro, che de li primi corpi, et de li principii cioè caldo freddo, humido, e secco. Dell'anima se ne dice pochissime cose. Et quelle sole, ch'appartengono alla materia delli principii, et all'anima sensitiva, et motiva [...]."

have been serious and might explain the long time, from 1570 to 1586, that Telesio took to elaborate the last version of his work, in which the distinction between the *spiritus e semine eductus*, i.e. the natural soul, and the *anima a Deo immissa*, i.e. the soul of divine origin, was emphasized. Nevertheless, this caution (or clarification) was not sufficient to avoid censure. The Clementine *Index librorum prohibitorum* of 1596 suspended *De rerum natura* and some of the small natural treatises with the clause *donec expurgetur*. As no expurgation was ever approved, Telesio's 'dangerous' works remained prohibited to orthodox Catholics up to 1900, when they were taken off Leo XIII's *Index*. Miguel Ángel Granada (Chapter 2) offers an accurate reconstruction of Telesio's evolving views on "spiritus" and "soul," the sources he referred to, and the contextual reasons for his intellectual engagement with this problematic.

Three further essays contribute to the comprehension of the cultural environment in which Telesio lived and was received. Drawing on Antonio Persio's *De natura ignis et caloris*, which is preserved in the *Biblioteca Corsiniana* in Rome, Martin Mulsow discusses (in Chapter 8) a manuscript that documents a philosophical discussion on warmth and light that occurred in Naples in the 1570s. It offers a unique insight into Telesio's dialogical attitude and his confrontation not with Aristotle in the abstract but with the views of his contemporaries adhering to the Peripatetic school. Alessandro Ottaviani's contribution to this volume (Chapter 9) is a discussion of early documents connected with the censure of Telesio's ideas. He analyses a rare copy of the 1565 edition of Telesio's *De natura* preserved in the Biblioteca Corsiniana, Naples, which includes significant marginal annotations revealing of a time of intellectual censure linked to Counter-Reformation policies. Ottaviani's philological analysis allows him to posit that the author of the annotations was probably the Friar Angelo Baronio. This chapter is a valuable contribution to the reception of Telesio's philosophy and would also be relevant for readers interested in the history of the book and in inquisitorial procedures. Finally, Giulia Giannini (Chapter 11) considers the development of Renaissance academies through an analysis of Telesio's connection with them. She discusses the historical attribution of Telesio as the founder of the *Accademia cosentina* in his hometown, and considers whether an *Accademia Telesiana* existed or whether it is just a historiographic construction.

Telesio's troubles with censorship should be read against the background of a climate of increased religious tensions and attempts at control and censure.[21]

21 In order to relate these documents to the evolution of Telesio's thought (and his work itself), see Ottaviani's introductions to his editions of *De natura iuxta propria principia* of 1565 and the new version of 1570.

In the same year in which Telesio wrote his self-exculpation to Cardinal Orsini (1570), the polymath Girolamo Cardano was tried by the Inquisition in Bologna for his heterodox views and astrology—sixteen years before this art was officially condemned by Sixtus V. He was subsequently forced to move to Rome, where he could be better controlled. The Neapolitan scientist Giambattista Della Porta was arrested in 1574 and condemned in Rome for necromancy. Many of his works, for instance the Italian version of *De humana physiognomonia*, could not be printed. Patrizi, who cherished Telesio's views and corresponded with him, was hired in Rome as a professor of Platonic philosophy and was immediately tried for the view on natural philosophy that he presented in *Nova de universis philosophia* (1591).[22] Besides the most severe theological allegations, he had to exonerate himself in 1592 from the suspicion of being a follower of Copernicus's system. As the Inquisition's documents read, "In *Pancosmia* [...] he [Patrizi] states 'that the motion of the Earth is by far in better agreement with reason than the motion of the heavens or the uppermost celestial bodies'. And he refers to Nicolaus Copernicus' sentence according to which the sidereal heaven is immobile, along with the stars, while the Earth moves."[23] Some of Patrizi's persecutors were later involved in Bruno's trial and condemnation to death, as well as in the Galileo *Affaire*. After a first imprisonment in 1592, Campanella was trialed in Padua and Rome (1594–1595) leading to his abjuration upon being vehemently suspected of heresy. His subsequent imprisonment in the Inquisition's jails of Naples and Rome lasted thirty years (1597–1634); he was spared the death penalty only because he resisted horrible tortures that deformed him permanently and he pretended to be insane.[24] Galileo and his Padua associate, the natural philosopher Cesare Cremonini, were investigated together by the Padua Inquisition (1604).[25] Telesio's works were publicly burned in front of the cathedral of Naples on St. Peter and Paul's day in 1610, together with other prohibited books. Hence it is no wonder that authors he inspired were reluctant to mention him directly, as was likely the case with Della Porta's views on heat and moving spirits in *De aeris transmutationibus* (Rome, 1610).[26] Campanella's Telesian work *De sensu*

22 Cf. Rotondò, "Cultura umanistica e difficoltà di censori."

23 Baldini and Spruit, *Catholic Church and Modern Science*, vol. 1, 3, 51, doc. 1, 2216: "Lib. 17 Pancosmias fol. 103, pag. 1, col. 2.a ait quod Terram revolvi longe videtur esse rationi consonantius, quam Coelum, vel suprema astra moveri. Et refert sententiam Nicolai Copernici dicentis Coelum sydereum stare simul cum stellis, Terram vero moveri."

24 Among others, see Del Col, *Inquisizione in Italia*, 552–565, and Black, *Italian Inquisition*, Chapter 7.

25 Poppi, *Cremonini e Galileo inquisiti.*

26 See *infra*, Borrelli, Chapter 4.

rerum et magia was confiscated by his oppressors, which forced him to rewrite it later by heart, in Italian, during his imprisonment, and then to translate it back into Latin.

Notwithstanding this climate of suspicion and censure, Telesio's ideas subtly entered the scientific culture of the seventeenth century. Bacon is perhaps the most evident instance of such influence, but other examples can be mentioned.[27] The English mathematician Henry Savile became acquainted with Telesio's philosophy during his continental *Bildungsreise*, discussed his ideas with Patrizi and Persio, and sent a copy of the 1570 edition of *De rerum natura* to the humanist Andreas Dudith-Sbardellati.[28] It has been shown that Telesian conceptions, particularly those relative to psychology, met with the keen interest of scholars linked to the Northumberand circle and were circulated among English philosophers.[29] In Italy, the founder of the *Accademia dei Lincei*, Federico Cesi, was profoundly influenced by Telesio's conceptions.[30] The philosophy of *De rerum natura iuxta propria principia* was held in great esteem and critically assessed by the Lynceans. Persio discussed Telesio's conceptions on the nature of light with them in the years following Galileo's astounding telescopic discoveries.[31] Further documentation of the circulation of Telesian views in the scientific debates of the seventeenth century is found in Galileo's references to his work. In the last part of the *Dialogo sopra i massimi sistemi del mondo* (1632) Galileo deemed it necessary to introduce his own 'proof' of the Copernican theory, a mechanical explanation of the sea tides, by distancing himself not only from astrological interpretations of the phenomenon but also from Telesio's 'thermo-dynamic' account.[32] In spite of his disagreement on specific points, Galileo fiercely reacted in defense of Telesio's memory when it was discredited by his opponents. In the course of their polemics over cometary theory, the Jesuit Orazio Grassi accused Galileo of following Cardano's and Telesio's "sterile and unfortunate philosophy" (*sterilem et infelicem philosophiam*).[33] Galileo's reply in *Il Saggiatore* (1623) in defense of the two natural thinkers was firm: "Does he [Grassi] not notice how impiously

27 See *infra*, Bondi, Chapter 1.
28 Iovine, "Henry Savile lettore di Bernardino Telesio."
29 Cf. Spruit, "Telesio's Psychology."
30 Galluzzi, *Libertà di filosofare*, 83–97.
31 Gómez López, "Telesio y el debate sobre la naturaleza de la luz."
32 See *infra*, Omodeo, Chapter 6.
33 Grassi, "Libra astronomica ac philosophica" [1619], 118.

he deprives them of all their reputation in order to hide a small blot on that of his master?"[34]

Telesio is emblematic of early-modern scientific culture in many ways. His attention to detail and experience and, at the same time, his aspiration to universality and all-comprehensiveness is typical of Renaissance science. His natural philosophy constituted a milestone in modern culture as a first systematic attempt at the foundation of the natural sciences, running counter to the Scholastic tradition. He is also representative of the ethical tensions affecting the natural debates of his age. His thought and its reception bear witness to the inseparability of the natural sciences and philosophy at a time when the natural disciplines underwent a process of rapid change, which led to an understanding of science and scientific methodology which is closer to our present understanding. This collection of essays is dedicated to him and the place of his thought at the crossroads of the natural sciences, philosophy, philology, and the life sciences. The contributors to the volume focus on the scientific-cultural context of this thinker as well as his scientific roots, and they deal with the question of his influence on the natural sciences of early modernity.

I would like to acknowledge the institutions which supported this project: the European Union's Horizon 2020 Research and Innovation Programme for the support of the research endeavor *Early Modern Cosmology* (GA n. 725883), of which this publication is part, the Max Planck Institute for the History of Science (Berlin), the Centro Internazionale di Studi Telesiani, Bruniani e Campanelliani "Alain Segonds" e "Giovanni Aquilecchia" (Cosenza), the Collaborative Research Centre 980 "Episteme in Motion" (Freie Universität, Berlin), and the Deutsche Forschungsgemeinschaft. I am thankful to Roberto Bondì for his valuable advice, to the blind referees for their precious suggestions, to Peter Christie for the translation of my chapter and the first stylistic revision of the volume, Louis Berger, Matteo Fornasier, Janna Katharina Müller, and Jason Terry for their support with the formatting and to Valentina Littera for her support with the indexing. I am very thankful to Lindsay Parkhowell for his invaluable support as copy-editor and the final stylistic revision of the entire volume.

34 Galilei, *Opere*, vol. VI, 236. On the reception of Telesio, also see Elisabetta Selmi, "'Formazione' e 'ricezione' del pensiero."

CHAPTER 1

The First of the Moderns: Telesio between Bacon and Galileo

Roberto Bondí

1 The Origin and Success of an Interpretation

From the nineteenth century to the present day the assessments of Telesio's 'modernity'[1] have been far from homogeneous. Like all the great classics, the work of the Cosentian philosopher has generated contrasting interpretations. There is Francesco Fiorentino's Telesio, who at times has a "hint of" and is the "true precursor of Galileo", and at other times is seen as the forerunner of the "Baconian contempt for sciences based on pure reasoning".[2] Then there is Ernst Cassirer's Telesio, who is far removed from the new mentality upon which the modern science of nature originates.[3] There is Erminio Troilo's Telesio the "radical empiricist" who is placed alongside modern thinkers,[4] and the "metaphysical materialist" Telesio of Giovanni Gentile, who argues that the philosopher only hinted at the modern age and essentially remained a Renaissance man.[5] There is Nicola Abbagnano's Telesio, who is critical of "all possible metaphysics" and of whom Galileo is the "natural successor",[6] and Giacomo Soleri's Telesio, the philosopher and non-scientist whose metaphysical interests prevail over his physical ones.[7] More recently, Eugenio Garin has stressed that in Telesio's *De rerum natura* the philosopher "precisely intended to seize the ideal moments, the first causes, which he understood as immanent and internal to nature itself", adding that:

1 See now Bondí, *Il primo dei moderni*.
2 Fiorentino, *Bernardino Telesio*, vol. 1, 111 and 304.
3 See Cassirer, *Das Erkenntnisproblem*, vol. 1, 238.
4 See Troilo, *Bernardino Telesio*, 71–73.
5 See Gentile, *Bernardino Telesio*, 64 and 81.
6 See Abbagnano, *Bernardino Telesio*, 245 and 246.
7 See Soleri, *Telesio*, 157.

 | DOI:10.1163/9789004352643_003

> Galileo, to whom Telesio has been compared, determining the laws of motion without looking out for essences, declared not to know or to care for Telesio and, in general, for the pettiness of philosophers. Telesio wanted to know what heat and cold were. He did not establish the rapport of motion and position of the Sun in respect to different bodies, but he was concerned with the relationships of the Sun with the vital and life-giving warmth that is the true substance of the universe.[8]

If one considers Telesio in the context of modernity, as one should in my opinion, one needs to take several specific points into account: the image of nature itself; the image of knowledge and of the learned; the contrast between the principles of authority and truth; and the assertion of the independence of scientific knowledge. Yet the question of Telesio's modernity arose long before it became the focus of Fiorentino's important monograph *Bernardino Telesio, ossia studi storici sull'idea della natura nel Risorgimento italiano* (Bernardino Telesio, or Historical Studies on the Idea of Nature in the Italian Renaissance) (1872–1874). Fiorentino asserted the anti-metaphysical significance of *De rerum natura*, claiming its author was in the vanguard of "all the moderns",[9] which was merely an amplification of Francis Bacon's representation of Telesio as the first of the moderns.[10] The interpretation of Telesio as one of the *novatores*, or the first of the *novatores*, as a staunch defender of philosophical freedom and as a fierce opponent of Aristotelian tyranny is omnipresent in seventeenth-century European culture: the writings of Tommaso Cornelio, Leonardo di Capua, Pietro Giannone, Gian Vincenzo Gravina, Daniel G. Morhof, Gabriel Naudé, and Charles Sorel all testify to this.[11] And this assessment was expounded again in the eighteenth century: see, for example, Johann G. Lotter's *De vita et philosophia Bernardini Telesii* (The Life and Philosophy of Bernardino Telesio) (1733) and Johann J. Brucker's *Historia critica philosophiae* (Critical History of Philosophy) (1742–1744).

Bacon's interpretation of Telesio as the first of the moderns is particularly present in *De principiis atque originibus secundum fabulas Cupidinis et Coeli, sive Parmenidis et Telesii et præcipue Democriti philosophia tractata in fabula de Cupidine* (On the Principles and the Origins According to the Fables of Cupid and Coelum, etc.), published in a posthumous collection (1653) that circulated

8 Garin, *History of Italian Philosophy*, vol. 1, 431–432.

9 Fiorentino, *Bernardino Telesio*, vol. 1, 244.

10 Bacon, *Works*, vol. 5, 495; Bacon, *Philosophical Studies*, 259. On Fiorentino's *Telesio*, see Savorelli, "Letture telesiane," 445–473.

11 See Agrimi, "Telesio nel Seicento napoletano"; Bianchi, "Des novateurs modernes en la philosophie."

widely in Europe. This text, which amounts to a monograph on Telesio and accounts for his success in Europe, presents a rather varied picture: Bacon captures key aspects of his work, alternating interpretations which are highly critical with others which are highly positive, and focuses on both specific points and general questions.

2 "... Calmly and as if in Idleness."

Much importance has rightly been given to the pages of *De principiis* in which Bacon presents the naturalism of *De rerum natura* as a "pastoral philosophy".[12] It is important to recount these passages in detail so that the reader has an adequate understanding of Bacon's judgement of Telesio. In *The Advancement of Learning* the Lord Chancellor proposes "to weigh the dignitie of knowledge in the ballance with other things" using "testimonies and arguments divine, and humane". This dignity can be sought "in the Arch-tipe or first plat-forme, which is in the attributes and acts of God". If we consider the work of the Creation, we observe "a double emanation of vertue from God": "The one referring more properly to power, the other to wisedome", "the one expressed in making the subsistence of the matter, and the other in disposing the beauty of the fourme".[13] When we turn to the "distribution of dayes", it is easy to understand that "the day wherin God did rest, and contemplate his owne works, was blessed above all the dayes, wherein he did effect and accomplish them":

> After the Creation was finished, it is sette downe unto us, that man was placed in the Garden to worke therein, which worke so appointed to him, could be no other than worke of contemplation, that is, when the end of worke is but for exercise and experiment, not for necessitie, for there being then no reluctation of the creature, nor sweat of the browe, mans employment must of consequence have ben matter of delight in the experiment and not matter of labor for the use. Againe the first Acts which man perfourmed in Paradise consisted of the two summarie parts of knowledge, the view of Creatures, and the imposition of names.[14]

12 Bacon, *Philosophical Studies*, 251. See Giachetti Assenza, "Bernardino Telesio: Il migliore dei moderni."

13 Bacon, *Advancement of Learning*, 33.

14 Ibid., 34.

Once this is clear, Bacon addresses "the first event or occurrence after the fall of Man". Here "wee see [...] an Image of the two Estates, the Contemplative state, and the active state".[15] These two ways of life are represented by Cain and Abel and in the "two simplest and most primitive Trades of life", the farmer and the shepherd. Unlike the farmer, who represents active life, the shepherd embodies the contemplative life and simply stands gazing at the sky. The pastoral life is totally absorbed in contemplation and idleness.

This is exactly what Bacon means when he speaks of Telesian philosophy as a pastoral philosophy in *De principiis*. According to Bacon, it is a philosophy of contemplation and idleness which should be rejected because the world is now something to be transformed and no longer an object to be contemplated. The human condition after the Fall is very different from before the Fall. Thanks to sin, writes Bacon in *Novum organum*, "man lost both his state of innocence and his command over created things". His task is now to try to recover both, albeit partially: the first through religion and faith; and the second by means of the arts and sciences ("the curse did not quite put creation into a state of unremitting rebellion ...").[16] Hence Bacon's insistence on the need to combine theory and practice. As he writes in *The Advancement of Learning*:

> but this [...] will indeed dignifie and exalt knowledge; if contemplation and action may be more neerely and straightly conioyned and united together, that they have beene; a Coniunction like unto that of the two highest Planets, *Saturne* the Planet of rest and contemplation; and *Iupiter* the Planet of civile societie and action.[17]

In *De rerum natura* Bacon instead finds a kind of pastoral philosophy which contemplates the world "calmly and as if in idleness". He finds assertions that are credible if, and only if, "man along with the mechanical arts which vex matter were removed from nature, and the fabric of the world were regarded artlessly".[18] It was a question of limits belonging to an entire tradition. In the words of Paolo Rossi, in Bacon's eyes, the natural philosophies

> have organized nature to perfection (and there are many new ways of organising it) but these arrangements are like well-acted plays that console or amuse the spectators for the time being but have no effect on

15 Ibid., 34.
16 Bacon, *The "Instauratio magna" Part II: "Novum organum"*, 447.
17 Bacon, *Advancement of Learning*, 32.
18 Bacon, *Philosophical Studies*, 251.

reality. Whereas the new philosophy, by acknowledging natural objectivity will attempt to adapt nature to the needs of man, for which purpose those technical tools—the mechanical arts—fashioned by man for such ends, cannot be overlooked.[19]

3 "Sunk in the Bottomless Pit of Peripatetic Darkness"

The copious passages that Bacon devotes to Telesio contain more than just a general critique as is found in the above passage; they also contain an interpretation of specific aspects of Telesian philosophy. In his *De principiis*, Bacon primarily emphasizes continuity with Aristotelianism. This is particularly clear in his analysis of the principle of matter, which had already been criticized by Francesco Patrizi of Cherso. In 1572, after having been invited to read *De rerum natura* by Telesio's disciple and friend, Antonio Persio, Patrizi set out several objections to which both Telesio and Persio then replied. Telesio posits three principles as the basis of natural phenomena: two incorporeal agents (heat and cold) and bodily mass, which is conceived of as a receptacle of opposites that neither acts nor operates and remains unchanged. Patrizi considers that matter thus conceived is unintelligible and impossible to understand with the senses.[20] According to Patrizi, this made Telesio inconsistent as he postulated a material principle that was a useless hypothetical pretense beyond the visible phenomena of the condensation and rarefaction of bodies.

In *De principiis* Bacon argues that Telesio, "having been corrupted [...] by Peripatetic notions", simply adds the principle of matter to the Parmenidean system, describing it incorrectly in terms of passivity. In this as in other cases,

---

19 Rossi, *Francis Bacon: From Magic to Science*, 54. In another passage, Rossi wrote (ibid., 38): "Classical philosophers: Plato, Aristotle, Galen, Cicero, Seneca, Plutarch, and those of medieval and Renaissance times: St Thomas, Duns Scotus, Ramus, Cardano, Paracelsus, and Telesius were not accused of committing theoretical errors. But their philosophies, all comparable to some extent, deserved the same condemnation and the same fate because their ethics were at fault. This seems so monstrous to Bacon that he says 'such profane and polluted' subjects cannot even be discussed without shame. He does not wish to replace them by a new philosophy of the same kind with identical principles, arguments, and aims but by an entirely new attitude to nature involving new principles and a different kind of argument, and different aims: in fact a new concept of truth, a new ethic, and a new logic."

20 See Telesio, *Varii de naturalibus libelli* (1981), 467 and Telesio, *Delle cose naturali libri due* (2013), 483–84: "Quae vero de mole remanente sub rerum immutatione affers, non videtur sensui respondere [...]. Certe sensus nullus, quo solo duce te in omnibus usurum initio es professus, materiam ulli unquam indicavit".

Telesio remains "sunk in the bottomless pit of Peripatetic darkness". The Cosentian philosopher is

> a mighty man furnished with the reasonings of the Peripatetics (if there were anything to them) which he has actually turned against them, but held back by his positive claims and better at knocking down than building up [...] [yet sometimes he] does not do his job very well but acts like his opponents, who, since they make up their minds before they try anything out, when they come to particular facts, abuse their minds and the facts, and wretchedly squander and torture both; and they are nevertheless busy and (if you believe the individuals themselves) victorious, and one way or another bubbling over with their own conceit.[21]

The context in which Bacon places this type of critique is clear. He is bent on championing the philosophy of the pre-Socratics, particularly Democritus, against Plato and Aristotle. For Bacon, Democritus "was held in high esteem among wiser men, and those who immersed themselves more profoundly in silent and demanding kinds of contemplation", whereas the founders of the Academy and the Lyceum "rang round the schools to great applause amid the pompous rumblings of professors". When matter is conceived as "despoiled and passive" it is no more than

> a figment, arising from the fact that as far as the human intellect is concerned, those things seem to have most reality which the intellect takes in most readily, and which affect it most.

Thus, forms end up having "more reality than either matter or action". After all, "it is from here that the reign of forms and ideas in essences seems to have originated, namely with the addition of a kind of fantasy matter". Bacon argues very clearly that Plato "made over the world to thoughts" and that Aristotle "made over thoughts to words". Instead, the Ancients "submitted their minds to the nature of things". Empedocles, Anaxagoras, Anaximenes, Heraclitus, and Democritus "were as one in maintaining that matter was active, had some form and imparted its form, and had the principle of motion within itself".[22]

21 Bacon, *Philosophical Studies*, 225, 247, 251, 261. On Bacon's critique of Telesio, see Pousseur, "Bacon, a Critic of Telesio;" Margolin, "Bacon, lecteur critique d'Aristote et de Telesio;" Pousseur, "La notion baconienne de principe." See also Whitaker, "Francesco Patrizi and Francis Bacon."

22 Bacon, *Philosophical Studies*, 205, 207 and 209.

4 Telesio: a Restorer of the Philosophy of Parmenides

Whatever their importance, the general and specific criticisms made against the arguments contained in *De rerum natura* in no way exhaust the association between Bacon and Telesio. Indeed, this relation risks being misunderstood if we do not bear in mind other important claims made by Bacon in *De principiis*. I have always believed that Bacon's basic judgment on Telesio was not so much a question of what has so far been discussed in this chapter so much as a question of the relationship between Telesio and Parmenides. On several occasions, Bacon claims that Telesio is the modern "restorer of the philosophy of Parmenides".[23]

Without playing down Bacon's interest in the heat/cold dialectic and the doctrine of the soul expounded and discussed in *De rerum natura*,[24] these arguments are not what makes Telesio so important for Bacon. The philosopher's great merit lay elsewhere, in having brought to light a part of pre-Socratic wisdom. In this respect Bacon felt himself to be—and was also in point of fact—fully congruous with Telesio. Pre-Socratic philosophy deserved to be rediscovered and set against the "false Philosophy" of Aristotelianism. In the first book of *Novum organum* Bacon had this to say about Aristotle's philosophy:

> Thus then the root cause of error is three kinds of *false Philosophy*: the *Sophistical*, *Empirical*, and *Superstitious*. The most obvious example of the first family is *Aristotle* who with his dialectic corrupted natural philosophy when he fashioned the world from categories; doled out to that most noble substance, the human soul, a genus of terms of the second intention; settled the business of *dense* and rare, whereby bodies take on greater and lesser dimensions or volumes, by the frigid distinction of act and potency; claimed that motion in particular bodies was singular and proper to them, and that if they shared in any other motion that circumstance came from an extrinsic cause; and forced on the nature of things countless other ideas as the whim took him. For he was everywhere more concerned about how someone replying to a debate might parry a thrust,

23 Bacon, *Philosophical Studies*, 259: "Now to some it may perhaps hardly seem worth while going to so much trouble to controvert *Telesio's* philosophy, a philosophy which is not of course very famous nor widely accepted. But I do not bother with such niceties. [...] our business is not with *Telesio* as such, but him as a restorer of the philosophy of *Parmenides*, to whom much respect is due."

24 See Walker, *Spiritual and Demonic Magic*, 189–202; Gemelli, *Aspetti dell'atomismo classico*, 104–31; Graham Rees, "Bacon's speculative philosophy."

> and retort with something positive and verbal, than about the inner truth of things.[25]

According to Bacon, to be convinced of the validity of this judgement it is enough to compare Aristotle with the pre-Socratics.[26] In Bacon's view, time is like a river that keeps what is "light and blown up" afloat, leaving what is "solid and grave" to sink. The most profound speculations on nature arise sporadically, but are soon broken up and snuffed out by the winds of popular opinion. In the midst of the array of opinions expressed by the pre-Socratics which have come down to us in a rather fragmented way, we can find viable claims regarding the observation of nature and the discovery of causes. In some cases, these claims have penetrated nature more acutely and deeply than Aristotle.[27]

When he presented Telesio as the modern restorer of the philosophy of Parmenides, Bacon was not raising the question of originality.[28] The great respect that he had for Telesio was due to the reaffirmation in *De rerum natura* of Parmenidean doctrines in a strongly anti-Aristotelian context. It was Telesio who had salvaged one of those "solid and grave" things often swallowed up by time.

5 Telesio: the First of the Moderns

We also find the juxtaposition of Telesio and Parmenides in *Sylva sylvarum*. Here it is accompanied by another challenging verdict that claims that Telesio

---

25 Bacon, *The "Instauratio magna" Part II*, 99.

26 Ibid., 99–101: "For the Homoeomera of *Anaxagoras*, the Atoms of *Leucippus* and *Democritus*, the Heaven and Earth of *Parmenides*, the Strife and Friendship of *Empedocles*, and *Heraclitus'* Resolution of bodies into the undifferentiated nature of fire and their Reconstitution as dense matter—all these have something of the natural philosopher about them and smack of the nature of things, experience and bodies, whereas *Aristotle's* Physics usually rings out with little more than the terms of dialectic, which he has dealt with again under a more imposing name, more as a Realist than a Nominalist, in Metaphysics. And no one should be impressed by the fact that in his books on Animals, in his Problems, and in other tracts of his he often deals in experiments. For he had made up his mind beforehand, and did not take experience into due account when he framed his decrees and axioms but, having made up his mind to suit himself, he bends experience to his opinions and drags it about in chains, so that in this respect too he is more blameworthy than his modern followers (the family of Scholastic philosophers) who have abandoned experience altogether."

27 Bacon, *Works*, 503, 565–71, 599–602.

28 For a contrasting opinion, see De Mas, "Introduzione," xxi, and "Bernardino Telesio e la falsità di Aristotele," 174.

is the best among those trying to renew philosophy, such as Bruno, Campanella, Cardano, Fracastoro, Gilbert, Paracelsus, Patrizi, and Severinus: "[Telesio] hath renewed the philosophy of Parmenides, and is the best of the novellists".[29] Bacon says "*and*", not "*yet*". In my opinion, whoever holds that Bacon is contrasting the two things is mistaken.[30] Telesio is instead—and in a way which is *not contradictory*—both the restorer of Parmenidean philosophy *and* the first of the moderns.

We also find the verdict on Telesio as first of the moderns in *De principiis*:

> I think well of *Telesio*, and recognize him as a lover of truth, a man useful to the sciences, a corrector of certain doctrines, and the first of the new philosophers (*novorum hominum primum*).[31]

De principiis is the text which is most helpful in clarifying Bacon's image of Telesio. If we ask what it was that led Bacon to reach this highly positive verdict we can find the reply primarily in Chapter XVII of Book I of *De rerum natura*. In Bacon's view Telesio had clearly grasped the knowledge-power nexus. Bacon completely transcribed the passage in question from Telesio:

> Further, what heat or how much, that is, what strength and what amount of heat, can change what earth and what entities into what, looks like an inquiry not to be pursued, and as something beyond (it seems to me) the bounds of human reason. For how can one divide either the powers of heat and heat itself into degrees as it were, or have a clear understanding of the amount and quantity of matter in which it is impressed, and allot to certain and determinate powers and amounts of heat a certain quantity, disposition and certain actions of matter, or, on the other hand, to a certain quantity and certain actions of matter a certain and determinate amount of heat? If only people who have the time and better minds, and the chance to investigate the nature of things in perfect tranquillity, would find this out, so that men may not only get to know all but pretty well become masters of all![32]

---

29 Bacon, *Works*, II, 370. See, for example, Bacon, *Advancement of Learning*, 92–93; Bacon, *Works*, III, 571, 603; Bacon, *The "Instauratio magna" Part II*, 174. See also, Bacon, *Works*, 13.

30 See Giachetti Assenza, "Bernardino Telesio," 76: "Come nel *De principiis*, Telesio è considerato il restauratore della filosofia di Parmenide, ed è tuttavia ritenuto il migliore dei moderni."

31 Bacon, *Philosophical Studies*, 259.

32 Ibid., 247. Bacon cites the passage from Telesio thus: "Qui porro calor vel quantus, hoc est, quod caloris robur, & quæ ejus copia, quam Terram & quæ Entia in qualia invertat,

Not only did Telesio not rule out the development of human knowledge, he also held out hope for the development of human knowledge. Bacon comments that Telesio is

> more candid in this than his opponents usually are, who insist that it is quite impossible for an art to find out whatever the arts they themselves have created cannot find out, so that no art can be condemned, since it is itself judge in its own cause.[33]

Reading *De rerum natura*, Bacon was able to appreciate Telesio's style, full of modesty and humility, which were the virtues that the Lord Chancellor deemed essential in the intellectual world. Telesio presents himself to the

---

minime inquirendum videtur, ut quod homini nulla (ut nobis videtur) innotescere queat ratione. Qui enim vel caloris vires, & calorem ipsum veluti in gradus partiri, vel materiæ, cui inditus est, copiam quantitatemque distincte percipere, & certis determinatisque caloris viribus copiæque, certam Materiæ quantitatem, dispositionemque, certasque actiones, aut contra, certæ Materiæ quantitati, certisque actionibus, certam determinatamque caloris copiam assignare liceat? Utinam id otio fruentes & perspicaciore præditi ingenio, & quibus in summa tranquillitate Rerum Naturam perscrutari licuerit, assequantur: ut homines non omnium modo scientes, sed omnium fere potentes fiant!". See Telesio, *De rerum natura iuxta propria principia* (2013), 40. The passage also occurs in the 1570 edition: Telesio, *De rerum natura iuxta propria principia* (2013), 30.

33 Ibid., 247. This is an assessment that recurs constantly through Bacon's works. See for example Bacon, *The "Instauratio magna" Part II*, 15: "Even those very authors who have elbowed their way into a kind of dictatorship in the sciences and have laid down the law with such assurance, when after a while they come to their senses again, they take to whining about the subtlety of nature, the inaccessibility of truth, the obscurity of things, the intricacy of causes, and the weakness of human wit. But in this they are not the least bit more modest, seeing that they put the blame on the common condition of men and matter rather than confess to faults which are all their own. In fact their way is that when any art fails to achieve something, they insist that such achievement is impossible on the authority of the same art. But an art cannot be condemned when it pleads and judges in its own cause"; Bacon, *The "Instauratio magna" Part II*, 119–20: "Another sign should also be taken into account [...], i.e. the admissions of the very authors whom men now follow. For these authors, who pronounce so confidently on the nature of things, still change tack in their more sensible moments to grumble about the subtlety of nature, the obscurity of things, and the weakness of human wit. Now if this were all they did, other less courageous people might perhaps be deterred from further investigation, while others again with quicker and bolder wits might be roused and incited to make further progress. But it is not enough for them to admit defeat on their own account, but whatever is unknown or untried by themselves or their masters, they set beyond the bounds of possibility, and decree, as if their art told them so, that the thing is impossible to know or achieve. This is the height of pride and malice, and turns the weakness of their own discoveries into a slander on nature itself, and reduces everyone else to despair".

reader contrasting these virtues with the pride and arrogance of the Ancients. He thought that they placed too much faith in themselves, had competed with God in wisdom, had claimed to use reason to discover the principles of things, and had ended up imagining on their own terms. The Ancients, indeed, described nature not on the basis of the proprieties that it actually possesses, but on the basis of the properties that it *should* possess, according to the tenets of a way of reasoning totally unconnected with reality. Rather than observing things close up, they all too often limited themselves to merely discussing them. And yet words are a continuous source of error, since language is ambiguous and inadequate, whereas things, when examined in the correct manner (*recte perspectae*), reveal their properties. Guilty of pride, the Ancients "preceded nature" (*veluti naturae praeeuntes*) and superimposed their voice on nature.

Bacon sets great store in this lesson of Telesian thinking, and traces of this can be found across his writings, including the *Historia naturalis et experimentalis* (Natural and Experimental History), in which the philosopher is considered one of the many who "have come upon the stage with fresh stories, neither honored by approbation nor elegant in argument":[34]

> For we copy the sin of our first parents while we suffer for it. They wished to be like God, but their posterity wish to be even greater. For we create worlds, we direct and domineer over nature [*naturae praeimus et dominamur*], we will have it that all things *are* as in our folly we think they should be, not as seems fittest to the Divine wisdom, or as they are found to be in fact; and I know not whether we more distort the facts of nature or our own wits; but we clearly impress the stamp of our own image on the creatures and works of God, instead of carefully examining and recognising in them the stamp of the Creator himself. Wherefore our dominion over creatures is a second time forfeited, not undeservedly; and whereas after the fall of man some power over the resistance of creatures was still left to him—the power of subduing and managing them by true and solid arts—yet this too through our insolence, and because we desire to be like God and to follow the dictates of our own reason, we in great part lose.[35]

In the *Novum organum* he writes that the "anticipations of nature" that indicate "human reasoning which we currently apply to nature" (and this is "an impetuous and premature proceeding"), must be carefully distinguished from

---

34 Bacon, *Works*, v, 131.

35 Ibid., v, 132.

the "interpretation of nature", that is instead "that reasoning elicited from things by proper means".[36] We should not forget that:

> words clearly force themselves on the intellect, throw everything into turmoil, and sidetrack men into empty disputes, countless controversies and complete fictions.[37]

Man does not study nature with sufficient care. In the *Redargutio philosophiarum* he writes that an inappropriate and abstract approach to nature is like observing and contemplating nature from a distant and lofty tower, while one should instead descend from the tower and approach things directly, without expecting nature to come to us: we must be content that nature allows itself to be examined when we approach it with respect.[38]

Man can indeed aspire to truth, but on the condition that he is willing to change his own attitude to nature completely. Neither Telesio nor Bacon had an instrumental vision of science, and both held, in contrast to the culture of magic, that truth should be a collective asset. Telesio was not only alien to the Hermetic tradition but had also taken a clear stand against it. From the first edition of *De rerum natura* he refused to conceal truth and regarded doing so an "iniquity".[39] In all three editions of *De rerum natura*, he felt the need to express himself with particular clarity:

> We certainly, drawn by love of the truth and worshipping only this, have long examined the nature of things, unable to satisfy ourselves with what had been handed down by the ancients. Eventually, after having perceived, unless we deceive ourselves, the truth, we wanted to reveal it to men, believing that we would not have done our duty of an honest and free man if, denying humankind or fearing men's envy, we had kept it hidden.[40]

Telesio always opposed any notion of the secret nature of knowledge, which, in Rossi's words, "appears as a sort of prevailing paradigm in European culture".[41] He never adopted the distinction—between the tight circle of learned men

---

36 Bacon, *The "Instauratio magna" Part II*, 75.

37 Ibid., 81.

38 See Bacon, *Works*, III, 581–82.

39 Telesio, *De natura iuxta propria principia* (2011), 13.

40 Ibid., 201–202; Telesio, *De rerum natura iuxta propria principia* (2013), 88.

41 Rossi, *Il tempo dei maghi*, 156.

and the mass of the unlearned—on which this secrecy was based. Instead, he insisted on its limitations and accepted, of his own accord, that his work contained "nothing divine and worthy of admiration", being satisfied that it was not contradictory with respect to reality.[42] These are positions which would certainly not have left Bacon unresponsive.

De natura opens with the polemical and disruptive statement "Bernardinus Consentinus haec cogitabat" (Bernardino of Cosenza thought this).[43] "Franciscus de Verulamio sic cogitavit" (Frances of Verulam reasoned thus) is the equally polemical and disruptive beginning of *Novum organum*.[44] In the *Cogitata et visa* (Thought and Seen), which not by chance begins with "Franciscus Bacon sic cogitavit" (Francis Bacon reasoned thus), we find that the first thing that a man wanting to discover something does is to search for and consult everything that others have said on the argument; only then does he add his own thoughts. But relying on authority is a procedure without grounds.[45] With Telesian expressiveness, Bacon writes in the *Praefatio* to the *Instauratio magna*:

> And the same humility that I use in discovering, I also use in teaching. For I do not seek either by victory in debate, appeals to antiquity, any arbitrary claim to authority, or even by cloaking myself in obscurity, to dignify or commend any of my discoveries with any majesty; which is the sort of thing anyone could easily do if he were trying to aggrandise himself rather than enlighten the souls of others. I have not (I say) sought nor do I plan to ensnare men's judgements by force or fraud; instead I want to lead them to the things themselves and their interconnections, so that they can see for themselves what they possess, what they may assert, and what they may add and contribute to the common good. For my own part, if I have wrongly given credit to anything, or grown sleepy and inattentive, or become weary on my way and left the investigation unfinished, I nevertheless make the things plain for all to see, so that my mistakes can be spotted and separated out before the body of science is further infected by them, and also so that my labors can be carried on easily and expeditiously.[46]

---

42 Telesio, *De rerum natura iuxta propria principia, libri IX*, 7, 14.

43 Telesio, *De natura iuxta propria principia* (2011), 45.

44 Bacon, *The "Instauratio magna" Part II*, 2.

45 Bacon, *Works*, III, 608–9.

46 Bacon, *The "Instauratio magna" Part II*, 21.

6 The Telesian Galileo

Bacon was certainly not the only great "modern" to grasp the explosive nature of the Telesian critique of the philosophical tradition in general and Aristotelianism in particular, just as he was not the only one to see Telesio as a figure of great importance. Obviously, Galileo denied ever having read Telesio, an author who had been placed on the *Index Librorum Prohibitorum*.[47] This denial is clear in both *The Assayer* (1623), and in one of the many marginal notes devoted to the *Ratio ponderum librae et simbellae* (1626) by the Jesuit Orazio Grassi who had made heavy insinuations about Galileo's work. It is nevertheless significant that the juxtaposition of Galileo and Telesio appears in both Aristotelian and anti-Aristotelian contexts. This is the case, respectively, with the Florentine philosopher and man of letters Ludovico Delle Colombe,[48] and with Tommaso Campanella,[49] to mention two examples. Furthermore, we know that Galileo had shown interest in one of Telesio's greatest pupils and followers, Antonio Persio, who was elected as a member of the Accademia dei Lincei *post mortem* with the approval of Galileo himself. And yet it was actually the founder and patron of the Academy, Federico Cesi, who showed more interest in Persio than Galileo did. He considered his work important and intended to publish it; he must have also judged the works of Persio's master important, as he owned both Telesio's *De rerum natura* (1570 and 1587) and *Varii de naturalibus rebus libelli* (1590).[50]

---

47 See Galilei, vol. VI, 236, 398: "Quello che abbiano scritto il Cardano e 'l Telesio, io non l'ho veduto"; "[…] per me, non ho mai fatto studio in tali autori".

48 See Galilei, *Opere*, vol. III, 253: "L'ambizioso animo umano, sospinto oltre ogni convenevole termine dal desiderio dell'immortalità, venutagli a stomaco la navigazione dell'oceano della veritade, s'ingolfa nel mar della bugia, sprezza le Colonne d'Ercole, schernendo Aristotele e beffeggiando Platone, grida *plus ultra*, in sin tanto che va a dare in non conosciute sirti, e romper in non antiveduti scogli. Imperò che alcuni, disperandosi d'intendere Aristotele, o di poter dir cosa che porti lor nominanza nella di lui filosofia, si risolvono, non potendo far pompa, come si dice, a far foggia, e di negare tutte le sue verità, altrettante menzogne opponendole, con ritrovare anzi sognar nuova filosofia e modo nuovo di filosofare. Tali già furono alcuni antichi, e a nostra etade i Telesisti. Altri, in niuna filosofia avendo fondamento, si danno alle matematiche, e quelle predican per sovrane sopra tutte l'altre facultà" (*Discorso apologetico d'intorno al Discorso di Galileo Galilei, circa le cose, che stanno su l'Acqua, o che in quella si muovono*).

49 Campanella, *Lettere*, 190: "Sic ergo stant principia doctissimi Telesii nostri" (*Letter to Galileo*, 13 January 1611).

50 See Galluzzi, *Libertà di filosofare in naturalibus*, 83–97.

The proximity between Galileo and Telesio is less surprising than it appears at first glance. To see how this is so it is worth first recalling some general theories. For example: faith in the uniformity of nature, which is clearly set out by Telesio in *De rerum natura*:

> I exclusively followed our senses and nature, which is always coherent to itself, always does the same things in the same manner, and always operates in the same way.[51]

In other words, Telesio embraced the theory of the progressive nature of human knowledge. In the same pages in which Telesio expressed his faith in the constancy of nature, he forcefully rejected the idea that knowledge was exhaustively summed up in Aristotelian texts. He thought there was still a great deal to be explored in the world, but that man had to be willing to reject the bookish culture and return to the observation of physical things. Aristotle had to be debunked. He wrote in *De rerum natura* that one should have the courage to dissent (*dissentire audeamus*) from the greatest interpreter of nature. We must realize and accept that respecting the greatest philosopher of antiquity does not mean accepting his doctrines as if they had come from the mouth of nature herself (*ex ipsius naturae ore*). Aristotle was not a god and did not deserve, as no man deserved, to be venerated more than the truth.[52] Delle Colombe spoke of Aristotle as one who held the *summa* of philosophical truth, adding that *natura locuta est ex ore illius*.[53] Benedetto Castelli replied that Galileo's philosophy did not mean attacking Aristotle but only speaking the truth. Galileo was driven solely by the "desire for truth" that "he places before all other things" as he thus "places nature itself before the authority of any famous writer".[54]

51 Telesio, *De rerum natura iuxta propria principia, libri IX*, 14: "Sensum videlicet nos, et naturam, aliud praeterea nihil sequuti sumus, quae perpetuo sibi ipsi concors, idem semper, et eodem agit modo, atque idem semper operatur". See also Telesio, *De rerum natura iuxta propria principia* (2013), 7–8. See Garin, *History of Italian Philosophy*, 429.

52 See Telesio, *De rerum natura iuxta propria principia, libri IX*, 88. See also Telesio, *De natura iuxta propria principia* (2013), 201; Telesio, *De rerum natura iuxta propria principia, liber primus, et secundus, denuo editi—Opuscula*, 87–88.

53 See Galilei, *Opere*, vol. IV, 317–318 (*Discorso apologetico d'intorno al Discorso di Galileo Galilei, circa le cose, che stanno su l'Acqua, o che in quella si muovono*).

54 See Galilei, *Opere*, vol. IV, 466–67 (*Risposta alle opposizioni del S. Lodovico delle Colombe, e del S. Vincenzio di Grazia, contro al Trattato del Sig. Galileo Galilei, delle cose che stanno su l'Acqua, o che in quella si muovono*). See Shea, *Galileo's Intellectual Revolution*, 33–34. The text was largely written by Galileo himself.

7 Like an Eagle

Historically, the most relevant encounter between the work of Galileo and Telesio took place during the famous debate on comets. It was, as we know, a very bitter debate, but one that in no way detracted from the validity of the encounter itself: the positions of Galileo and those of Telesio were in fact particularly close to one another. The relevant Telesian text is the *De cometis et lacteo circulo* (*On Comets and the Milky Way*) (written around 1580), which, as Miguel A. Granada has observed, must be considered "an important contribution" to the debate "on heavenly novelties and their cosmological implications".[55]

In the nine chapters of this *libellus* Telesio ended up rejecting the Aristotelian theory of comets completely, with serious consequences at the cosmological level. Telesio discussed the *nova* of Cassiopeia—visible between 1572 and 1574, and which he believed to be a comet, along with many other authors of the time—and discussed the comet of 1577, recognizing in both cases their heavenly location. Hence, contrary to Aristotle, the comets were not exclusive sublunary phenomena (*quae cometis attribuit Aristoteles loca, et non recte ea attributa esse*);[56] and not only that, they were once again seen as the opposite of what Aristotle claimed: comets were not real objects, but a phenomena produced by the reflection of the sun's light on exhalations that rose up to the sky (*cometem solis lucem esse a vaporibus relucentem*).[57] This discussion gave rise to a subversive vision of cosmology, which had already been variously defended long before 1580, that rejected the traditional distinction between the celestial sphere and the sublunary sphere in favor of the existence of a unitary and homogeneous universe. It was a total subversion of Aristotelian cosmology and one that would inevitably invite both attacks and enthusiasm.

Here it is worth remembering that the theory of comets expounded in *De cometis et lacteo circulo*, a theory then shared by several authors, coincided with what Galileo had expounded "to an audience of over a thousand" during three lessons held at the University of Padua from late November to early December 1604 on the *nova* which had appeared some weeks earlier.[58] In his view, the aim of these lessons, of which unfortunately only a few fragments remain, was to demonstrate that "the place of the new star" was "far above the lunar orb".[59]

55 Granada, "Introduzione," xx. See also Lerner, "La physique celeste de Telesio."

56 Telesio, *Varii de naturalibus rebus libelli* (2012), 28–32.

57 Ibid., 12.

58 See Galilei, *Opere*, vol. II, 267–84. See Camerota, *Galileo Galilei e la cultura scientifica*, 118.

59 Galilei, *Opere*, vol. X, 134.

Galileo defended the odds that the Earth produced an evaporation which formed a huge mass, that of the *nova* in question, and that its brilliance was due to the reflection of the sun's rays.[60]

It is also worth recalling that Galileo explained comets years later in the same way. It is difficult to imagine that he was surprised by the comparison between his theories and those of Telesio (and Cardano), which is what the Jesuit Orazio Grassi did with "malice". In *Discorso delle comete* (1619) by Galileo-Guiducci, he assimilated comets to "simulacra" and defined the "distinction, or rather contrast, between the elements and the heavens" as "vain". He also emphasized the likelihood that the sublunary comet matter reached the heavens spreading out "in the immense spaces of the universe".[61] Comets end up representing "a material bond between the terrestrial and celestial region".[62] Galileo was in fact affirming the theory of the homogeneity of the cosmos, while Grassi, in *De tribus cometis anni MDCXVIII disputatio astronomica* (Astronomical Disputation on the Three Comets of the Year 1618) (1619), had emphatically rejected the likelihood that comets came from "ex huius Terrae sordibus" (from the mud of this earth).[63]

In the *Libra astronomica ac philosophica* (1619), Grassi provocatively likened the positions taken by Galileo to those of authors who—not by chance—had been condemned by the Church, such as Telesio and Cardano;[64] followers of a "sterile and unhappy" philosophy, who had left for posterity not "sons" (*libe-*

---

60 Ibid., vol. II, 282, 283. Among the most interesting passages, we find the following: "Stellam novam non fuisse incendium, patet ex eo quod quae citissime incenduntur, brevi quoque extinguntur: materia, enim, quae statim incenditur, est admodum combustibilis; quod vero maxime est combustibile, statim comburitur. Exempla de hac re habentur innumera" (280); "Quod circa terram eleventur vapores, qui, ascendentes, solis lumen reflectant, saepissime apparet" (281); "Alcuni fuochi, che da lontano appariscono splendentissimi, da vicino non si veggono niente, per la loro tenuità: così la Stella nuova può essere una esalazione illuminata; e chi vi fosse vicino non la vedrebbe, ed apparirebbe solo come i vapori elevati ed illuminati la notte" (282); "Non esse absurdum, talem condensationem ponere in caelo, cum et circa Lunam, veluti circa terram, consimilem videamus" (282); "Ut luceat haec nova Stella ut reliquae, non infert ut debeat esse solidissima substantia, veluti creduntur illae: eadem, enim, fit reflexio et a solidissimo corpore atque a tenuissimo, ut a nubibus, etc." (282); "Qua ratione possit a terra maxima fieri evaporatio, ut moles adeo immensa, quali erit Stella nova, conficiatur, nullam habet impossibilitatem" (283).

61 Galilei, *Opere*, vol. VI, 71, 93. See Galilei, *Opere*, vol. VI, 278: "Quello che ha scritto il Sig. Mario è, che non ha per impossibile che tal volta possano elevarsi dalla Terra essalazioni ed altre cose tali, ma tanto più sottili del consueto, che ascendano anco sopra la Luna, e possano esser materia per formar la cometa" (*The Assayer*).

62 Heilbron, *Galileo*, 238. See also Bucciantini, *Galileo e Keplero*, 270–271.

63 See Galilei, *Opere* VI, 33.

64 See Baldini and Spruit, *Catholic Church and Modern Science*, vol. 2, 1033–1472, vol. 3, 2415–2425. See Redondi, *Galileo Heretic*.

ros), but "books" (*libros*).[65] Galileo commented on this passage concentrating on the question of the lack of followers, noting the absurdity of thinking that the secrets of philosophy, which are only completely known by God, could be understood by many, even by unlearned men.[66] He only focused on Grassi's heavy insinuation later, however, when Grassi reinforced the criticism in his *Ratio ponderum librae et simbellae* (1626).[67] Galileo was cautious, stating that he had never studied Telesio or Cardona, but still defended their philosophical views against allegations of impiety. He questioned Grassi's affirmation that seeing a comet as an appearance also implied denying divine providence or even God himself.[68]

In *The Assayer*, however, Galileo did not back down from defending these two condemned authors that he insisted he had never read. Galileo wrote:

---

65 See Galilei, *Opere*, vol. VI, 118. See also references to Galileo and to Telesio in Fortunio Liceti, *De novis astris et coometis libb. sex*, 47–48. On relations between Galileo and Liceti, see Guerrini, *Galileo e gli aristotelici*.

66 This is the passage in full: "Et hoc praestantiae et dignitatis maximum est argumentum. Videmus enim et leonum et generosissimorum animalium minus numerosas esse proles; sed murium et fatuorum piscium, and contra, etc. Et quis unquam credat, vere sapientum maiorem esse numerum quam insipientum? [P]hylosophiae autem, quam inte[g]re solus Deus intelli[g]it, arcana erunt [p]opularia? Etc" (Galilei, *Opere*, vol. VI, 118–19).

67 See Galilei, *Opere*, vol. VI, 397–398: "At quae porro pietas est, Galilaee, eorum partes suscipere (dicam multo quam antea liberius) quorum damnata multis partibus, ut parum catholicis rebus amica, philosophia aeterna potius oblivione sepelienda iam fuerat? En, ut illos ingenii illa libertas toties decantata, commendata toties, sine lege excurrentes ac vagos tandem devolvit in praeceps! Hae sunt doctrinae hominum et daemoniorum, ut cum Tertulliano loquar, prurientibus auribus natae, de ingenio sapientiae saeculi, quam Dominus stultitiam vocans, stulta mundi in confusionem etiam philosophiae ipsius elegit: ea est enim materia sapientiae saecularis temeraria interpres divinae naturae et dispositionis. Cardanum igitur ac Tilesium laudent qui volent, ego laudationum religiosius argumentum exposco."

68 See Galilei, *Opere*, vol. VI, 398: "Poco appresso spacciate me per Epicureo, con la giunta (per ben bene specificar la vostra intenzione) "il quale o totalmente negava Dio, o almeno la sua Provvidenza"; altrove mi fate seguace di Telesio e di Cardano, con la dichiarazione, per chi non lo sapesse, che la loro filosofia e dottrina è parimenti dannata. Le proposizioni poi per le quali voi volete registrarmi tra i Telesiani e Epicurei, sono (per quel che dite voi stesso, ché io, per me, non ho mai fatto studio in tali autori) che la cometa è una pura apparenza e che il moto è causa di calore, intendendo che il moto sia in quella materia che ha da scaldare, e non in quella che deve essere scaldata, come intendono i Peripatetici. Adunque, se voi non volete confessare d'esser mosso da altro che da un puro zelo a darmi simili note, è forza che diciate di tenere veramente, sinceramente e internamente, che l'affermare che la cometa è un puro simulacro, e che il moto nel riscaldante è causa di calore, sieno 2 proposizioni empie e dannande, come il negar la Provvidenza divina o l'istesso Iddio. Ah, che dovreste vergognarvi! E sotto qual maschera volete nasconder la vostra arrabbiata malignità?".

> Now Sarsi rises up in arms and passionately strives to prove that this suggestion is beside the point and false to boot. Yet in order to be prepared for anything (lest the idea appear worthy of some consideration), he robs me of any possible credit by calling this "an ancient notion of Cardano and Telesio", which his teacher disparages as a fantasy of feeble philosophers who had no followers. And under this pretense, without the least shame for his disrespect, he robs those men of their reputations in order to cover up a slight oversight of his teacher's.[69]

Telesio and Cardano are two *venerandi padri* whose value could certainly not be measured by the number of their followers:

> Perhaps Sarsi believes that all the host of good philosophers may be enclosed within four walls. I believe that they fly, and that they fly alone, like eagles, and not in flocks like starlings. It is true that because eagles are rare birds they are little seen and less heard, while birds that fly like starlings fill the sky with shrieks and cries, and wherever they settle befoul the earth beneath them.

Grassi should know that "the crowd of fools who know nothing is infinite", that "those who know very little of philosophy are numerous", that "few indeed are they who really know some part of it, and only One knows all". Galileo insisted:

> I consider it not very sound to judge a man's philosophical opinions by the number of his followers. Yet though I believe the number of disciples of the best philosophy may be quite small, I do not conclude conversely that those opinions and doctrines are necessarily perfect which have few followers, for I know well enough that some men hold opinions so erroneous as to be rejected by everyone else. But from which of those sources the two authors mentioned by Sarsi derive the scarcity of their followers I do not know, for I have not studied their works sufficiently to judge.[70]

For reasons that are obvious Galileo once again emphasizes that he is unacquainted with the works of Telesio (and Cardano), but is courageous enough to make his thoughts clear; he who Francis Bacon had defined as the first of moderns was an eagle and a *venerando padre* for Galileo.

---

69 Galilei, *Discoveries and Opinions*, 239.

70 Ibid., 236–237.

8 Concluding Remarks

Bacon and Galileo both held Telesio in high esteem despite the distance that they had established between themselves and him. As for Bacon, we are well acquainted with his critique of Telesio: he blamed him for developing a "pastoral philosophy", a philosophy of contemplation and idleness to be rejected, and for having not entirely shaken off Aristotle's principles. As for Galileo, we can imagine the effect that reading works that attached no importance to mathematical knowledge would have had on him.

Nonetheless, both Bacon and Galileo showed an interest in many themes expounded and discussed in Telesian works, such as the heat/cold dialectic, the doctrine of the soul, the theory of comets, and the nature of heat and its relation to motion. More crucially, both of them considered Telesio as the first or at least one of the main *novatores*. They appreciated his break with tradition, his empiricism, and his anti-intellectualism. According to Bacon, Telesio was, above, all, a restorer of the philosophy of Parmenides. Bacon and Galileo understood one key issue, namely that Telesio had made a decisive contribution to changing the old way of thinking.

Acknowledgements

I wish to thank Clare Tame for her translation of the Italian text and for her patience, and Pietro Daniel Omodeo for his advice and suggestions.

CHAPTER 2

"*Spiritus*" and "*anima a Deo immissa*" in Telesio

Miguel Ángel Granada

According to Telesio, man has a *spiritus*, just like the animals do, and it is derived from the semen (*e semine eductus*). Its origin is thus completely natural, since it appears during the generative process of the individual. The human *spiritus* is not essentially or qualitatively different from that of animals; at most, it shows a difference of degree and is responsible, through sensation (*sensus*), for knowledge and the preservation of life. As in animals, the human *spiritus* perishes along with the decomposing body that it animates. Yet in man (and only in man) there is also a soul infused by God (*anima a Deo immissa*), and thus created immediately by God. This soul is immaterial and immortal.

The doctrine of the soul was a decisive factor in the ecclesiastical condemnation of Telesio's works, insofar as the mortal *spiritus* seemed to clash with the Christian concept of the immortal soul; and the *anima a Deo immissa* was perceived as a subterfuge intended to conceal real unbelief. In 1596, the great treatise *De rerum natura iuxta propria principia*, whose third (considerably expanded) edition had appeared in 1586, and two pamphlets published posthumously in 1590 (the ponderous *Quod animal universum ab unica animae substantia gubernatur. Contra Galenum*, and the short satellite-pamphlet *De somno*) were included in the *Index librorum prohibitorum* with the clause "*donec expurgetur*". And yet this expurgation, requested in 1601 by the philosopher's nephew, was declared "*impossibilis*" by the commission set up for this task.

Our intention is to analyze Telesio's conception of the soul as he presented it in complex and various ways in the three very different editions of *De rerum natura* (published in 1565, 1570, and 1586), as well as in the opuscules cited, and to examine the diverse and conflicting interpretations of this conception left to us by twentieth-century historiography. Among these interpretations we find, at one extreme, Telesio's sincere Catholicism, defended by Luigi De Franco (possibly the most devoted and unremitting Telesio scholar of the past century), and at the other the veiled heterodoxy advocated by Roberto Bondì in later years. We intend to explain Telesio's position as objectively as possible, without ignoring the possible intended ambiguities in the presentation of his doctrine. We will then draw some conclusions concerning the coherence and substance of his doctrine of the two souls.

 | DOI:10.1163/9789004352643_004

1 The Soul in *De natura iuxta propria principia* (1565) and in *Quod animal universum*

A first reading of *De natura* (1565) reveals that the question of the soul is all but absent, since Telesio focuses on the fundamental question of the constitution of the universe. In the few places where the topic is addressed, the soul is always understood as *spiritus* produced by the semen (*e semine eductus*). At no time does Telesio consider it necessary to introduce a clarification or an explanation indicating the existence of another soul (in this case immortal and immaterial, *anima a Deo immissa*). Indeed, he proceeds as if the term *anima* is unequivocal and simply indicates the *spiritus*.[1] Thus, in the only chapter in which the *spiritus* is the subject of a very detailed analysis (Book I, Ch. 8), Telesio states that it is situated in the ventricles of the brain and spreads throughout the body via the nerves. Sensation (*sensus*) takes place through it.

Despite its brevity, the description of *spiritus* given in 1565 clearly indicates its inclusion in the medical tradition but also reflects a complete departure from the astrological connection of *spiritus* (revived by Ficino and Renaissance Platonism), consistent with Telesio's disdain for astrology.[2]

We also find a presentation of *spiritus* in *De natura* [DN] II, 49, where Telesio refers to the broader explanation in his treatise against Galen. The reference proves that the anti-Galen (and anti-Plato) pamphlet already existed in 1565. What is of interest to us here is that in 1565 the only soul that Telesio contemplated was the *spiritus* produced by the semen, whose location is the brain, from which it spreads via the nerves throughout the entire body of the animal and man. The extensive treatise *Quod animal universum ab unica animae substantia gubernatur* provides full confirmation of this. I will merely

1 *Anima* also indicates, logically, the Aristotelian soul in the places in which Telesio describes or criticizes the peripatetic doctrine of the soul.

2 See, for example, the disparaging attitude he shows in the opusculum *De cometis et lacteo circulo*. Cf. Telesio, *Sobre los cometas y la Vía Láctea*, 23. On the astrological conception of the *spiritus* in Ficino as *vehiculum animae* or astral body of the soul (enveloped by it in its descent to the body via the celestial spheres), see Ficino, *Commentaire sur Le Banquet de Platon*, 134–135. Ficino provided a development extended to the astrological-cosmological dimension of the *spiritus* in his *De vita coelitus comparanda*, the third of the *Libri de vita* dedicated to the treatment of the melancholic *spiritus* of intellectuals. Cf. Ficino, *Three Books on Life*. As we shall see, although Telesio is completely alien and indifferent to Ficino's concept of *spiritus*, his conception of the soul *immissa a Deo*, as formulated in the third edition (1586) of *De rerum natura*, appeals to the theory of the rational soul expounded by Ficino in his *Theologia platonica de immortalitate animorum*. For a brief summary of the complexity of the concept of *spiritus* in the Renaissance, cf. Garin, "Il termine '*spiritus*'." Walker, *Spiritual and Demonic Magic*, 189ff., remains essential.

say that the objective of the treatise is to refute the Platonic and Galenic conception of the tripartite soul in favor of a single soul or *spiritus* deriving from the semen, located in the brain and operating via the nervous system. We see this is so in the title of Chapter 6, which states that "spiritum in cerebri ventriculis contentum animalium animam esse" (the *spiritus* that is contained in the brain ventricles is the animals' soul)[3] and insists that the *spiritus* carries out all the functions that "the soul seems to act and put to completion", from which it can be inferred that the latter is reduced to the former.[4] In logical consonance with this doctrine, which is the cornerstone of the whole treatise (the only soul that governs man is the *spiritus* residing in the brain), a parenthesis at the beginning explicitly states the proposition with an additional set of problems.

This parenthesis clarifies that the soul discussed throughout the pamphlet is the *spiritus*, which is also the soul mentioned occasionally in *De natura* (as we have seen). How then is it possible to state that "in the comments themselves [*De rerum natura*, in this case], what has been presented more fully [is] the doctrine of that other soul that is its [of the *spiritus*] form, as well as of the whole body, [and] whose substance is infused in men by God?"[5] This new soul is obviously the *anima a Deo immissa*, but the "comments" in which it is explained in detail can only refer to the third edition (1586) of *De rerum natura iuxta propria principia* because the version of the topic that appears is the same as the one that appeared in 1565. Hence, everything suggests that the parenthesis is an addition subsequent to the original drafting of *Quod animal universum*. Naturally, this must be an addition by Telesio himself,[6] introduced when the "comments" in question (actually the theory of the *anima a Deo immissa*) were written. However, we know (albeit without much precision regarding the actual content and the date) that in the 1560s this part of the final work contained a text that is unknown to us. It is therefore equally possible

---

3 *Quod animal universum ab unica animae substantia gubernatur. Contra Galenum* [QAU], in Telesio, *Varii de naturalibus rebus libelli* [1590] (2012), 150. Cf. ibid., VII, 154: "Itaque animae substantiam spiritum esse liquido patet, quocum scilicet et sensus et motus et ipsa recedit vita, et quocum et sensus et motus et ipsa redit vita."

4 QAU, 154: "qui [*spiritus*] igitur ad motus quosvis aedendos promptissimus sit et qui vel a languidissimis rerum viribus immutari, et vel angustissimos nervosi generis poros subire inque iis inhabitare et per eos, ut libet, meare queat, ad ea omnino omnia, quae operari praestareque anima videtur, operanda praestandaque aptissimus cum sit *spiritus*, quin animae is substantia et ipsa omnino sit anima, nulli dubium esse potest."

5 QAU, I, 144: "cuius [spiritus], corporisque universi, ut in iisdem amplius expositum est commentariis, quae a Deo in homines immittitur substantia, forma sit." Cf. also Bondì, *Introduzione a Telesio*, 6–7.

6 Or (though this is not very probable) an addition by Antonio Persio on the occasion of the publication of *Libelli* in 1590.

that the doctrine of the *anima a Deo immissa* may or may not have been present in 1565.

Just to complicate matters further, towards the end of the *De natura* of 1565 a passage introduces a discordant note in the melody, which until then had consistently equated *anima* and *spiritus*. In the middle of the refutation of Aristotle's separate motive intelligences, Telesio now introduces a short chapter (56) in which he forcefully affirms that the ordered structure of the world "demonstrates both that there is God and that He is the creator of all things, as said, while it seems arrogant more than foolish to investigate His substance or others of His activities, as these are the sort of things that cannot in any way be known by [any] human being".[7] This profession of metaphysical skepticism allows Telesio to relate the epistemological limits of our "soul" to the immortal destiny reserved for it:

> God has revealed to us what he wished that we know of Him. Let us be content with this and not dare inquire, through our own forces and reasons, about anything else that far surpasses the acuteness of our mind and is much more brilliant than what the human eye can perceive. All this will be seen, I *hope*, when our soul [animus], free from the bonds of the body, shall fly to the very same God who created it.[8]

It is easy to see that Telesio is not speaking here in terms of experience or of an inference based on experience but is expressing a "hope", that is, he has made the leap toward the profession of his (presumably sincere?) religious faith. Therefore, the discussion is heterogeneous or of another order with respect

7 *De natura* [DN], II, 56, in Telesio, *De natura iuxta propria principa* [1565] (2006), 186–187: "Deum esse et rerum omnium conditorem non caeli motus, at is caeli motus, eaque caeli a terra manifestet distantia, quibus, ut dictum est, provisum videtur, ut longissime etiam exuperans terram sol, nullam ipsius partem exurat tamen et nullam non illustret vivificetque et nulli non sua largiatur bona. Et ipsum per se caelum, longe maximum longeque conspectum pulcherrimum. Et paulo reliqua entia quaevis minus, animalia praesertim eorumque constructio et partium singularum usus. Haec quaecunque [sic; read 'quicunque'] intuitus fuerit, quantumvis ferus impiusque et stupidus etiam, nequaquam fortuito, aut casu constructa suspicari aut queat aut velit, sed vere divino omnia consilio. Haec et Deum esse et rerum omnium conditorem, ut dictum est, manifestare queant, ipsius substantiam aut operationes alias inquirere, non arrogans magis, quam stultum videtur, ut quae homini innotescere nullo queant pacto."

8 "Quae sui sciri voluit, ea ipse nobis manifestavit Deus, iis contenti nequaquam nostris alia viribus nostrisque inquirere audeamus rationibus, longe *animi* nostri aciem exuperantia et longe lucidiora quam quae humanus intueri queat oculus, ibi *spero* intuenda omnia, ubi corpore exolutus *animus*, ad ipsum, a quo est factus, evolarit Deum", ibid., 187; translations and italics are always mine unless otherwise indicated.

to the naturalistic and sensualistic discussion developed up to that moment. The problem that arises is whether we are dealing with a merely precautionary, rhetorical declaration meant to confront and neutralize possible charges of impiety[9] or with the realization of the limit of natural philosophy, which is superseded either by the metaphysical discipline (which Telesio does not intend to discuss but does not deny as a possibility)[10] or simply by faith and revelation.[11] We cannot support Giovanni Di Napoli's hypothesis that Telesio does not deny the possibility of a "scientific" metaphysics but simply confines himself to the physical plane, leaving the superior discipline to those who feel they have the strength to realize it. Rather, we believe that passages such as the above-mentioned *De natura* II, 56 clearly affirm Telesio's full conviction that metaphysics is an impossible task[12] and that beyond correctly performed natural philosophy (i.e. according to the naturalistic and sensualistic perspective) there is space only for religious revelation, in so far as it is to be accepted.

2 The First Accusations of Impiety and the Second Edition of *De rerum natura* (1570)

In April 1570 Telesio sent a very significant letter to Cardinal Flavio Orsini, Archbishop of Cosenza, reporting some unfounded accusations of impiety levelled at him on the basis of some passages supposedly contrary to religion in *De natura*. One of these accusations referred precisely to the doctrine of the soul: "in my work, printed in Rome five years ago, [...] there are other statements contrary to religion. From them one may infer that I intend that the soul is mortal and deny that the heavens are moved by intelligences."[13] In his letter, Telesio defends himself against the accusations, noting that his (notably

9 This is the interpretation seemingly proposed by Roberto Bondì. See Bondì, "'*Spiritus*' e 'anima' in Bernardino Telesio", 413 and 417. Cf. also Bondì, *Introduzione a Telesio*, 36–39 and Bondì, "Dios, naturaleza y alma en Telesio."

10 This is clearly the interpretation proposed by Di Napoli in *L'immortalità dell'anima*, 389–403.

11 Telesio's sincere religious conviction is the hypothesis vigorously supported by De Franco, according to whom Telesio is "un cattolico osservante, il quale non ebbe mai a nutrire dubbio alcuno sulle sue credenze religiose." Cf. De Franco, "Telesio e la *libertas philosophandi*", 113. Cf. also De Franco, *Introduzione a Bernardino Telesio*, 143f.

12 Therefore, Telesio shows total scepticism about metaphysics or rational theology. On the sceptical component of his thinking, cf. Mulsow, *Frühneuzeitliche Selbsterhaltung*, 179–183.

13 "Nell'opra mia stampata già cinque anni in Roma [...] ci son altre propositioni contra la religione. Et dalle quale si può cavar ch'io metto l'anima mortale, et che negho che'l Cielo sia

enlarged) work is being reprinted at that very moment ("avendola io repolita alquanto, et ampliatola assai [...] già si stampa") after undergoing ecclesiastical examination and after being "approbata senza difficoltà", just as the Roman edition of 1565 had been approved by the theologians who examined it. Telesio reassures the Cardinal of what we already know: that the work deals with cosmology and that the problem of the soul is addressed only occasionally and always in reference to the "sensitive" and "motive" soul (i.e. the *spiritus*).[14] Telesio adds: "the entire doctrine of it [the soul] is presented in the following books, already finished, and [...] they should be revised and completely ready for print; with God's help they will be printed around October, I think in Rome. And it is my hope that they shall not be found contrary to religion, but most conform to it."[15]

According to Telesio's account, the work was now finished and ready for printing, which the philosopher expected to occur in October of the same year. At that time, the full doctrine of the soul would be offered to the public and it would be possible to ascertain what his real position was in the 1565 edition and in the reprinted version which was in progress.[16]

The complete edition of the work was postponed until 1586. Nevertheless, there is no doubt that, whatever the doctrine of the soul in 1565, it had been put down in writing in 1570 in what Telesio considered to be a sufficient manner. Meanwhile, in 1570 the second edition of his work was published in Naples under the title *De rerum natura iuxta propria principia, liber primus et*

---

mosso dall'intelligentie." I quote the letter from the complete transcription in De Franco, *Introduzione a Bernardino Telesio*, 66.

14 Ibid., 67: "veramente in questi doi libri [of the edition in print and of that from 1565] non si tratta d'altro, che de li primi corpi et de li principij cioè caldo freddo, umido e secco. Dell'anima si dicono pochissime cose. Et quelle sole, ch'appartengono alla materia delli principi, et all'anima sensitiva et motiva."

15 "L'intera dottrina d'essa se tratta nelli libri, che sequeno, quali son già finiti, et [...] già sarebbono revisti, et posti in ponto per posserse stampare, ma con l'aiuto di n(ostro) S(igno)re Dio stamparono verso Ottobre, et penso in Roma. Et spero non solo non parranno [contrari] alla religione, ma conformissimi", ibid. Telesio again adds a profession of sincere Catholic faith and submission to the Church, on which we should not exclusively base our judgment if we bear in mind the normal cautionary function of such declarations by authors clearly suspected of heresy. Cf. De Franco, *Introduzione a Bernardino Telesio*, 68: "la mente mia [...] sarà sempre soggettissima et inchinatissima alla vera et cattolica religione, et sarei prontissimo ad abbruggiar tutte le mie opere, quando mi fusse mostro che non siano piene di pietà christiana."

16 "Vers'Ottobre [...] disegno venirci, et per [...] farci stampare il restante delle cose mie et tanto miglior gioditio si potrà far di questi libri stampati, quanto che se potrà largamente vedere la mente mia in quelli ristampati", De Franco, *Introduzione a Bernardino Telesio*, 68.

secundus, denuo editi.[17] This was a largely new work, not only because it had been corrected in terms of style and expanded but also because the material itself had undergone a profound redistribution. With regard to the doctrine of the soul, there is no novelty except for the clarification (Book I, Chap. 31) that the work is limited whenever the soul is mentioned (in truth, only a few times), to treating the soul *e semine*, as distinct from the soul infused by God in men's bodies ("a Deo ipso in singula hominum corpora infusam"), whose doctrine is presented "in its proper place", which is not the present edition. Here is a passage that provides an interesting consideration on the origin of the *spiritus* in the action of the Sun on the Earth:

> Therefore, the Sun (which is hot to the highest degree and acts upon the Earth, which is cold and heavy to the highest degree, being far removed from [the Earth] and acting only for a short time upon each part of it [the Earth] since it [the Sun] is carried with a swift, continual, and always varying motion) does not transform into itself the parts of the Earth that it changes, nor into something very similar or akin to it, just as fire does not change into fire everything upon which it acts. Nevertheless, the Sun still removes them from the [nature of the] Earth, and renders them after its own nature, that is, it gives them all some heat and a certain tenuity (meaning that it makes them share in its own substance and form). And some of these parts are made very hot and very thin, such as those transformed into extremely hot and tenuous exhalations or into the very souls of animals. But let nobody think that I say this with respect to the human soul infused by God himself into the individual bodies of men, such as we are taught not only by Sacred Scripture but also by human reason (as set forth in further detail in its proper place).[18] Rather, I say this only with respect to that soul which is produced from the semen and can be considered a work of nature. I will always speak here only of the latter.[19]

In addition to this, the second edition presents the aforementioned passage of *De natura* II, 56 without any substantive change but with a careful revision and refinement of style. Precisely for this reason it is surprising that Telesio retained the imprecise term *animus* and did not replace it with the

17 I refer to this edition (hereafter DRN[2]) in the publication by Bondì, *La natura secondo i suoi princìpi* (2009). Cf. the reprint: *De rerum natura iuxta propria principia* (2013).

18 Perhaps Chapter V, 2 of the 1586 edition.

19 DRN[2], 96–98.

more technical and precise phrase *anima a Deo immissa*.[20] But now the passage becomes an independent chapter, the very one that concludes the work (II, 60), giving it added value and emphasizing Telesio's declaration of skepticism about metaphysics even more strongly, at least in relation to the question of the origin of the universe.

3 The Doctrine of *anima a Deo immissa* in *De rerum natura* of 1586

Finally, in 1586, two years before his death, Telesio published the definitive edition of *De rerum natura* [DRN³] in Naples. In five additional books, Telesio presents the natural philosophy of the "minor bodies" of the universe, more specifically living beings (plants and, above all, animals and man), retaining its empiricist methodology and including a considerable amount of material from contemporary biology and medicine.

Plants, animals, and humans are explained as compounds formed by two substances: a body, whose generation, structure, and vital course are studied extensively in Book VI, according to the medical-physiological knowledge of the time in the case of man,[21] and a *spiritus e semine eductus* (material soul), the subject of a broad presentation in Book V. The *spiritus* is not a form of the body, with which it composes the individual human substance, but a substance distinct from the body, a substance that is also material.[22] The *spiritus* is matter possessed by heat, which has made it very thin[23] and therefore maximally mobile. It is, as Telesio had stated in previous works, a unique substance that resides in the brain and spreads from it via the nervous system in order to

20 No doubt this responds to a precise intention and certainly contains a message for the reader. It is perhaps a hope which cannot have a theoretical demonstration.

21 Interestingly, in explaining the anatomy of animals and in particular man, Telesio systematically makes use of a teleological explanation of the configuration and arrangement of the different parts and various organs of the body; they are so according to the performance they must undertake for the *spiritus* or soul they serve. Cf. De Franco, *Introduzione a Bernardino Telesio*, 279–280. Hence the perspective agrees with Aristotelianism and strongly evokes Galen's paradigm in *De usu partium corporis humani*.

22 Any analogy with the subsequent Cartesian dualism disappears when we realize that the soul (*spiritus*) is not an immaterial *res cogitans*, provided with a reflective capacity independent of the senses, but also a corporeal substance or *res extensa*, according to the Cartesian terminology.

23 *De rerum natura* [DRN³], V, 13: "tenuissimum [...] et levissimum", in Bernardino Telesio, *De rerum natura/Intorno alla natura, libri IV–V–VI* (1974), 286. On the origin of the *spiritus* from celestial heat and the connection of Telesio's conception with Aristotle's doctrine of *pneuma* related to the celestial substance (*De generatione animalium*, II, 3) and Hippocrates' *De carnibus*, see Hirai, "Il calore cosmico di Telesio," 72 and 76–77.

move the body[24] and returns to the brain during sleep to regain strength and be reconstituted, with the consequent immobility of the body parts.[25] For the *spiritus*, the body is merely an integument protecting it from external aggression (thereby preventing the volatilization to which it is exposed as *spiritus*) and which it uses as an instrument or organ[26] for its preservation. Initially, therefore, man is not—as in Platonism—a soul exiled from its celestial homeland and imprisoned in the body but basically a corporeal spiritual vapor, the result of the action of the Sun's heat on inert matter. The corporeal instrument of this spiritual vapor is a protective armor and, at the same time, the organ which, through the inhalation of air[27] and through the vapors deriving from the digestion of food,[28] facilitates the regeneration which it always requires. However, the *spiritus* is mortal, since it is unable to survive the dissolution of the corporeal instrument that protects it, and in turn is a prisoner of its own body without which it cannot maintain its existence.

In Chapter V, 2, Telesio affirms the existence in man of the second soul:

> Let it be allowed in advance that, even if it appears (as has been shown) that the spirit derived from the semen is, in those things produced from the semen, the substance of the soul in other animals, by no means should we believe that it is also the substance of the human soul and realizes by itself everything realized by man, even if we have seen that it may be found in men and works in the animals in the same way as it works in men. For not only does the Sacred Scripture teach us that there is in man another substance that is completely divine and infused by the Creator himself but we can also understand this through human reasons.[29]

24 DRN^3, V, 15, 301–302: "Quomodo *spiritus* corpus commovet", with the theological inference based on intelligent design: "rerum nimirum omnium conditoris Dei sapientia suspicienda est admirandaque, qui ita animalium corpora construxit [...]".

25 Thesis developed in the opuscule *De somno*.

26 DRN^3, V, 1, 208: "[*spiritus*] substantia per se existens et corpori, ut proprio tegumento proprioque inexistens organo, quaecumque operatur animal, ipsa ea operetur omnia, corpore universo singulisque ejus artibus, ut organis, usa."

27 A reasoning also developed in the opuscule *De usu respirationis*.

28 DRN^3, VIII, chapters 30–36, in Bernardino Telesio, *De rerum natura libri VII–VIII–IX* (1976), 302–328. The original or natural composition of the *spiritus* and the different materials (air, food) by which it is regenerated explain the differences between men.

29 DRN^3, V, 2, 210: "At illud praefari liceat, quod si spiritus e semine, ut dictum est, eductus rebus e semine confectis inexistens reliquis in animalibus animae substantia visus est, nequaquam tamen propterea et in hominibus, vel si iis ille inesse, eaque quae in animalibus operatur, eodemque operari visus sit modo, humanae etiam animae substantiam illum esse, et per se quae homo operatur operari omnia, existimandum est. Siquidem aliam homini inesse substantiam, et penitus eam divinam, et ab ipso immissam Creatore

The third edition peremptorily introduces the distinction between the two souls but in such a way that there are no doubts about Telesio's orthodoxy and about the unity of the soul in man. At this point, it should be noted[30] that no outlines or alternative handwritten drafts of Chapter V, 2 and 3 survive, and above all that the outlines of other chapters generally use the term *anima* in the sense of *spiritus*. In addition, according to De Franco, Telesio never adds "the expression 'e semine educta', as instead he would always do in the final edition, in order to differentiate it from that 'a Deo immissa'."[31] For Roberto Bondì, this fact confirms that from the 1570 edition onwards, and clearly in the edition from 1586, Telesio accepts a "compromise" with orthodoxy in contrast to his "genuine" thinking, a compromise he accepts with "displeasure".[32] For De Franco, on the other hand, the distinctions reveal the greater care and attention the philosopher gives to his way of expressing himself, the "greater caution" in the presentation of his thesis, without it being possible to infer any change in his doctrine or an insincere genuflection to Catholic orthodoxy.

In any case, the *anima a Deo immissa* is introduced in Chapter V, 2 with the aforementioned schematic formulation. Telesio states there that it is accredited by Scripture (however, he offers no reference)[33] and by "ragioni umane".[34] Yet here again these arguments receive a schematic and brief presentation. The explanations or arguments for the soul *immissa a Deo* have usually been considered Platonic and linked to Marsilio Ficino but without any precise textual references.[35] The succinct explanations adopted by Telesio undoubtedly

non sacrae tantum, divinaeque litterae nos docent, sed humanis etiam intelligere licet rationibus."

30 This was candidly recognized by Luigi De Franco, the most enthusiastic defender of Telesio's Catholicism and sincere orthodoxy who was, at the same time, the only scholar who seems to have thoroughly examined the Roman manuscripts which present a complex and highly varied editorial phase of *De rerum natura* between 1570 and 1586.

31 "Telesio non usa aggiungere, come invece farà sempre nell'ultima redazione a stampa, l'espressione '*e semine educta*' per poterla così differenziare da quella '*a Deo immissa*", De Franco, *Introduzione a Telesio*, 188.

32 Bondì, "'*Spiritus*' e 'anima' in Bernardino Telesio", 416–417.

33 Later Francis Bacon, who follows Telesio on this point and calls the created soul *spiraculum*, will refer to Genesis 2, 7: "Formavit igitur Dominus Deus hominem de limo terrae, et inspiravit in faciem eius spiraculum vitae, et factus est homo in animam viventem." Cf. Bacon, *De dignitate et augmentis scientiarum* (1623), IV, Chapter 3, in Bacon, *Works*, vol. I, 605.

34 DRN^3, V, 2, 210: "aliam homini inesse substantiam, et penitus eam divinam et ab ipso immissam Creatore, non sacrae tantum divinaeque litterae nos docent, sed humanis etiam intelligere rationibus licet."

35 Cf. Corsano, "La psicologia di Telesio", 9–10; Kristeller, *Eight Philosophers*, 137; Di Napoli, *L'immortalità dell'anima*, 402; Vasoli, Introduction to the 1971 reprint of DRN^3, xviii–xix.

recall Ficino's argumentation and the reader is led to think that Telesio is not only inspired by it but also cites passages of Ficino's discussion verbatim in Book XIV of *Theologia platonica de immortalitate animarum* (Platonic Theology on the Immortality of Souls), in which the Florentine Platonist shows that the soul is immortal, insofar as it pursues the twelve attributes of God and naturally strives in a way to equal God. This impression is strengthened by the fact that Antonio Persio, Telesio's faithful disciple, reveals a similar tendency (in an even more marked manner) to combine Telesio's naturalist and empiricist perspective with Platonism in his *Trattato dell'ingegno dell'huomo* (A Treatise on Human Wit) published in 1576. In this case, Persio combines the Telesian doctrine of the *spiritus* with the doctrine of the *spiritus* that Ficino had presented in his *Libri de vita* (Books on Life) and, to a lesser extent, with the arguments on the dignity of man set out in Book XIII of *Theologia platonica*.[36] A thorough comparison with Ficino's text fails to reveal a literal repetition of the passages in Telesio; nevertheless, it cannot be denied that the *four* arguments that we can distinguish in Telesio's text are a synthetic paraphrase—with frequent lexical coincidences—of the lengthier presentation we find in Ficino, particularly in Book XIV of his *Theologia platonica* but also in other places.

These four arguments by Telesio establish the existence and immortality of the soul created and infused by God based on the following considerations: 1) man harbors contempt for the purely natural and an anxiety to know what is higher, even to know God himself, entrusting his desired and highly sought-after happiness to this end; 2) man is not content with the purely natural goods that are present but disdains them in the search and desire for future goods and for God; 3) man admires his virtuous fellows (over and above and contrary to the mere preservation of his own being), even if they are poor and wretched, and despises the vicious and evil, even if they are rich and content; 4) God would be unjust (*iniquus*) if He did not give another life to men after this one, in which they receive just remuneration for virtue and for vice which often is not given in this life. In the four cases, these behaviors go beyond

Nor does Pupo provide references in his study *L'anima immortale di Telesio*, 92–93 on the Platonism of the conception of the soul *immissa*. It should be noted that I only had access to Pupo's work after finishing this essay; Pupo's goal is similar to mine, but his conclusions are quite different.

36 Cf. Persio, *Trattato dell'ingegno dell'huomo*. In his introduction, in addition to mentioning the theoretical continuity with Telesio in the conception of the *spiritus*, Artese indicates that Persio's work "sempre in bilico tra naturalismo e neoplatonismo [...] finisce per suggerire una prospettiva metafisica al naturalismo telesiano", 10–11. See also Garin, "Nota telesiana."

knowledge and action by the *spiritus* or *anima e semine*, limited to that which is purely natural.[37]

Below in parallel columns is the correspondence that Telesio's arguments find in Ficino:

DRN[3], V, 2

Non animalium reliquorum ritu, in earum rerum sensu cognitioneque ac fruitione, e quibus nutritur servaturque et voluptate afficitur, acquiescere homo videtur; sed aliarum quarumvis, vel earum, quae nullo ipsi usui esse, quin quae nullo prorsus comprehendi possunt sensu, et divinorum etiam entium Deique ipsius substantiam operationesque summa inquirit cum anxietate; quin in horum contemplatione intenta beataque rerum reliquarum cognitionem, vel quae se ipsam servandi oblectandique modum afferat facultatemque, et ipsas etiam res, quarum ope servari oblectarique queat, et corporis etiam, cui indita est, salutem negligit contemptui habet, similium nimirum entium, quorum summo tenetur desiderio, intuitu consuetudineque, qua potest ratione, frui appetens quaerensque (pp. 210s.);

Theol. platonica, XIV, 2 and 8[38]

Animus noster communem veritatis bonitatisque concipit rationem, per quam commune verum quaerit et commune appetit bonum. In communi veritate omnia vera, in communi bono bona omnia continentur. Hinc fit ut cuncta vera naturaliter quaerat, cuncta naturaliter appetat bona. Quod ex eo patet quod cognita una quadam alicuius rei veritate non quiescimus, sed aliam inquirimus rursusque aliam, quamdiu putamus veritatem aliquam superesse noscendam. Idem in bonis comparandis inspicitur. Omne autem verum et omne bonum Deus ipse est, qui primum verum est primumque bonum. Ergo Deum ipsum appetimus. Sed quid in eo potissimum affectamus? Illi similes fieri. [...] Finis ergo noster est per intellectum Deum videre, per voluntatem viso Deo frui, quia summum bonum nostrum est summae potentiae

37 DRN[3], V, 2, 210: "Quaedam seorsum a reliquis animalibus operari patique et appetere homines videntur, quae sublimiori omnino substantiae, quam qualis *spiritus* e semine eductus, videri potest, assignanda sunt" [Unlike the rest of animals, men seem to operate, suffer, and wish certain things which must be assigned to a more exalted substance than spirit can appear].

38 Ficino, *Théologie platonicienne*, vol. II. Cf. Ficino, *Platonic Theology*, vol. 4, Books XII–XIV, 226–228, 230–232, 280 and 288.

[anima humana] anxia semper semperque remotis futurisque prospiciens appetensque, nequaquam e semine educta nec mortalis certe videri potest [...] Propterea scilicet numquam in praesentibus neque in iis plane bonis, quibus, praesentem degens vitam, potiri fruique potest, acquiescit homo, sed veluti remotum appetens quaerensque bonum, et aliam longeque beatiorem praesagiens vitam, remotis futurisque assidue prospicit appetitque, quod substantia ei inest, vere Dei ipsius opus, et quae sui ipsius opificem parentemque Deum, ut supremum propriumque ipsius bonum, anxie appetit quaeritque, et illius orbata intuitu rebus aliis nullis adeo deliniri potest, ut vel illius oblivione capiatur vel desiderio non teneatur (pp. 212s.).

nostrae obiectum summum sive actus perfectissimus circa ipsum. Summae autem potentiae nostrae sunt mens mentisque caput atque voluntas. Summum harum obiectum est commune verum bonumque commune et integrum, id est autem Deus (pp. 250s.);
Dedit [natura] brutis cibum et coitus appetitum, addidit membra ad esum et coitum praeparata. Dedit naturae dux animo universalis et totius veri bonique votum, quod quidem eo naturalius est quam cibi et coitus appetitus quo illud est magis continuum. Comedere quidem raro corpus exigit. rarius vero coire. Verum autem bonumque singulis optamus momentis. Semper enim novarum rerum imaginationumque et rationum cupidi sumus. Semper oculos patefacimus ad quaelibet occurrentia et longissimo amplissimoque prospectu nimium delectamur, immo solo contenti sumus immenso (p. 252)

Pravos praeterea homines, vel in summa rerum copia summaque positos fortuna, non odio modo habens, sed despiciens etiam longeque eos infelicissimos judicans humana anima; probos contra non amans modo colensque et venerans, sed beatos etiam statuens, manifeste aliam a spiritu, e semine educto, substantiam homini indit (p. 214).

Cuncti denique homines excellentissimos animos atque optime de humano genere meritos in hac vita ut divinos honorant, solutos a corporibus adorant, ut deos quosdam Deo summo charissimos, quos prisci heroas nominaverunt. Tanta vero ad se et ad suos reverentia non apparet in bestiis, nedum vilioribus, sed neque etiam in maioribus (p. 274)

Forte etiam iniquus nonnullis Deus videri queat, si una cum corpore ipsa itidem dissolvatur intereatque anima et non aliam post hanc

Itaque homines multi omnia, omnes certe quam plurima vitae commoda temporalis abiiciunt Dei amore vel suspicione vel metu. Deus igitur vivat vitam: utpote qui pravissimos saepe homines summis affluere bonis, optimos contra malis miseriisque obrui omnibus et pessimis etiam deterrimisque subdi subicique et divexandos praeberi sustineat (p. 214).

debet pro temporalibus aeterna tribuere. Nullum vero animalium reliquorum praesentibus abstinet bonis propter futurorum aviditatem. Quo fit ut homo stultissimus sit animalium omnium atque miserrimus, si neque praesenti fruitur vita neque futura (p. 277).

4 Conclusion

Telesio's initial text (1565) assumes that the soul is a natural product generated by the semen, and is ultimately the result of the possession of a portion of matter through the Sun. It is certainly the soul that can be known naturally, by means of sensation and within the framework of a program limited to the explanation of nature *iuxta propria principia*. The terminological precision and distinction between *spiritus e semine* and *anima a Deo immissa* comes after the accusations of impiety (of which Telesio was aware in 1570); indeed, the slight shift in terminology mirrors a defensive strategy. The available evidence does not necessarily indicate that this acceptance of an immortal soul is a uniquely defensive move concealing disbelief and heterodoxy. It could also be the demonstration of sincere faith and belief, over and above his philosophy, which emerge even though they do not yet have a theoretical foundation in the 1570 edition. This absence could be due to the fact that the second edition is a repetition of the first one without the study of nature *iuxta propria principia* extending to the field of living beings and man.

That moment arrives with the 1586 edition and the broad study of animals and man. But even in this case, the naturalist premise of his method prevents Telesio from giving an extensive presentation of the "nature" and the operations of the *anima immissa*. The *spiritus* clearly appears as the substance that sufficiently explains (within the framework of the naturalist perspective) human behavior in the biological, epistemological, and ethical context; in the last case, *conservatio sui ipsius* is the ruling principle. Telesio also recognizes that the divine soul never acts autonomously and independently in this life but always acts through the *spiritus*, with which it forms a single thing.

On the one hand, Telesio reduces the operation of the higher soul (intellection) to a process of association by similarity with previous perceptions whose memory is preserved, a process realized by the *spiritus* that is paralleled among animals.[39] This might lead us to think that the belief that the most perfect human intellection is a result of its most perfect spirit[40]—and thus something taking place within the framework of nature—is a logical step in Telesio's philosophical system. However, Telesio states that the superiority of human intellection is not the work of the spirit alone but rather the result of its union with the infused soul,[41] a soul introduced by an appeal to Scripture and to "human reasons" of Platonic origin, and thus extraneous to the methodological principles of his philosophy. Simply put, the *anima immissa a Deo* is less the result of a final stage in the methodical construction of Telesio's philosophy of nature and more the affirmation of a religious postulate whose otherness is manifested in its positioning in relation to the dogma of the resurrection of bodies and the reunion of the soul with the glorious body. In the same way, the "human reason" demonstrating the immortality of the created soul, based on the altruistic behavior that causes a person to sacrifice his life for a reward in the afterlife, as opposed to the *conservatio sui ipsius* (the only thing that moves the spirit), is weakened by Telesio himself when he appeals to the (in his view) verified testimony of such behavior in animals such as the lion and the horse.[42]

All this does not necessarily mean that Telesio's appeal to the created soul and his profession of orthodoxy are a pretense, and thus that he is, if not a hypocrite, at least a philosopher far removed from Church doctrine who actually professes that the human soul is nothing but mortal *spiritus*. Kristeller observed that Telesio may be sincere.[43] Certainly, we do not find elements of a philosophical tradition which, in the same way as Averroism, places philosophy above religion or which considers religion a pragmatic fiction (a "noble lie"

39 DRN³, VIII, 15, 233.

40 Ibid.: "the human spirit (since it is purer and more copious than that of the animals and has obtained a place much more suitable to recall the motions and, as a consequence, the motions in it are never interrupted or confused by any sort of smoke, and thus the spirit preserves them entirely pristine and pure) reasons more excellently than the spirit of animals to such a degree that, compared with the human [spirit], the latter may seem incapable of reasoning." Bondì has insisted on this naturalization of the intellective process in man and on the extrinsic and disjointed character of the appeals to transcendence with regard to the theoretical development of Telesio's philosophy; cf. Bondì, *Introduzione a Telesio*, 103. De Anna recently recognized this point in a nuanced analysis whose conclusions I share; see De Anna, "Telesio e il naturalismo: le critiche alle tesi aristoteliche sull'immortalità dell'intelletto", 94.

41 DRN³, VIII, 15, 235.

42 DRN³, V, 40, 449–451.

43 Cf. Kristeller, *Eight Philosophers*, 141.

in the sense of Plato's *Republic*) devised by the legislator for the benefit of the ignorant common people naturally inclined to vice in order to make their virtue possible—and this as the basis for a political community. In short, Telesio is neither a Pietro Pomponazzi nor a Giordano Bruno.

Nevertheless, it remains true that Telesio abandoned the traditional line of a "Christian philosophy", which in concordist Aristotelianism (Thomas Aquinas) and in Platonism (Ficino, Giovanni Pico della Mirandola) straightforwardly established the immortality of the rational soul starting from its intellectualistic inception. Rather, Telesio embarked on a rigorously naturalistic and sensualistic way of thinking, which (as would emerge from the developments of the subsequent philosophy) blocks and excludes any appeal to a supernatural and metaphysical supposition, denying the immortality of the soul or at least rejecting the possibility that this hypothesis can be the subject of philosophical affirmation. In summary, the considerations Telesio uses to introduce the *anima immissa a Deo* are extraneous to the methodological principles supporting his philosophical system,[44] and this casts a shadow of suspicion on the sincerity and validity of his expansion of the doctrine of the soul.

In the dedication to the third edition of his work, addressed to his patron the Neapolitan patrician Fernando Carafa, Duke of Nocera, Telesio indicates that Aristotle's philosophy is contrary to the experience of the senses, to itself, and to God, whereas his own philosophy is free from these three sins and is thus true.[45] For this reason, since the truth cannot be in conflict with itself, in the preface to the first book Telesio confidently expresses his submission to the Church and his complete willingness to recant his philosophy: "*if* any of the things that we have stated does not agree with Sacred Scripture or with the decrees of the Catholic Church, it cannot be held but must be rejected completely. Not only should any human reason be suspended before these things but sense itself; and the very same sense must be rejected entirely *if* it is not in accordance with them."[46] This is undoubtedly a very strong declaration, a rather theatrical act of submission—like the frequent *protestationes*

44 As reported by Giglioni in his Introduction to the very recent reprint of DRN³ *De rerum natura iuxta propria principia* (2013), xxii, "è chiaro che l'inserzione di una dimensione sovrannaturale nel centro dell'anima umana introduce un elemento di possibile incoerenza nella compattezza del sistema filosofico telesiano, un elemento che rischia di sovvertire il senso dell'originario progetto diretto a investigare la natura in piena autonomia, *iuxta propria principia*."

45 DRN³, 20: "Nostra siquidem doctrina quoniam nec sensui nec sibi ipsi nec sacris etiam litteris repugnat unquam, quin adeo his et illi concors est, ut ex utrisque enata videri possit; quoniam omnino vera est."

46 DRN³, 28: "Nec tamen, *si* quid eorum, quae nobis posita sunt, sacris litteris catholicae ecclesiae decretis non cohaereat, tenendum id, quin penitus reiiciendum asseveramus,

fidei in contemporary authors—which professes the captivity not so much of the intellect[47] as of the very principle of philosophical authority according to him: *sensus*. *Libertas philosophandi*, another great sign of Telesio's philosophical identity, would be freedom of thought against Aristotle and all philosophical authorities but not against the Church and Scripture; the latter constitute its limit. Yet Telesio establishes this point only here (a very important point, in our opinion, which lends strength to the thesis of disbelief), clearly in conflict with the set of programmatic and methodological declarations present in explanatory and neutral places in his writings. Those declarations differ from the present thesis, which is clearly defensive and meant to provide security to the reader and immunity to the author.

According to De Franco, Telesio's declaration does not prevaricate, and is merely an expression of his sincere religious faith and his conviction (note the conditional nature of the prefatory statement) that there can never be a contrast between his true philosophy and Holy Scripture.[48] As a consequence, recantation is impossible.[49] The documentation available to us (and primarily the work of Telesio, an intelligent author) does not allow us to repudiate these declarations of orthodoxy, nor does it allow us to confirm them with any proofs or arguments other than these very same declarations.[50] They are simply external to the logic of his theoretical itinerary. With respect to his metaphysical skepticism and his naturalist philosophical system, Telesio's affirmation of the

contendimusque. Neque enim humana modo ratio quaevis, sed ipse etiam sensus illis posthabendus; et si illis non congruat, abnegandus omnino et ipse etiam est sensus."

47 Cf. 2 Corinthians, 10: 5: "in captivitatem redigentes omnem intellectum in obsequium Christi."

48 Cf. De Franco, *Introduzione a Bernardino Telesio*, 226–229. Similar conclusions were made by Spruit, "Elementi aristotelici e polemica anti-peripatetica", 370 and De Anna, "Telesio e il naturalismo," 96: "dal fatto che i riferimenti di Telesio all'anima creata da Dio manchino di giustificazione filosofica non possiamo dedurre immediatamente che allora dipendono certamente da motivazioni estrinseche, pratiche o 'politiche'."

49 However, where there is a seemingly insoluble contrast (as in the question of the indefinite duration of the world in its current and permanent configuration; cf. DRN[3], Book I, Chap. X, 15–17), Telesio adapts the Scriptures to the conclusions of philosophy. On this question see Bondì, "Dios, naturaleza y alma en Telesio," 28–32.

50 I believe that we can agree with the recent conclusion of Giglioni in the already cited Introduction to DRN[3], xxxi: "l'annosa questione della presunta sincerità o meno delle opinioni di Telesio a proposito dell'anima immateriale [...] è in gran parte irrisolvibile, visto che comunque la sfera delle intenzioni rimane irriducibilmente privata e si sottrae per sua natura a qualunque intervento inquisitorio, anche quello animato dalle migliori intenzioni investigative, come può essere l'indagine di uno storico."

divine and immortal soul (if it is not to be considered a defensive strategy[51]) certainly does not exceed a declaration of faith and hope[52] and of rhetoric borrowed from a Platonic philosophy extraneous to his own.

51 This position is based on a fragile hermeneutic (*à la* Leo Strauss) since, as Giglioni states (ibid.), "un possibile modo di risolvere l'*impasse* è allora quello di ipotizzare in Telesio una deliberata strategia di dissimulazione in virtù della quale, mentre egli dice una cosa, in realtà ne intende un'altra. Quest'approccio presuppone un lettore altrettanto abile in strategie oblique di interpretazione e che sia in grado di vedere cosa l'autore volesse davvero dire al di là delle tecniche depistatorie da lui messe in atto per sfuggire al controllo del censore di turno. In questo caso, si istaura una complicità tra lettore e autore che va ben al di là del testo, e che è spesso basata su tenui indizi, ammiccamenti e preferenze ideologiche." For this reason, I have tried to avoid this path in this essay.

52 See De Anna, "Telesio e il naturalismo," 91: "Il suo naturalismo ha forti connotazioni empiriste e il suo sensismo riduce l'esperienza alle proprietà sensibili esperite. In questo modo, però, preclude possibili aperture alla trascendenza della realtà rispetto all'esperienza. Ogni nozione della divinità dovrà essere di tipo fideista", 93: "le sue affermazioni relative a una finalità soprannaturale dell'uomo e a un desiderio umano trascendente la natura non trovano alcuna motivazione intrinseca nella sua psicologia o nella sua antropologia."

CHAPTER 3

Telesio, Aristotle, and Hippocrates on Cosmic Heat

Hiro Hirai

1 The Cosmological Passage in Aristotle's *Generation of Animals*[*]

The enigmatic passage given below, which is from Aristotle's *Generation of Animals*, II 3, was one of the most commented lines from his biological works during the Middle Ages and Renaissance. It refers to an invisible and hidden component residing in the seeds of living beings:

> The faculty of the soul of every kind has to do with a matter different from and more divine than the so-called elements. But as one soul differs from another in the scale of value, so does the nature of the corresponding matter. In every seed there is that which causes it to be fertile, that is, what is called "heat." This heat is neither fire nor any such faculty but the *pneuma* which is enclosed in the seed and a foam-like body. Nature in this *pneuma* is analogous to the element of the stars. That is why fire does not generate any animal, and we find no animal taking shape in either fluids or solids under the influence of fire; whereas the heat of the Sun and that of animals do generate them. Not only the heat residing in the seed but also whatever other natural residue, which there may be, has in itself a vital principle too. Considerations of this sort clearly show us that the heat contained in animals neither is fire nor draws its origin from fire.[1]

The seed contains the *pneuma*—which is translated as *spiritus* into Latin—and its heat. Aristotle suggested that the nature of this heat corresponds by

* This chapter is based on my article, "Il calore cosmico in Telesio fra il *De generatione animalium* di Aristotele e il *De carnibus* di Ippocrate," in *Bernardino Telesio tra filosofia naturale e scienza moderna*, ed. Giuliana Mocchi, Sandra Plastina and Emilio Sergio (Rome: Fabrizio Serra Editore, 2012), 71–83. My warmest thanks go to Clare Hirai, Elisabeth Moreau, and Pietro Daniel Omodeo for their help in its preparation.

1 Aristotle, *Generation of Animals*, II 3, 736b29–737a5. On this passage, see Solmsen, "Vital Heat"; Preus, "Science and Philosophy in Aristotle's *Generation of Animals*," 35–38; Gad Freudenthal, *Aristotle's Theory of Material Substance*, 107–119; Lennox, *Aristotle's Philosophy of Biology*, 229–249.

 | DOI:10.1163/9789004352643_005

analogy to the "element of the stars," which is the celestial substance, *aether*. A cosmological dimension is thus introduced in the middle of an embryological discourse. The aim of the present article is to examine the impact of this passage on Telesio's natural philosophy.[2] But it is necessary above all to address its historical and intellectual context.

Aristotle's Greek commentators such as Alexander of Aphrodisias (fl. ca. 200) and Simplicius (fl. 529–?) seem to have paid no particular attention to the interpretative potential of the passage. We have to wait a long time for Michael of Ephesus's (fl. ca. 1138) commentary on Aristotle's same treatise. One rare exception in Antiquity was a paraphrasing of Aristotle's *On the Soul* by Themistius (ca. 317–ca. 388), who made a brief reference to the work in question.[3] It was probably following Themistius's text, which was widely read in the Arabic world, that authors such as Avicenna (980–1037) and Averroes (1126–1198) seized the value of the cosmological passage in Aristotle's embryological work. Simultaneously philosophers and physicians, they used it to explain the theory of "formative power" (δύναμις διαπλαστική), which was believed to reside in the seed. They went even further and connected it to the notion of "intellect" (νοῦς).[4] Indeed two passages in the same chapter from Aristotle's *Generation of Animals* address the origin of the intellect. The first of them affirms:

> It remains, then, that the intellect alone comes from the outside and that it alone is divine; for the bodily actuality has nothing to do with its actuality.[5]

The second argues:

> But the body of the semen, in which there also comes the portion of the soul's principle—partly separable from body in all those in which something divine is included (and such is what we call the "intellect") and partly inseparable.[6]

2 On Telesio, see De Franco, *Bernardino Telesio: La vita e l'opera*; De Franco, *Introduzione a Telesio*; Bondì, *Introduzione a Telesio*; Mulsow, *Frühneuzeitliche Selbsterhaltung*. See also the conference proceedings dedicated to him: Santoro, *Bernardino Telesio nel 4° centenario*; Galasso, Sirri and Torrini, *Bernardino Telesio e la cultura napoletana*.

3 Themistius, *In de anima*, I 3 = Heinze, *Themistii in libros Aristotelis de anima paraphrasis*, 19 = Todd, *Themistius: On Aristotle's* On the Soul, 35.

4 See Hirai, *Medical Humanism*.

5 Aristotle, *Generation of Animals*, II 3, 736b27–29. Cf. Moraux, "À propos du νοῦς θύραθεν chez Aristote." See also Themistius, *In de anima*, I 5 = Heinze, 37 = Todd, 54.

6 Ibid., II 3, 737a7–12.

These two passages were often employed to confirm the incorporeal and incorruptible, and thus immortal, character of the human soul. Following their Arabic forerunners, Western intellectuals developed their interpretations not only from a philosophical and theological point of view, as in Albertus Magnus (ca. 1193–1280) and Thomas Aquinas (ca. 1225–1274), but also with a medical dimension as in Pietro d'Abano's (1257–ca. 1315) work.[7] All three passages from Aristotle's *Generation of Animals*, II 3, were often evoked by theologians, philosophers, and physicians during the late Middle Ages.[8]

In the Renaissance, the French physician Jean Fernel (1497–1558) provided a new impulse to the humanist tendency of harmonizing Plato and Aristotle by developing a particular notion of *spiritus* in his dialogue *De abditis rerum causis* (*On the Hidden Causes of Things*) (Paris, 1548). Frequently calling upon the cosmological passage of Aristotle's embryology, Fernel linked it with the idea of the "World-Spirit" (*spiritus mundi*) advanced by the Florentine Platonist Marsilio Ficino (1433–1499). Fernel transformed this passage into a *topos* of Renaissance natural philosophy and medicine, which were permeated by Platonic philosophy.[9]

In terms of natural philosophy and medicine, the cosmological passage of Aristotle's *Generation of Animals* involves another important element since it refers to some "vital principle" (ἀρχὴ ζωτική) as the origin of life. This principle is meant to be included in the heat of the Sun and in that of living beings. Thus the seed possesses a heat which holds the secret of life's origin. To help illustrate this dimension, another passage from Aristotle's *Generation of Animals*, III 11, was often invoked in the Renaissance. It addresses the spontaneous generation of lesser living beings:

> Animals and plants are generated in earth and water because there is water in earth, and *pneuma* in water, and in all *pneuma* is soul-heat, so that all things are in a sense full of soul.[10]

This connection offered a formidable occasion for Girolamo Cardano (1501–1576) of Milan to present his pan-vitalistic world view in a widely-read treatise.

---

7 Pietro d'Abano, *Conciliator*, differentia 48, f. 72*r*. Cf. Paschetto, *Pietro d'Abano, medico e filosofo*, 199–202; Jacquart, "Médecine et astrologie à Paris," 121–134.

8 See Nardi, *Problemi d'embriologia umana*; Nardi, *Studi di filosofia medievale*; Martorelli Vico, *Medicina e filosofia*.

9 On Fernel, see Hirai, *Medical Humanism*, 46–79. On Ficino's idea of the World-Spirit, see Hirai, "Concepts of Seeds and Nature," 273–274.

10 Aristotle, *Generation of Animals*, III 11, 762a18–21. Cf. Hirai, *Medical Humanism*, 42, 95, 112, 124, 148 and 164.

In *De subtilitate* (*On Subtlety*) (Nuremberg, 1550), he defended the idea of the soul's omnipresence in the universe along with his theory of cosmic heat.[11] For Cardano, this special heat was the source of all other sorts of terrestrial heat. As the efficient cause of all kinds of generation, it played the role of the active principle in the formation of natural beings, including minerals and metals. He went even further and argued that cosmic heat was the soul itself or its essential instrument:

> [...] it is evident that Hippocrates correctly said: the soul is nothing but that celestial heat. This also corresponds well to the opinion of Aristotle since he wants the heat of *spiritus* to have a certain analogy with the element of the stars. Indeed, whether the heat is the soul or its first instrument, wherever there is this [kind of] heat, it is evident that the soul itself should also be present; therefore [there should be] life too. For life is nothing but the work of the soul.[12]

Here Cardano refers to the cosmological passage of Aristotle's *Generation of Animals*.[13] He also relies on the passage on spontaneous generation from the same treatise to defend the soul's omnipresence. But there is yet another crucial element in this quotation. Cardano mentions Hippocrates's idea in order to reinforce his argument although he does not reveal his source. Actually, this idea was drawn from the treatise *On Fleshes*, in which the Hippocratic author proposed a cosmological dimension of the soul's origin:

> [...] I should say nothing about heavenly and sublime things, unless to the extent that human beings and other animals that live and are begotten on earth have their principle and origin there; and that the soul comes from heaven. In my view, what we call 'heat' is immortal, perceives all, and sees, hears and knows all that is and all that will be.[14]

---

11 See Hiro Hirai, *Le concept de semence*, 135–156: 140–148; Hirai, *Medical Humanism*, 110–114.

12 Cardano, *De subtilitate*, lib. 5 = Nenci, *Girolamo Cardano: De subtilitate*, 451–452: "[...] manifestum est quod recte dixit Hippocrates, animam nihil aliud esse quam coeleste illud calidum. Quod et Aristotelis opinioni adhaeret, dum vult spiritus calorem proportionem habere quandam cum elemento stellarum. Nam seu calor sit anima, seu illius primum instrumentum, ubi calor hic erit, palam est oportere adesse etiam ipsam animam, quare et vitam; est enim vita nihil, aliud quam opus animae."

13 Cf. Bianchi, "Scholastische Motive im ersten und zweiten Buch des *De subtilitate*."

14 Cf. Hippocrates, *On Fleshes*, 2 = Littré, *Œuvres complètes d'Hippocrate*, VIII, 584. Cf. Spoerri, "L'anthropogonie du *Peri sarkon*."

Fernel built his argument in favor of the celestial origin of the human soul and its immortality precisely upon these lines.[15] For his singular reading of the Hippocratic text, he followed the Latin translation by Marco Fabio Calvo (ca. 1440–1527).[16] Cardano, in his turn, went even further to affirm that cosmic heat is "animated" and "endowed with intelligence" according to the teachings of the Greek physician. After Cardano, this connection with Hippocrates's *On Fleshes* became very popular in the Platonizing reading of the cosmological passage of Aristotle's embryology during the Renaissance.

Likewise, the Flemish physician with Platonic tendencies, Cornelius Gemma (1535–1578), adopted this connection, which in his eyes successfully explained the secret of life's principle and the soul's origin in a cosmological dimension. In his early work, *De arte cyclognomica* (*On the Cyclognomical Art*) (Antwerp, 1569), this great admirer of Fernel and Cardano naturally connected the idea of cosmic heat to the Ficinian theory of the World-Spirit:

> Above quality, I place the *spiritus* which, in my opinion as well as according to the view of Hippocrates in the treatises *On Regimen* and *On Fleshes*, does not really differ from innate heat, just as that universal *spiritus* of the world does not differ from the element of the stars. Although many people often speak of it, very few [of them] really understand it. Here the former kind of *spiritus* is the first instrument of a future form or soul. It connects the form to bodies as the *spiritus* is tied to these bodies by a carrier quality which intervenes. It is the same *spiritus* as that which perfects, connects, sees and understands everything according to Hippocrates.[17]

---

15 Fernel, *De abditis rerum causis* II 4 = Forrester, *Jean Fernel's* On the Hidden Causes, 445.

16 This was published in *Hippocratis octoginta volumina* (*The Eighty Treatises of Hippocrates*) (Rome, 1525) and in a Hippocratic collection (Paris, 1527), although another translation by Janus Cornarius (1500–1558), published in *Hippocratis opera omnia* (*The Complete Works of Hippocrates*) (Venice, 1546), was closer to modern readings; on Calvo, see Gualdo, "Marco Fabio Calvo." On Cornarius, see Mondrain, "Éditer et traduire les médecins grecs."

17 Gemma, *De arte cyclognomica*, II 3, 42: "Supra qualitatem constituo spiritum, qui meo judicio, et secundum mentem Hippocratis libro *De diaeta*, *De carnibus*, a vere innato calido minime differt, uti nec ille communis spiritus mundi ab elemento stellarum, quod multi in ore saepius habent, intelligunt sane perpauci. Hic ille est spiritus adventantis formae, vel animae primarium instrumentum; connectit formam corporibus, uti alligatur iisdem spiritus intercedente vehiculo qualitatis. Ille est idem spiritus quem censet Hippocrates cuncta perficere, nectere, videre et intelligere."

2 Telesio on Cosmic Heat

Now let us turn to Bernardino Telesio (1509–1588). In his major work, *De rerum natura iuxta propria principia* (*On the Nature of Things according to Their Own Principles*) (Naples, 1586) or even its first version, *De natura iuxta propria principia* (*On Nature according to Its Own Principles*) (Rome, 1565), Telesio—probably following Cardano—reduced the number of elements to two: "heat" (*calor*) and "coldness" (*frigus*). These two elements held the active role in Aristotle's conception of the four elementary qualities.[18] To these Telesio added the third principle: "matter" (*materia*), which was purely inert and formless.[19] For him, heat is represented by heaven and the Sun as the active and positive principle of everything in the universe, while cold is the active but negative principle, symbolized by the earth.[20] He justified his choice of heaven and earth in the name of the Bible. Needless to say, *Genesis* teaches that God created heaven and earth at the beginning of the world. Thus, we can see that the authority of the Bible was more important for Telesio than it might appear.[21]

As the active principle of the universe, cosmic heat is responsible for all the terrestrial phenomena and acts on matter with the help of coldness. In one passage, Telesio justifies his principles by calling upon Aristotle's *Metaphysics*, XII 4, which states: "The elements of perceptible bodies have heat as their form and, in another sense, cold, which is privation; and as matter, that which is in potentiality."[22] This passage is very important for Telesio in his revision of Aristotle's theory of the elements. Indeed, what his particular naturalism presented was not a pure and simple rejection of Aristotle but a radical reform of Peripatetic philosophy. This reform was supported by his remarkable knowledge of Aristotle's Greek commentators.[23]

Let us take a closer look at Telesio's idea of cosmic heat. In the second book of his *On the Nature of Things*, dedicated to the principles of the universe, he

18 I have used *De rerum natura iuxta propria principia*, ed. De Franco (Cosenza, 1965–1974 and Florence, 1976); *La natura secondo i suoi principi*, ed. Bondì (Florence, 1999); *De natura juxta propria principia*, ed. Ottaviani (Rome, 2008).

19 Schuhmann, "Telesio's Concept of Matter."

20 On his idea of coldness as the principle, see De Franco, *Telesio*, 143–158.

21 Telesio, *De rerum natura*, I 3 = ed. De Franco, I, 48. Cf. De Franco, *Telesio*, 203–212. See also Telesio, *De natura*, I 43 = ed. Ottaviani, 164.

22 Telesio, *De rerum natura*, III 4 = ed. De Franco, I, 414. Cf. Aristotle, *Metaphysics*, XII 4, 1070b11–12; Averroes, *Commentarium magnum in libros metaphysicarum*, XII 22 = Martin, *Averroès: Grand commentaire de la* Métaphysique *d'Aristote*, 150.

23 On his attitude toward Aristotle, see Lerner, "Aristote 'oublieux de lui-même' selon Telesio."

presents the argument of some Aristotelians. According to them, heat and coldness are not the substances which "feel and know what they do"; they are the instruments of another principal and higher substance, that is the "form" (*forma*) or the "soul" (*anima*). Telesio rejects this idea by responding that heat cannot be the instrument of another principle because it acts according to its own natural disposition, which he calls *ingenium*:

> For heat never acts in everything according to the *ingenium* of another substance but always according to its own [*ingenium*]. Evidently, natural things are produced by heat alone, which acts according to its own *ingenium* and its forces without the help of another active substance [...].[24]

To reinforce this argument, Telesio proceeds to explain the constitution of the four elements. Using the expressions found in Aristotle's *Generation and Corruption*, he tries to show that two "contrary agents" (especially heat and coldness among the four elementary qualities) and matter are sufficient for the formation of the elements. Telesio deduces that these qualities are the forms of the elements since everything is composed of form and matter.[25]

To justify his reasoning, Telesio appeals to Aristotle's Greek commentators, who were, according to him, "the best and most excellent of the Peripatetics."[26] His method is typical of Renaissance humanists. Telesio first quotes Alexander of Aphrodisias, who, in his commentary on Aristotle's *Metaphysics*, XII 4, affirmed that: "heat is a substance since it is the form of fire."[27] Then he calls upon Ammonius (ca. 435/45–517/26) and Galen (129–216).[28] The reference to Ammonius, rare in the discussions on natural philosophy of the time, attests to his profound knowledge of Aristotle's Greek commentators.[29] Next, Telesio returns to his favorite, Alexander, according to whom the qualities are the

24 Telesio, *De rerum natura*, II 22 = ed. De Franco, I, 334: "Siquidem calor numquam substantiae alterius sed iuxta proprium ingenium in omnibus perpetuo agit. Et manifeste uno a calore, suo ingenio ac viribus, non alterius substantiae ope agente, entia prorsus naturalia constituuntur [...]." Telesio's notion of *ingenium* is complex and requires further research. Here I leave it untranslated in Latin.

25 Ibid., II 24 = ed. De Franco, I, 342–344.

26 Ibid., II 24 = ed. De Franco, I, 344. Cf. Telesio, *De natura*, I 56 = ed. Ottaviani, 76.

27 Cf. (Ps.-)Alexander of Aphrodisias, *In Aristotelis metaphysica*, XII 4 = Hayduck, *Alexandri Aphrodisiensis in Aristotelis metaphysica commentaria*, 680.

28 Cf. Ammonius, *In Porphyrii isagogen* = Busse, *Ammonii in Porphyrii isagogen*, 113; Galen, *De elementis ex Hippocrate*, 9 = Kühn, *Galeni opera omnia*, I, 482 = De Lacy, *Galen: On the Elements*, 128.

29 On Ammonius in the Renaissance, see Lohr, "Renaissance Latin Translations of the Greek Commentaries on Aristotle," 27–30.

substantial forms of the four elements.[30] Then he adds Philoponus (490–570), who adhered to this thesis by affirming that the qualities are the formal principles.[31] After referring to Olympiodorus (495/505–after 560), Telesio finally reaches Hippocrates:

> Did not that divine father of medicine, Hippocrates, say these wonderful things on heat in his treatise *On Fleshes*? It is my pleasure to quote his words in Greek because they contain things worthy of admiration: "In my view, what we call 'heat' is immortal, perceives all, and sees, hears and knows all that is and all that will be." Is heat an accident for Hippocrates? [Of course not.] If [the Peripatetics] paid attention to all these, they would affirm that heat is not only the form of the elements but also that of all other entities produced by heat and that coldness [is] the form of those which are produced by coldness.[32]

Here, as the decisive piece of evidence, Telesio adopts the same passage from Hippocrates's *On Fleshes* as Fernel, Cardano, and Gemma did before him. Moreover, this is one of the rare occasions in Telesio that a Greek text is quoted at length.

Telesio then emphasizes that Aristotle himself acknowledged no other substance than that which comes from heaven for the formation of natural things:

> For, according to Aristotle, all other entities, including plants and animals, are produced not from other things or by other natures but from simpler bodies and by their forms after being mixed and united to each other or broken up by mutual actions. He never announces another substance for their formation. Even if he wished [to do so], he never considers where to find it. In fact, for Aristotle, this [substance] can derive only from heaven, and heat alone flows down from it or, at least, emanates from the substance of heaven […].[33]

---

30 Cf. Alexander of Aphrodisias, *De anima*, II 8 = Bruns, *Alexandri Aphrodisiensis praeter commentaria scripta minora*, I, 127 = Sharples, *Alexander of Aphrodisias: Supplement to* On the Soul, 86.

31 Philoponus, *In de generatione et corruptione*, II 2 = Vitteli, *Ioannis Philoponi in Aristotelis libros de generatione et corruptione commentaria*, 215 = Williams, *Philoponus: On Aristotle's On Coming-to-Be and Perishing*, I 6–II 4, 125.

32 Telesio, *De rerum natura*, II 24 = ed. De Franco, I, 348 = Appendix 3. Cf. Hippocrates, *On Fleshes*, 2 = ed. Littré, VIII, 584.

33 Ibid., II 24 = ed. De Franco, I, 348–350 = Appendix 3.

Telesio clearly affirms that according to Aristotle's teachings what emanates from heaven is nothing but cosmic heat. However, he does not interpret this in Platonic dimension—as Cardano and Gemma did—but within a very physicalistic perspective:

> [...] maybe [such heat] could be seen to be nobler than an elemental [entity] or different from it. But it is produced in the fire, moved and robbed by the movement of the Sun without receiving anything else, and is brought here from that fire. Thus this [heat] is not nobler than the fiery one or different from it for any other reason than [that it is different] not by its nature, namely by the faculties of acting and operating, but by its strength.[34]

Telesio concludes that if the Aristotelians acknowledge heat and coldness as the forms of the elements they must accept them as the forms of all other beings which are produced from the elements. For him, cosmic heat does not act under the control of another substance but according to its own *ingenium*.[35] Further, Telesio refuses to consider the soul of living beings to be the cause of their generation as some Peripatetics did. Indeed, he knows Aristotle's words, which clearly show that heat and the *spiritus* produce the substance of everything, even the very soul of living beings.[36] To confirm this, he quotes the passage on spontaneous generation from Aristotle's *Generation of Animals*, III 11, and turns again to the cosmological passage of the same treatise.[37]

On the basis of the same two passages, Cardano tried to emphasize the omnipresence of the soul resulting from cosmic heat. To this end, he called upon Aristotle's conception of "soul-heat" (θερμότης ψυχική), which is said to operate in spontaneous generation and hold the secret of life's principle.[38] By contrast, Telesio used the same passages to emphasize the role of cosmic heat and the *spiritus* more than the soul itself. Indeed, for him, the *spiritus* played the role

34 Ibid., II 24 = ed. De Franco, I, 350 = Appendix 3. In the 1570 version of *De rerum natura*, I 53 = ed. Bondì, 176, this part also comes after the quotation of the cosmological passage from Aristotle's *Generation of Animals*, II 3.

35 Cf. also Telesio, *De rerum natura*, II 25 = ed. De Franco, I, 352: "Quin, et quae e semine enascuntur animalia, si eorum constitutionem proprius Peritatetici intueantur, uno et ipsa a calore, et formae nullius recto habenis, sed iuxta proprium agente ingenium effecta esse decernant oportet."

36 Ibid., II 25 = ed. De Franco, I, 354 = Appendix 4.

37 Ibid., II 25 = ed. De Franco, I, 354–356 = Appendix 4. Telesio identifies the heat hidden in putrefied matter with the soul. Cf. Telesio, *De natura*, I 15 = ed. Ottaviani, 31 = Appendix 1.

38 On the soul-heat, see Aristotle, *Generation of Animals*, II 1, 732a18–19; II 4, 739a11; III 1, 752a2–3; III 11, 762a20. Cf. Freudenthal, "Medieval Astrologization of Aristotle's Biology."

of the soul of living beings. It is the portion of matter in which cosmic heat is highly concentrated.[39] The *spiritus* is, so to speak, a quasi-material carrier of cosmic heat or even its material representation. That is why he frequently speaks of the *spiritus* "drawn from the seed" (*eductus e semine*).[40] Indeed, this is nothing but a clear allusion to that special life-giving *spiritus* enclosed in the seed of living beings according to the cosmological passage of Aristotle's embryology.

3 Conclusion

The cosmological passage of Aristotle's *Generation of Animals*, II 3, stands at the origin of Telesio's conception of cosmic heat.[41] In 1565 his *De natura* had already justified the choice of replacing the soul with cosmic heat on the basis of that very passage, by arguing that Aristotle identified the soul with heat. In another place in the final version of his work, Telesio also compared the *spiritus* with seminal heat by linking the cosmological passage with the lines on spontaneous generation from the same treatise.[42] He advanced the omnipresence of heat as the principle of the universe instead of universal animation advanced by Cardano on the basis of the omnipresence of the soul derived from cosmic heat.

However, it is not sufficient to understand Telesio's thesis on the basis of Aristotle's cosmological passage alone, or with the help of other passages from the same treatise. According to him, cosmic heat as the principle of the universe (or its material representation, *spiritus*) "feels and knows what it does" and is endowed with its own *ingenium*. Although this thesis is enigmatic at first glance, it reveals its real meaning when the passage from Hippocrates's *On Fleshes* is rightly taken into account. According to the latter, cosmic heat is "endowed with intelligence" and "sees and knows everything" and thus even "the future." Most probably, Telesio first learned of Hippocrates's words from Cardano's work. But he interpreted it in a very naturalistic way after his own standard. Behind this radical naturalism, it is possible to detect the shadow of

39 On his notion of *spiritus*, see De Franco, *Telesio*, 167–176; Badaloni, "Sulla costruzione e sulla conservazione della vita in Bernardino Telesio."; Ingegno, "Corpo, spiritus, anima," Bondì, "*Spiritus* e *anima* in Bernardino Telesio."

40 See the article by Granada in this volume.

41 Telesio, *De natura*, I 15 = ed. Ottaviani, 30–31 = Appendix 1. Cf. Telesio, *De natura*, I 56 = ed. Ottaviani, 76–77 = Appendix 2. See Bianchi, "Scholastische Motive," 128–129; Mulsow, *Frühneuzeitliche Selbsterhaltung*, 234.

42 Telesio, *De rerum natura*, 6.40 = ed. De Franco, II, 670.

Averroes, who developed a very physicalistic interpretation of cosmic heat as the generative principle of the soul in the sublunary world.[43]

The theme of cosmic heat under the authority of Hippocrates's *On Fleshes* became a *topos* in the natural philosophy and medicine of the late Renaissance and beyond. It was to reappear especially in the discussions about the notion of nature among seventeenth-century intellectuals such as Justus Lipsius (1547–1606), Ralph Cudworth (1617–1688), Robert Boyle (1627–1691), and Gottfried Wilhelm Leibniz (1646–1716).[44]

Some decades ago, Alfonso Ingegno masterfully showed the Hippocratic foundation of Cardano's natural philosophy.[45] Without a doubt, the imprint of this particular Hippocratism of the Renaissance can also be found in Telesio, who was an attentive reader of his precursor from Milan.[46]

Appendices

In the following selection I provide four chapters that are crucial for Telesio's notion of cosmic heat: *On Nature* (1565), I 15 and I 56; *On the Nature of Things* (1586), II 24 and 25. These chapters show the textual evolution between the two treatises and the context of Telesio's argument.

1. Telesio, *De natura*, I 15 = ed. Ottaviani, 30–31:[47] At et sensus vitaque cum motu a calore constitui Aristoteli videntur et proprii esse caloris: "Generantur, inquit, in terra humoreque animalia quoniam humor in terra, spiritus in humore, calor animalis in universo inest, ita ut quodammodo plena animae sunt omnia, quoamobrem consistunt celeriter, cum calor ille comprehensus sive exceptus est, comprehenditur autem et humoribus corporeis incalescentibus efficitur velut bulla spumosa."[48] Calor igitur anima Aristoteli videtur, at non quilibet, sed in tenui, in proprio nimirum existens subiecto, qui igitur, ubi in humido efficitur, retineturque corpore, animal constitutum apparet; et omnia animae plena, quod spiritus, calidae nimirum tenuisque rei. Id enim

43 See Freudenthal, "The Medieval Astrologization," 128–137. I presented my preliminary study "Telesio's Hidden Use of Averroes" at the conference on which this volume is based. I plan to publish a more complete version in the near future.

44 See Lipsius, *Physiologia stoicorum*, I 6, 13–14; Cudworth, *A True Intellectual System of the Universe*, I 3.7, 109; Boyle, *A Free Inquiry into the Vulgarly Received Notion of Nature*, 43; Leibniz, "Natura Ipsa (1698)," 499.

45 Ingegno, *Saggio sulla filosofia di Cardano*, 226–234. See also Siraisi, *The Clock and the Mirror*.

46 On Telesio's debt to Cardano, see Bianchi, "Scholastische Motive," 129.

47 See also the 1570 version of *De rerum natura*, I 53 = ed. Bondì, 174–176.

48 Aristotle, *Generation of Animals*, III 11, 726a18–24.

spiritus Aristoteli etiam videtur, et non hoc modo, sed quod caloris animalis, propterea, reor, quod ipse innuere videtur, non tota spiritus moles, sed calor tantum spiritui inexistens, videri anima poterat. Et non e putri tantum enatorum animalium anima calor Aristoteli videtur, sed semine etiam constitutorum maximeque perfectorum: "Inest, inquit, in omnium semine quod facit ut foecunda sint semina, quod calor vocatur. Id autem non ignis, non talis aliqua facultas sed spiritus, qui in semine spumosoque corpore continetur, et natura, quae in eo est spiritu, proportione respondens elemento stellarum. Quamobrem ignis quidem nullum animal generat, at solis calor, et animalium, non modo qui semine continetur, verum etiam siquid excrementi sit, quamvis diversum a natura. Tamen id quoque principium habet vitale, animalem nimirum calorem."[49] Alibi itaque: "Putrefactis, inquit, facillime ingenerantur animalia, quod excreta caliditas constituit et componit corpora,"[50] ipsa videlicet anima: ipsa enim constituit Aristoteli corpora, sive igitur semini inclusus et a sole ingeneratus spiritus animae sit substantia, sive ut Aristoteli forte placet, spiritui inexistens calor e menstruo foemineo sanguine et e putri animam educat, quoniam calorem modo e cuiusvis materiae sinu quivis educere potest calor, aliud nihil praestantius praesertim quod sit diviniusque, nequaquam animale principium atque animam ipsam calorem esse negare aut queat aut velit Aristoteles; at igneum esse qui nihil illi generare omnino videtur uspiam [...].

2. Telesio, *De natura*, I 56 = ed. Ottaviani, 76–77:[51] Aliam agentem substantiam nulla agnoscit Aristoteles simplicia constituens et neque mista etiam componens, ex elementorum ipsa constituens differentiis, non quidem integris, at mutuo passis imminutisque: "Elementa, inquit, omnia contrarietatem habent, quoniam ipsorum differentiae contrariae existunt."[52] Differentiis porro a quibus constituuntur immutatis, immutantur et ipsa, ex aqua igitur aer fit, frigidate in caliditatem versa: "At si neutra, inquit, differentia neutram vicerit, mutuo autem in se ipsas agentes et mutuo patientes, sese mutuo imminuerint debilitaverintque, utriusque igitur retusa appareat vis et refracta utrumque immutatur, non igitur aqua amplius remanet nec aer et neque aer constituitur nec aqua, sed tertium quoddam, veluti ex utroque compositum. Quoniam autem virium inaequalitas non in indivisibili, non unum modo generetur, sed infinita propemodum, carnes, inquit, et ossa."[53] Et rerum generationem constitutionemque edocens Aristoteles: "Est, inquit, simplex et naturalis generatio transmutatio facta ab activis qualitatibus, a caliditate videlicet et firigiditate, e subiecta materia."[54] Non igitur simplicium modo formae caliditas Aristoteli videtur et firigiditas, sed compositorum

49 Ibid., II 3, 736b34–737a6.

50 Aristotle, *Meteorology*, IV 1, 379b7–9.

51 See also the 1570 version of *De rerum natura*, I 53 = ed. Bondì, 174.

52 Aristotle, *Generation and Corruption*, II 4, 331a15–16.

53 Ibid., II 7, 334b7–27.

54 Aristotle, *Meteorology*, IV 1, 378b33–35.

ex ipsis omnium, refractae illae quidem et mutuo a se ipsis passae imminutaeque, at ipsae eaedem tamen et neque aliae videri queant; neque enim, ut dictum est, aliud a se ipsis constituant illae; at neque aliunde compositis forma alia superveniat ulla, quae nimirum e caelo defluere queat tantum. At e caelo nihil praeter calorem Peripateticis manat, et non ille a propria caeli substantia, sed in sublunari aere a caeli motu factus, qui igitur non alius ab elementari videri queat; at quicunque sit calor, certe superaddi e caelo mistis potest tantum. Non igitur alia insit mistis forma, at calor vel frigus tantum; et non caetera omnia a calore constitui Aristoteli videtur, sed animalium etiam animae, et non e putri tantum enatorum, sed e semine etiam constitutorum. Neque igitur in illius neque in huius constitutione aliam affert substantiam agentem ullam, at illam a solis, ut dictum est, calore incorporeo humido, hanc a semini incluso spiritu, a spiritui inexistente natura quae, quid sit, aliud quam calor effingere nullus queat. In menstruo foemineo sanguine generari constituique affirmat Aristoteles: animam itaque calorem esse.

3. Telesio, *De rerum natura*, II 24 = ed. De Franco, I, 346–350: Quoniam igitur ignis aerisque et aquae generatio, eorumque omnium et terrae etiam conditiones quae non penitus a solis calore, in suprema ipsius superdicie, quae nobis conspicua est, corruptae sunt; prerterea et actiones operationesque uno illa omnia a calore et uno terram a frigore constitutam, et non aliam illorum ulli substantiam sed unum illis omnibus calorem et unum terrae frigus inesse declarant; nihilque in illorum ullo spectatur, quod alii naturae merito attribuatur et quod calori frigorive attribui non possit; ad illorumque constitutionem naturam aliam nullam sed unum modo calorem unumque Aristoteles affert frigus; utique, nisi positionis tenaces esse velint calorem frigusque non amplius ignis reliquarumque simplicium formarum organa, sed terrae quidem et aquae frigus, aeris vero et ignis calorem formam ponant. Quam rem bene aperte scribit Galenus, dum elementa non aliud esse quam materiam primam et solas qualitates docet, ex quibus habent quod elementa sint. Alexander, praeter adducta superius loca, clarissime libro secundo *De anima*, qualitates formas substantiales elementorum, ac etiam libro primo *Naturalium quaestionum* appellat. Philoponus itidem huic assentitur sententiae, secundo *De ortu et interitu*: inquit enim, ex qualitatibus et prima materia simplicia constare corpora qualitatesque principium esse formale. Hoc idem habet Olympiodorus, quarto *Meteorologicorum*. Sed quid multa? Divinus medicinae parens, Hippocrates in libro *De carnibus* quae mirabilia de calido narrat? Libet ejus verba, quia admiratione digna continent, graece referre: "Νῦν δὲ ἀποφαίνομαι αὐτὸς ἐμεωυτοῦ γνώμας. Δοκέει δέ μοι ὃ καλέομεν θερμόν, ἀθάνατόν τε εἶναι, καὶ νοεῖν πάντα καὶ ὁρῆν καὶ ἀκούειν καὶ εἰδέναι πάντα καὶ τὰ ὄντα καὶ τὰ μέλλοντα ἔσεσθαι."[55] Numquid Hippocrati calidum est accidens? Heac omnia si attente animadvertent, non elementorum tantum sed reliquorum itidem entium, quae a frigore constituta apparent, frigus, quae vero

55 Hippocrates, *On Fleshes*, 2 = ed. Littré, VIII, 584.

a calore, calorem formam asserent. Nam entia reliqua omnia et ipsae etiam plantae et animalia non aliis e rebus neque aliis a naturis, sed e simplicioribus ab eorumque formis, sibi ipsis commixtis complicatisque et mutua actione refractis, Aristoteli efficiuntur: et substantiam, aliam nullam ad eorum constitutionem is affert usquam, nec si afferre velit, unde illam eruat habet. E caelo enim, e quo solo defluere illa queat, calor tantum Aristoteli defluit, minimeque is a caeli substantia emanans, qui elementari forte praestantior vel ab eo diversus videri queat, sed ipso in igne, a solis motu commoto contritoque, aliud nihil passo, factus et ab ipso huc delatus igne; qui igitur nequaquam igneo praestantior vel alterius ab illo rationis, non natura videlicet agendique et operandi facultatibus sed robore tantum differens est. Quamobrem calorem frigusque simplicium formas esse si, quod oportet faciant, Peripatetici assentiant, reliquorum itidem entium quorumvis, vel calor vel frigus vel huic commixtus ille, forma necessario ponendus est; nihilque nobis reprungnandum, calorem frigusque substantias et rerum omnium principia ponentibus.

4. Telesio, *De rerum natura*, II 25 = ed. De Franco, I, 354–356: Aristoteli porro non simplicium modo corporum sed mixtorum etiam omnium et animalium animarum insuper substantiam calorem et spiritum visum esse, non ex ipsius dictis colligere sed ipsum apertissime id enunciantem audire licet: "Elementa, inquit, omnia contrarietatem habent, quoniam ipsorum differentiae contrariae sunt."[56] Differentiis porro a quibus constituuntur immutatis, et ipsa itidem immutantur. Frigore igitur in calorem immutato, aqua in aerem transit et aer fit. "At si neutra, ait, differentia neutram vicerit, mutuo autem in se ipsas agentes et mutuo a se ipsis patientes, mutuo sese imminuerint ac debilitaverint, et utriusque vis retusa appareat refractaque, utrumque immutatur: et nec aqua amplius remanet nec aer, neque aer constituitur nec aqua, sed tertium quoddam veluti ex utroque compositum. Quoniam vero virium inaequalitas non in indivisibili, non unum modo constituitur sed infinita propemodum, carnes, inquit, et ossa."[57] Tum propius eorum generationem edocens, "Simplex, ait, et naturalis generatio transmutatio est, facta ab activis qualitatibus, a calore videlicet et frigore, e subiecta materia."[58] Praeterea animalium sponte nascentium exortum explicans, "Generantur, inquit, in terra humoreque animalia, quoniam humor in terra, spiritus in humore, calor animalis in universo inest, ut omnia quodammodo animae plena sint. Quoamobrem consistunt celeriter, cum calor ille comprehensus sive exceptus est; comprehenditur autem, et humoribus incalescentibus efficitur veluti spumosa bulla."[59] Et alibi, "Putrefactis, dicit, facillime ingenerantur animalia; quod excreta caliditas constituit et

56 Aristotle, *Generation and Corruption*, II 4, 331a15–16.

57 Ibid., II 7, 334b7–27.

58 Aristotle, *Meteorology*, II 1, 378b33–35.

59 Aristotle, *Generation of Animals*, III 11, 726a18–24.

componit corpora,"[60] ipsa nimirum anima: siquidem anima corpora Aristoteli constituit. Tum perfectorum generationem tradens, "Inest, inquit, in omnium semine quod calor vocatur: id autem non ignis, non talis aliqua facultas sed spiritus, qui in semine spumosoque corpore continetur, et natura, quae in eo est, proportione respondens elemento stellarum. Quamobrem ignis nullum animal generat: at solis et animalium calor generativus, neque is modo, qui in semine continetur, verum etiam si quid excrementi sit, quamvis a natura diversum."[61] Clarum est, animalium animae generationem comminus intuito Aristoteli, non igneus quidem at calor certe ea esse visa est.

---

60 Aristotle, *Meteorology*, II 1, 379b7–9.

61 Aristotle, *Generation of Animals*, II 3, 736b34–737a5.

CHAPTER 4

Heat and Moving Spirits in Telesio's and Della Porta's Meteorological Treatises

Arianna Borrelli

1 Introduction

In the year 1610, at the festivity of St. Peter and Paul, forbidden books were burned in front of the cathedral of Naples. According to an extant list, among them were "several volumes of Bernardino Telesio's *De rerum natura* [On the Nature of Things]" and "the works of Telesio".[1] In the same year, the meteorological treatise *De aeris transmutationibus* (On the Transmutations of Air) by the Neapolitan scholar and experimenter Giambattista Della Porta was published in Rome with the *imprimatur* of the Court of the Holy Office of the Inquisition.[2] Given these premises, it would be very surprising to find Telesio's name mentioned in Della Porta's treatise, and indeed that was not the case. Moreover, given Della Porta's experimental interests and syncretic style and Telesio's more systematic approach to natural philosophy, it might appear that their writings would hardly display points of similarity. Yet I will argue that, at least as far as meteorology is concerned, their works have more in common than one might expect.

In his short booklet *De iis quae in aere fiunt et de terraemotibus* (On What Takes Place in the Air and on Earthquakes, 1570) Telesio used meteorology to offer an example of how his natural philosophy fared in practice, but the text was also part of a broader contemporary trend.[3] In the Renaissance there was a growing interest in describing, explaining, and predicting meteorological phenomena. This was spurred by factors ranging from reports of exotic weather from the East and West Indies to the revival of Stoic philosophy and the increasing social and economic significance of winds and water flows. The explanation of weather phenomena also had a political significance because extraordinary "meteorological" events, which at the time also included

1 "Berardini [sic] Telesii de rerum natura, più volumi", "Telesij opera", see Agrimi, "Telesio nel Seicento napoletano," 345.

2 Paolella, "Prefazione," xiv.

3 Telesio, *De iis quae in aere fiunt et de terraemoribus* (1570).

 | DOI:10.1163/9789004352643_006

earthquakes and comets, could be seen as announcements/messengers of divine punishment for lay and religious leaders. Telesian meteorology builds upon the Aristotelian framework regarding "vapors" emitted by earth and water under the influence of the Sun as the material out of which all weather events are made. However, Telesio holds very different views from Aristotle on how the Sun acts upon bodies and disagrees with the Stagirite because he conceives that vapors are capable of moving of their own will, prompted by a desire to escape condensation or rarefaction. Telesio's meteorological explanations are based on his own general view of natural phenomena, but they also reflect notions that were starting to shape meteorology in his age and which were central to Della Porta's *On the Transformations of Air*. Della Porta's treatise is longer and of broader scope than Telesio's booklet and combines detailed discussions of how the various meteorological phenomena are to be explained, with descriptions of their role in everyday life and of the various traditions and superstitions associated with them.[4]

In his treatise Della Porta is not primarily interested in explaining weather and climate according to some predetermined natural philosophical framework, and he is happy to combine ideas from different world views. He deploys a bottom-up approach to investigating nature, regarding each phenomenon in its empirical details and trying as far as possible to illustrate the origin of wind, rain, or thunder by means of simple experiments, which he usually states he has performed personally in front of an audience. However, Della Porta also presents his meteorology as embedded in an overarching natural philosophical context, albeit a rather sketchy one, and assumes that all meteorological phenomena result from the transmutations of the element "air". For Della Porta there are no vapors coming from the earth or the sea, and this point sets him in contrast to both Aristotle and Telesio. Yet he regards the heating action of the Sun as the main motor of weather and climate, as Telesio does, and also shares Telesio's notion that the matter out of which winds, rains, and other phenomena are made is not passive but displays a degree of independent activity. Although these general similarities between the two authors can easily be attributed to more general trends of thought in their time, there are also more specific parallels between their meteorological works, and I believe that locating and analyzing them will contribute to a better understanding and contextualization of the development of natural philosophy in southern Italy around 1600. This paper aims to elaborate such a comparison between the two authors. Following a short overview of the many facets of Renaissance meteorology (section 2), I will discuss the contents of Telesio's meteorological

4 Della Porta, *De aeris transmutationibus* [1610] (2000).

booklet (section 3), and then go on to compare its main features with some aspects of Della Porta's meteorological treatise (section 4), closing with some concluding remarks in section 5.

2 Renaissance Meteorology

Today weather and climate are still fundamental aspects of human life, society, and culture, but this was even more true in pre-modern and early modern times.[5] Evidence of systematic observations and records of local and seasonal weather variations is extant from earliest antiquity, and meteorological predictions were among the main tasks of ancient prophecy. In the Graeco-Roman and Judeo-Christian traditions, discussion of meteorological phenomena contributed to shaping philosophy and theology. At the same time, at a less scholarly level, local weather knowledge, both with or without practical interest, was collected and formed the basis for proverbs and legends. In this sense, meteorology was part of each epoch of Western culture, and its subject matter ("meteors") comprised not only weather and climate but also topics from what we would today call geophysics such as earthquakes or rivers. However, in the European Renaissance the different dimensions of meteorology combined and resonated with each other: popular lore with scholarly reflections, practical information with tales of real or alleged weather marvels, religious interpretations with astrological predictions.

This trend had already started in the late Middle Ages, but grew in the Renaissance thanks to a number of factors supporting and strengthening it: the rediscovery of ancient philosophical traditions offering explanations of weather and climate which differed from the dominant Aristotelian world view, reports on exotic weather phenomena from the oceans and the East and West Indies, the increasing development of (and scholarly interest in) pneumatic, thermodynamic, and chemical technologies (e.g. pumps, ovens, windmills, gunpowder), and the relevance of weather "miracles" in the

5 A more detailed discussion of Renaissance meteorology than the following short overview can be found in: Borrelli, "The Weatherglass and its Observers," and "Pneumatics and the Alchemy of Weather," where a complete list of references on the subject is given. The main works on the history of pre-modern and Renaissance meteorology are: Bos and Burnett, *Scientific Weather Forecasting in the Middle Ages*, Casali, *Le spie del cielo*, Hellmann, "Wetterprognosen und Wetterberichte," "Meteorologische Beobachtungen," and "Entwicklungsgeschichte des meteorologischen Lehrbuches," Kahn Heninger, *Handbook of Renaissance Meteorology*, Martin, *Renaissance Meteorology*, Schneider-Carius, *Wetterkunde—Wetterforschung*, Taub, *Ancient Meteorology*, Vanden Broecke, *Limits of Influence*, and Vermij, "A Science of Signs."

context of the political and religious conflicts of the Reformation and Counter-Reformation. The printing press was a further element reinforcing this trend and allowing an increasing number of people to take part in it. In the sixteenth century an increasing number of almanacs with weather predictions were produced, as well as broadsides with images, descriptions, and interpretations of extraordinary weather occurrences. Especially from the middle of the century onward, treatises on weather and climate aimed at a learned but not necessarily scholarly audience also appeared, some written in Latin and some in vernacular languages, among them Antoine Mizauld's *Meteorologia* (Meteorology, 1547), Cesare Rao's *I meteori* (The Meteors, 1582), and William Fulke's *A Goodly Gallerye with a Most Pleasant Prospect, into the Garden of Naturall Causes of All Kind of Meteors* (1563).[6] Simple manuals explaining how to predict weather were also quite successful, such as Antoine Mizauld's *Le Mirouer du temps* (The Mirror of Weather, 1547) or the German *Bauernpractick* (1508).[7] Probably the best known example of how meteorological predictions and political and religious struggles could spark mass panic when combined with the power of the press was the prediction of a flood for the year 1524.[8]

Besides these new texts, classics of meteorology were also reprinted and translated into vernacular languages, including Aristotle's *Meteorologia* (Meteorology), whose principles had dominated reflection on the subject since the thirteenth century. Aristotle also remained a key reference point during the Renaissance, but even authors clearly identifying themselves as Aristotelian often proposed revisions of his views, while outside the Aristotelian community a number of works in Latin and vernacular appeared in which Aristotelian meteorology was not only criticized and revised but also thoroughly rejected.[9] Among the most innovative treatises were Paracelsus' *De meteoris* (On Meteors, written before 1541 and printed 1566), the first meteorological treatise written in the German language, Della Porta's *On the Transmutations of Air*, the widely read *Ein kurzer Tractat von der Natur der Elementen* (Short Treatise on the Nature of the Elements, 1608) by the Dutch scholar and engineer Cornelis Drebbel, and eventually also René Descartes' *Les Météores* (The Meteors, 1637).[10]

6 Mizauld, *Meteorologia*, Rao, *I meteori*, Fulke, *A Goodly Gallerye*. For more titles see Borrelli, "Pneumatics", 44–46, 55–59.

7 Hellmann, "Die Bauern-Praktik (1508)," Mizauld, *Mirouer du temps*.

8 Zambelli, "Fine del mondo o inizio della propaganda?"

9 For a discussion of the Aristotelian tradition in meteorology during the Renaissance see Hellmann, *Geschichte des meteorologischen Lehrbuchs*, Martin, *Renaissance Meteorology*, Thomas, "Experience of the New World."

10 Della Porta, *Aeris*, Descartes, *Les Méteores/Die Meteore*, Drebbel, *Ein kurzer Tractat von der Natur der Elemente*, Paracelsus, "De meteoris," in Paracelsus, *Sämtliche Werke*, 125–206.

Telesio's meteorology is to be placed in this cultural context. As Nicola Badaloni noted, Telesio had a special interest in Aristotle's *Meteorology*, and not only published the booklet *On What Takes Place in the Air and on Earthquakes* but also separate short treatises *De mari* (On the Sea, 1570), *De iride* (On the Rainbow), *De cometis et lacteo circulo* (On Comets and the Milky Way) (both published posthumously in 1590), as well as a text *De fulmine* (On Lightning), which was preserved in manuscript form and only printed in 1874.[11] In these works Telesio appropriated and modified Aristotle's meteorological principles, according to which all meteors are due to two kinds of exhalations drawn by the Sun from the earth: a dry one giving rise to winds and 'fiery' meteors (lightning and thunder, comets, earthquakes) and a humid one causing 'watery' phenomena (rain, snow, hail). Weather and climate are determined by the interplay of two pairs of opposite qualities: hot-cold and dry-humid, but for Aristotle the heat which appears to derive from the Sun has nothing to do with its light, and is actually due to the friction of the rotating celestial sphere on the higher levels of the air.

Since the early Renaissance, however, schools of ancient philosophy were being rediscovered which provided alternative templates to explain meteorological phenomena. In this context the Stoic views on nature in general and weather in particular were important, especially as expounded in Lucius Annaeus Seneca's *Naturales quaestiones* (Natural Questions).[12] According to the Stoics' views, all forms of life and motion on Earth were due to the activity of a "spirit" or "pneuma" which is conceived as a mixture of air and fire emanating from the celestial regions and bringing warmth and life to the terrestrial sphere.[13] Stoic *pneuma* could be seen as a subtle material substance or as a form of immaterial divine emanation, and in the Renaissance it was one form of a conceptual structure which had emerged from the interplay of various traditions; it is best exemplified by the triad body-spirit-soul, where the term

---

11 Telesio, *De mari* (1570), id., *De iride* and *De cometis* in *Varii de naturalibus rebus libelli* (1590). A more recent edition and discussion of the treatise *On lightning* is found in Delcorno "Il commentario 'De fulmine' di Bernardino Telesio," while the booklet on comets has been recently edited anew on the basis of additional manuscript evidence in Telesio, *Sobre los cometas y la Vía Láctea*. On Telesio's interest in meteorology see Badaloni, "Sulla costruzione e la conservazione della vita," 27.

12 Seneca, *Naturales quaestiones/Natural Questions*.

13 On Stoic philosophy in the Renaissance see Barker, "Stoic Contributions," Verbecke, *L'evolution de la doctrine du pneuma*. On the various Renaissance notions of "spirit" discussed in the following pages see: Bianchi and Fattori, *Spiritus*, Debus, "Chemistry and the Quest for a Material Spirit of Life," Debus, "The Paracelsian Aerial Niter," Klier, *Die drei Geister des Menschen*, Putscher, *Pneuma, spiritus, Geist*, Walker, *Spiritual and Demonic Magic*, Walker, "Astral Body."

"spirit" (*spiritus, Geist, spirito, esprit, pneuma*) indicates a middle and mediating instance between incorporeal soul and material body. Besides Stoic philosophy, other traditions which involved a similar structure were Neoplatonism, Hippocratic-Galenic medicine, and of course the Christian doctrine of the holy trinity. A more recent trend contributing to the theme was Paracelsian alchemy, with its three principles "salt", "mercury", and "sulphur". Notions of spirit in these schools of thought were very different in their features and functions, but could be related to the general tripartite scheme and so could combine and resonate with each other. Any attempt to discuss the complexity and variety of Renaissance notions of spirit would go well beyond the limits of the present paper, but it is important to note that in the course of the late sixteenth and early seventeenth century a tradition of meteorological thought emerged in which the fundamental explanatory pattern was the activity of a spirit-like substance—usually air—which was made alive by solar heat.[14] Della Porta's treatise belonged in this tradition and, as we will see, some similarities between Telesio's and Della Porta's meteorologies can be attributed to the common influence of the broader trends sketched above. However, more specific analogies will emerge upon closer inspection of the writings.

3 Telesio's Treatise *On What Takes Place in the Air and on Earthquakes*

Telesio's booklet *On What Takes Place in the Air* was published for the first time in 1570, at the same time as the second edition of his magnum opus *De rerum natura iuxta propria principia* (On the Nature of Things according to Their Own Principles, 1570) and two further booklets: the treatise *On the Sea* mentioned above and a text *De colorum generatione* (On the Generation of Colors).[15] The meteorological booklet comprises eighteen chapters and is only twenty-one pages long. The first eleven chapters are an exposition of Telesio's own views on weather and climate, while the last seven are a critique of some key aspects of Aristotelian meteorology. The first four chapters are devoted to vapors and their different behavior depending on their rareness or denseness, while Chapters 5 and 6 discuss the heating action of the Sun. In Chapters 7 to 9 the principles expounded in the previous sections are used to account for local and seasonal weather variety, while Chapters 10 and 11 are devoted to thunder and lightning and to earthquakes, respectively. In the

14 Borrelli, "Pneumatics".

15 Telesio, *De iis quae in aere fiunt*, Telesio, *De mari*, Telesio, *De colorum generatione opusculum* (1570), Telesio, *De natura secondo i suoi principi* [1570] (1999).

following pages I will offer a brief overview of the contents of the individual sections, expanding more on those points which are particularly relevant for the later comparison with Della Porta.[16]

3.1 *Chapters 1–4: Vapors and Their Properties*

The first four chapters of Telesio's booklet introduce the vapors from which all meteors are made, and discuss their properties. The titles of the chapters are:

> Chapter 1: Which vapors primarily become hail, rain, and snow, and which instead spread as winds; and we should investigate the way in which the vapors become these things, and what moves them.[17]
>
> Chapter 2: Which vapors primarily become snow, rain, and hail, and which flow in the air and appear as winds.[18]
>
> Chapter 3: How vapors transform into waters, hail, and snow.[19]
>
> Chapter 4: Why vapors move obliquely, and how long.[20]

Telesio begins his treatise by stating that meteorological phenomena are caused by vapors emitted by the earth under the action of the Sun: some of these vapors condense into rain, hail, and snow, while others flow around as winds. However, he continues, the question is why the same vapors can lead to such different weather manifestations. The answer given in Chapter 2 is that some vapors are unable to give rise to rain, hail, or snow because of their "extreme rareness" (*summa tenuitas*) which makes it impossible for them to condense when they are in an open space or to be condensed by cold. Therefore, the vapors giving rise to rain, snow, and hail have to be "somehow

16 As a basis for my discussion I have used the edition Telesio, *De iis quae in aere fiunt*, as well as the Italian translation made by Francesco Martelli (1534–1587), which is preserved in a manuscript printed in 2013. Cf. Telesio, *Delle cose naturali libri due* (2013), 433–478, Puliafito, "Introduzione."

17 "Qui vapores praecipue in grandines, pluvias et nives aguntur, quive in ventos diffunduntur; et modus, quo in ea quae dicta sunt illi coguntur et quo hi moventur, inquirendus videri debet—Caput 1" (Telesio, *De iis quae in aere fiunt*, 2r).

18 "Qui praecipue vapores in nives, pluvias grandinesque aguntur; et qui per aerem fluunt ventique apparent.—Caput 2" (Telesio, *De iis quae in aere fiunt*, 2r–2v).

19 "Quomodo vapores in aquas, et grandines, et nives aguntur.—Caput 3" (Telesio, *De iis quae in aere fiunt*, 3v–4r).

20 Telesio, *De iis quae in aere fiunt*, 4r–4v: "Cur vapores in obliquum moventur, et quamdiu—Caput 4."

dense" (*crassiuscolos*) and not able to move fast. In short, the quality pair rareness-denseness determines the behavior of the vapors.

After this introduction, the causes of condensation (Chapter 3) and rarefaction (Chapter 4) of vapors are explained. According to Telesio, vapors can condense without needing any external causes, such as cold or compression: "nothing happens to them [i.e. the vapors] other than that they simply appear to condense in themselves."[21] Thus, it may rain or hail even when it is very hot, although snow requires cold. This explanation is particularly important, as it is based on the assumption that vapors have some kind of independent agency and somehow tend to condense. This notion is in line with Telesio's views on the activity of nature and is the starting point for explaining wind motion as well. Rare vapors, Telesio explains, "diffuse obliquely of their own motion and out of their own free will, as they are naturally mobile, but do not have the same rareness as pure air, so that they can be neither moved upwards, nor tend at all to do so".[22] Once again, the degree of density of the vapors is the main characteristic determining their behavior, and Telesio states that the vapors, despite being rarer than rain-vapors, are nonetheless too dense either to be moved upwards by some external cause or to move upward by their own will. However, they can and do tend to move obliquely, to avoid being made denser:

> Being by nature rare, they hate and greatly resist being compressed in themselves, or condensing into a completely different substance. Sometimes it happens that they have increased so much that the space in which they are cannot contain a larger number or a greater volume of them—either because new ones are coming up under them, constantly emerging from earth and sea, or because the same vapors need a larger space, having been made rarer by the Sun, or in any circumstance in which they cannot preserve their rareness, but are compressed and made smaller. In these cases, they move horizontally with a light, peaceful motion, if by doing so they can avoid compression. [...] but it may happen that the air is completely full of vapors and they are so pressed in themselves that they cannot stand further pressure, and then they are all made larger at the same time by the Sun, and other vapors are added to them from earth or sea, so that they are all steadily further compressed, but cannot escape anywhere. In those cases they will steadily increase their motion, become very fast, destroy everything standing in their way and

21 Ibid., 3r: "nihil omnino aliud pati, nisi in se ipsos conspissari videntur."

22 Ibid., 4v: "proprio forte motu, sponteque sua in obliquum feruntur, bene propria natura mobiles, nec dum aeris nacti tenuitatem, nec ad superiorem efferri possunt locum, nec efferri omnino appetuntur."

> remove all obstacles, so as to avoid compression and move away. [The vapors] will then continue to flow until they arrive in an open space, or are reduced to water, or become so rare that they attain the same nature as pure air and become one with it, and never again move downward, but tend to move up and do so.[23]

This passage offers a good example of how the variety of weather phenomena resulted from the interplay between the (possibly changing) tendency of vapors to keep a certain (rare or dense) density and the more or less accidental external causes (e.g. heat, cold, obstacles) trying to change it. Rare vapors tend to remain rare and escape compression by gently flowing away, but if they cannot do so they generate a storm to remove obstacles. Depending on circumstances, the same vapors may in principle flow like wind, turn into water or become air. It must also be noted that once the vapors have attained the same rareness as air, they also desire to move upward like air does.

3.2 *Chapters 5–6: The Heating Power of the Sun*

The following two chapters deal with the way in which the Sun acts on vapors, the earth, and the sea. Their titles are:

> Chapter 5: Why vapors move in the way they seem to do.[24]

> Chapter 6: What kind of heat emanates from which things, and in what state of rareness.[25]

---

23 Ibid., 4v: "At vel nihil moveri appetentes a seipsis moveri tamen, impellique videri possint. Quoniam enim tenues sui natura in seipsis spissari, densarique in alienam omnino agi substantiam, summe odio habent summeque aversantur, ubi tanti facti sunt, ut locus in quo continentur plures, aut ampliores eos capere non possit, si vel novi, qui assidue a terra, marique emergunt, eos subeunt, vel amplius ipsi a sole attenuati ampliore indigeant loco, quacunque omnino occasione in propria ibi servari tenuitate non possint, sed comprimantur, inque angustius agantur, id ut vitent in obliquum feruntur, et levi quidem, placidoque motu, si eo compressionem vitant. [...] Nam ubi aer universus vaporibus oppletus est adeoque ii in se ipsos coacti ut cogi amplius non sustineant et universi simul ampliores a Sole fiunt, aliique e Terra marique educti illos subeunt, assidue nimirum magis undique comprimuntur, itaque commoti compressionem non effugiunt, assidue motum magis concitant et rapidissimo eo tandem et quo quae obstant omnia amoliantur feruntur, ut quibus quidvis sustinendum moliendumque sit compressionem ut vitent; et tamdiu per aerem fluunt, dum vel apertum liberumque nanciscantur locum vel in aquas cogantur, vel etiam longe a motu tenuiores facti aerisque nacti naturam et unum omnino aeri facti nequaquam in inferiore acquiescant loco, sed superiorem appetant et ad superiorem ferantur."

24 Ibid., 4v–5r: "Cur vapores eo motu moveantur, quo moveri videntur.—Caput 5."

25 Ibid., 5r–6r: "Qui calor quali e re, qua elabatur in tenuitate—Caput 6."

Chapter 5 explains how the Sun stimulates winds by making vapors rarer: whether winds rise or not depends on the interplay between solar heat, vapors, and the sea. Here Telesio makes use of a theory of the heating action of the Sun which he discusses in more detail in his main philosophical work. Unlike Aristotle, he thought that sunlight is a direct source of heat that works more effectively when the Sun is higher above the horizon and its rays hit the earth's surface perpendicularly and less effectively when the Sun is lower and its rays are inclined with respect to the ground. The strongest heating action occurs when the Sun stands perpendicular above deep sea, since "the light is reflected back onto itself at length by the solid, impenetrable sea bottom and so becomes very strong".[26] The minimum heating effect is instead generated by inclined solar rays on shallow water. What does this imply for wind production? Here, once more, the complexity of weather phenomena comes to the fore: neither direct Sun on deep sea nor low Sun on shallow sea produce winds, in the first case because the vapors drawn up are too rare, and rise up like air, and in the second case because the vapors are too dense and not mobile enough. It is only with an overall intermediate heating effect (high Sun on shallow sea or low Sun on deep sea) that winds arise. The explanation of why high Sun heats more than low Sun is only given in the meteorological treatise as an aside but is discussed in more detail in *On the Nature of Things*, as we shall see later on.

In Chapter 6, Telesio expounds his views on the action of heat on different bodies. The premise is that heat has goals to attain through its interactions with bodies, and that bodies in turn may resist its actions. Once again, the fundamental quality in this interplay is the rareness (*tenuitas*) or denseness (*crassitia*) of bodies. Heat, states Telesio, moves and penetrates into things similar to itself, making them rarer and mobile. As long as it can do this, it keeps on making them rarer, but "as soon as both tasks are completed, even if the things could become even rarer, the heat stops rarefying them further and expires away, as though its aim had not been to make things rarer, but rather to escape its opposite and move itself".[27] As with denseness and rareness, heat also has different levels with different behavior: maximum heat never emanates from the rarest bodies, but always from bodies which are not usually rare yet have been made to reach their maximum rarefaction by the action of heat, such as a burning piece of wood. Heat of the lowest level also comes from bodies at their maximum rarefaction, while medium heat propagates from one body to a thicker and denser one, although it never emanates from things which are

26 Ibid., 5r: "in se ipsam et a solido renitenteque maris fundo reflexa lux longe fit robustissima."

27 Ibid., 5v: "at utroque peracto, vel si amplius illa attenuari possint, nihil eas attenuat amplius, sed statim evolat, ut eius non attenuandi, sed e contrariis abeundi et sese commovendi propositus finis sit."

extremely thick and dense. These rather complex principles allow Telesio to make sense of a series of meteorological phenomena, e.g. that rare mountain air is due to an oblique Sun, while dense air is produced by high Sun on grassy landscapes.

3.3 *Chapters 7–9: The Variety of Weather Explained*

> Chapter 7: In which seasons of the year and in which lands rain, snow, and hail occur, and winds flow, and which winds flow with good weather and with rain.[28]

> Chapter 8: How it was correctly understood that water and rain derive from thoroughly dense vapors, and hail from the densest vapors of all.[29]

> Chapter 9: Why the vapors flowing in the air are correctly held to be rarer.[30]

After listing various seasonal combinations of rain, wind, and other weather phenomena in Chapter 7, Telesio uses the interplay of heat and cold and rare and dense vapors to explain them. Chapter 8 discusses watery meteors, stating that rain is more frequent in spring and autumn in humid and "soft" (*mollis*) lands, when solar heat is not very strong. In summer, however, when solar heat is strong and the land is dry, no rain occurs. However, exceptions are possible under specific circumstances. In Arabic lands, for example, the high Sun draws rare vapors from dry land, and these flow like wind until they reach very high mountains: at that point, even those very rare vapors are forced to condense into heavy rain. Chapter 9 explains that winds are made out of rare vapors moving obliquely which are usually raised by strong heat from hard, dry land. However, due to the differences in seasonal heat and local terrain, different vapors can be raised from the same land according to the season, or in the same season depending on the local features of the land. Here too Telesio makes clear that there is no strict difference between vapors which flow like wind and those which condense into rain: "those of them [vapors] which are condensed into water or even hail are not very different from those which flow through

28 Ibid., 6r–6v: "Quibus in anni temporibus, quibusque in terris pluviae, nivisque, et grandines fiunt, ventique spirantes, et qui venti quando sereni, quandove pluvii—Caput 7."

29 Ibid., 6v–7v: "Aquas, nivesque e crassis omnino vaporibus, et grandines e crassissimis constitui iure visas fuisse—Caput 8."

30 Ibid., 7v–8r: "Iure qui per aerem fluunt vapores, tenuiores visos fuisse—Caput 9."

the air, and both can, with little changes, transform into each other, and are forever mixing with each other".[31]

3.4 *Chapters 10 and 11: the Origin of Thunder and Lightning and of Earthquakes*

> Chapter 10: How thunder and lightning happen and why lightning moves downward and comes with various colors and strength.[32]
>
> Chapter 11: From which causes and how earthquakes are produced.[33]

Telesio explains fiery meteors largely along Aristotelian lines. Thunder is due to clouds colliding and lightning to rare vapors escaping from clouds. During storms, the upper part of the clouds becomes colder and denser than the lower one and, since the vapor in the lower clouds cannot stand becoming denser and colder, it throws itself in the opposite direction, that is towards the earth. The vapor emits light and heat because it contracts into itself and moves very fast. Telesio's explanation of earthquakes is more original. Initially, he follows Aristotle in stating that vapors are drawn by the Sun not only above the earth's surface, but also below it, so that empty caves underneath the ground become filled with vapors which cannot escape and eventually cause an explosion. However, Telesio describes earthquakes by using an analogy with gunpowder explosions:

> [Vapors] open up the earth—which is, in a sense, keeping them bound—as violently as they can, more or less in the same way in which the powder made out of sulphur and saltpetre, when it is given fire and thus made larger than what fits in the weapons or in the underground mines where it was set, sends balls of lead very far and sometimes even blows up the cannons and the mountain with great violence.[34]

31 Ibid: "[vapores] qui eorum in aquas, aut etiam in grandines coguntur, non valde ab iis, qui per aerem fluunt, differre, at utrique paululum quid immutati in alteros agi videantur, et perpetuo sibi ipsis commisti esse."

32 Ibid., 8r: "Quomodo tonitrua fulguraque fiunt et cur fulgura deorsum ferantur, variusque eorum fit color et vis—Caput 10."

33 Ibid., 8r–8v: "A quibus et quomodo terraemotus fiant—Caput 11."

34 Ibid., 8v: "[vapores] Terram, a qua veluti coercentur, quam possunt vehementissime amoliuntur; eo ferme pacto, quo pulvis e sulphure, salnitroque confectus in ignem actus, et propterea amplior factus, quam ut tormenta in quae adactus est, cuniculique, quibus inditus est, capere eum possint, pilas plumbeas quam longissime abigit, et ipsa interdum tormenta et montes et ipsos dirumpit, summaque in sublime attollit vi."

Using gunpowder to explain weather phenomena either by analogy or in real terms was an idea introduced by Paracelsus and further developed in the Renaissance in a theory of "aerial nitre", a substance responsible for cold and fire in the air.[35] However, Telesio's use is rather early and does not appear to be directly linked to a Paracelsian template, as he only used gunpowder to explain earthquakes and not thunder and lightning. It is nonetheless interesting to note that the technological innovation of gunpowder also found its way into Telesian meteorology.

3.5 *Chapters 12–18: Telesio's Critique of Aristotelian Meteorology*

In his works Telesio often goes on to criticize Aristotle's opinions after having expounded his own views, and this is what he does in this case, too. The titles of the relevant paragraphs are:

> Chapter 12: For Aristotle rain and hail derive from humid vapors and winds from dry ones, and the humid vapors are drawn out of waters and the dry ones out of earth, and both at the same time.[36]
>
> Chapter 13: Aristotle wrongly claimed that vapors drawn from earth are dry, and that cold is necessary for the production of waters.[37]
>
> Chapter 14: Aristotle incorrectly believed that the substance of winds could be drawn only from earth and that the two exhalations cannot be produced independently of each other.[38]
>
> Chapter 15: Which arguments convinced Aristotle that the exhalations from which winds are made cannot be produced in a different way than the other ones, and can be produced only from humid earth.[39]
>
> Chapter 16: It is not true, as instead Aristotle claims, that winds occur after rains, that humid earth emits exhalations, that the winds Auster and

---

35 Debus, "Aerial Niter".

36 Telesio, *De iis quae in aere fiunt*, 8v–9v: "E vaporibus humidis pluviae grandinesque, venti vero e siccis Aristoteli fiunt et humidi quidem ex aquis, sicci vero e Terris, at utrique perpetuo simul educuntur. Caput 12."

37 Ibid., 9v–10r: "Perperam vapores, qui e Terra educuntur, siccos Aristoteli visos fuisse, et ad aquarum constitutionem frigus allatum—Caput 13."

38 Ibid., 10r–10v: "Non recte ventorum materiam e Terra modo, et neutram exhalationem seorsum ab altera educi posse Aristoteli visam fuisse—Caput 14."

39 Ibid., 11r: "Quae exhalationem, e qua venti fiunt, seorsum ab altera et e Terra omnino non madefacta educi non posse Aristoteli persuasere—Caput 15".

Borea mostly blow because they come from places where there has been much rain.[40]

Chapter 17: Etesian winds are not due to the melting of subpolar snow, as Aristotle claims, but come from the sea.[41]

Chapter 18: Aristotle did not give a correct account of the oblique motion of vapors.[42]

The titles of these chapters already state the subject they deal with. One of Telesio's main points of critique is that the traditional Aristotelian distinction between dry and humid vapors is linked to secondary qualities (dry and humid) and not to the primary ones (rare and dense). The distinction is also too severe, since in principle both kinds of vapors can give rise to all weather phenomena. Finally, Chapter 18 is devoted to disproving Aristotle's classical explanation of the oblique motion of winds, which states that the dry vapors, when they reach a certain height, are carried away by the circular motion of the heavenly spheres.

3.6 *Telesio's Meteorology and His Theory of Light and Heat*

Telesio expounded his natural philosophical views in *De rerum natura iuxta propria principia* (On the Nature of Things According to Their Own Principles) (1st ed. 1565, 2nd ed. 1570, 3rd ed. 1586), a text whose originality and complexity have been noted by many early modern and modern scholars.[43] Without in any way attempting to offer an overview of Telesio's general philosophy, I would like to briefly discuss how the issues he dealt with in the meteorological treatise relate to the broader picture and, in some cases, may have provided the starting point for developing it further.

As we have seen in the previous section, Telesio thought that all meteorological phenomena are due to vapors drawn up by solar heat from the earth

40 Ibid., 11v–12r: "Nec post pluvias ventos fieri, quod Terra madefacta exhalationem emittat, nec Austrum Boreamque plurimum spirare, quod a locis oriantur, in quibus pluviae multae fiunt, ut Aristoteli placet."

41 Ibid., 12r–12v: "Etesias non subpolaris nivis liquefactione, ut Aristoteli placet, sed e mari oriri—Caput 18."

42 Ibid., 12v–13r: "Non recte obliqui vaporum motus causam ab Aristotele traditam fuisse—Caput 17."

43 Telesio, *De natura iuxta propria principia* (1565), *De rerum natura* (1570), Telesio, *De rerum natura: libri I–II–III* [1586] (1965) and *De rerum natura: libri VII–VIII–IX* [1586] (1976). For an overview of Telesio's life and work, see Bondì, *Introduzione a Telesio*.

and sea, and the only relevant quality whose variations determine the behavior of the vapors is their rareness or denseness. For Telesio there are not two distinct kinds of vapors but a continuous scale of rareness and denseness according to which vapors give rise to different phenomena. The changes in rareness or denseness of the vapors are in turn due to the interplay of three factors: (1) the varying action of solar heat, (2) the agency and goals of the vapors, which may tend to remain rare or dense and are able to move themselves and other bodies, and (3) contingent circumstances, such as the features of the local land and sea. This picture broadly fits some main principles of Telesio's natural philosophy: the role of heat and cold in driving natural phenomena, rareness, and denseness as the primary qualities of bodies which are changed by heat and cold, and the activity and sensitivity of all kinds of bodies which blur the distinction between animate and inanimate matter. A fundamental component of Telesio's philosophy is the notion of a corporeal but rare spirit (*spiritus*) present in humans, in addition to the incorporeal soul (*anima*), but also existing in animals and to some extent in all bodies. It is not possible for me to discuss Telesio's notion of spirit in its complexity and transformations, but various historians have noted how it can be connected to the different notions of spirit which mutually resonated in the Renaissance.[44] Roberto Bondì analyses the different versions of Telesio's notion of *spiritus*, which in his opinion originally combined different traditions, including the magical-hermetic one.[45] According to Martin Mulsow, Telesio's notion of spirit is closely linked to the "innate heat" (*calor innatus*) of the medical tradition, but is at the same time analogous to solar heat and to air as a principle of life, and Mulsow suggests that Telesio followed a "strategy of naturalization of *calor coelestis* [celestial heat]."[46] Mulsow also notes that Telesio's reflections were initially motivated by an interest in meteorology, and Raffaele Cirino, in his discussion of *On What Takes Place in the Air*, remarks on how Telesio establishes a connection in *On the Nature of Things* between the spirit of human bodies on the one hand and heat and air on the other by referring to his meteorological treatise.[47] Hiro Hirai views Telesio's heat as a universal life-giving principle in the tradition of "cosmic heat" linked to the interpretation of a specific passage of Aristotle's *De generatione animalium* (On the Generation of Animals) which had already been at the core of Girolamo Cardano's concept of cosmic heat.[48]

44 See *infra*, Granada, Chap. 2.

45 Bondì, *Telesio*, 3–39.

46 Mulsow, *Frühneuzeitliche Selbsterhaltung*, 304: "eine Strategie der Naturalisierung des *calor coelestis*." Cf. 234–246, 251–305

47 Cirino, "Bernardino Telesio e 'delle cose che in aria si fanno,'" 69–70.

48 Hirai, "Il calore cosmico di Telesio." See *infra*, Hirai, Chap. 3.

More generally, Roberto Bondì underscores the importance of reading Telesio's magnum opus in connection with his shorter booklets; indeed in Chapter 6 of his meteorological treatise, as we saw, Telesio refers to *On the Nature of Things* when explaining how heat penetrates bodies, making them rarer and giving them the ability to move.[49] In this sense the dynamics of meteorological vapors reflects the complexity of the Telesian interplay between heat, spirit, and matter at a cosmic level. For our purposes it is not very important whether Telesio was influenced primarily by the medical tradition or if he also had Stoic or alchemical views in mind: what is relevant is that his picture of meteorology is very compatible with those views. This fact potentially makes his interpretations of weather and climate of interest for Della Porta, as we shall see.

In his meteorological treatise, Telesio not only discusses the effect of heat on bodies but also the reason why solar heat works differently depending on the season, the hour, and the location: the heating power of the Sun is linked to its inclination above the local horizon. The exact way in which the connection occurs is only hinted at in the meteorological treatise but is explained in more detail in *On the Nature of Things*. The most detailed treatment of the subject is given in Book IV of the third and final edition of the work, which was published in 1586, but the key points are already found in both its first and second editions (1565, 1570).[50] Since the meteorological booklet was published in 1570, the same year as the second edition of *On the Nature of Things*, in my discussion I will refer to that edition of the work, where the question of how the Sun heats the earth and the relationship between light and heat are dealt with in Chapters 43 to 48 of Book II.[51]

Chapters 43 and 44 of Book II of the second edition of *On the Nature of Things* (1570) are devoted to criticizing the Aristotelian claim that the Sun heats the sublunary world by way of the friction between the rotating celestial spheres and air, a theory which is, according to Telesio, incapable of explaining the local and seasonal differences in climate.[52] Chapter 45 offers Telesio's view of the matter, which is already summarized in its title: "Light is hot and has the ability of multiplying and reflecting itself from solid objects, and heat becomes most intense where light is most collected in itself and becomes more

49 Bondì, *Telesio*, 4–5.

50 On Telesio's theory of light and heat, see De Franco, "La teoria della luce," Maierù, "Alcune riflessioni sul contesto in cui leggere il *De rerum natura*," Mulsow, *Selbsterhaltung*, 104–139.

51 I have made use of the following edition and Italian translation of Telesio, *De rerum natura* [1570], *La natura secondo i suoi principi* (1999).

52 Ibid., 361–375.

abundant".[53] Thus, solar light and heat are intimately related and, as explained in the meteorological treatise, the heat of the Sun is more intense where its light hits the ground perpendicularly and is therefore reflected onto itself and concentrated.[54] Telesio illustrates his view by describing a simple experiment with a concave mirror, and it is worth quoting the passage in full:

> Light reflected from concave mirrors becomes so strong for no other reason that it arrives in great quantity in those mirrors, and is all reflected at the same point. And this can be seen to happen with one's own eyes, if one sets a sheet of paper with many holes near a [concave] mirror; when the paper is at a shorter distance from the mirror, one sees light coming through all holes from each point of the mirror on which it is reflected, but when you move the paper away from the mirror bit by bit, you will see the light come out from less and less holes, as it increasingly concentrates more and more, until in the end all light comes out from only one hole, since it flows into one point. This point is called the point of combustion, because any body which is exposed to light only and precisely there will burn. [...] But now let us return to what we were discussing, and let us demonstrate that heat is produced according to the reflection and condensation of light described above. It would not even need to be proven, since it is evident in all places and at all times that heat is greater where the light is reflected and brought together into itself. For example, because of this effect, southern lands are hotter than northern ones, and the summer Sun at midday is hotter than the winter Sun when it is rising or setting, because the nearer the Sun is to the vertical, the more its light will reflect onto itself and unite to itself.[55]

53 Ibid., 375: "Lucem calidam esse et sese multiplicandi et a solidis resiliendi facultate praeditam, et ibi omnino maiorem fieri calorem, ubi magis in se ipsam colligitur et copiosior fit lux."

54 Ibid., 375–381.

55 Ibid., 376–379: "Nec alia de re quae e speculis concavis resilit lux adeo robusta fit, nisi quod multa in specula huiusmodi incidit et quae incidit universa in unum resilit punctum; et id fieri ipsis intueare oculis, si chartam speculo apposueris foraminibus multis distinctam; dum enim parum ea a speculo amota est, e foraminibus omnibus egredientem aspicies lucem ab universis resilientem speculi partibus foraminibus appositis, at quo chartam a speculo amoveas magis, eo a paucioribus, assidue magis videlicet sese colligente luce, et ab uno tandem universam, universa tandem confluente in unum qui combustionis dicitur punctus, quod ibi modo quodcunque fere luci exponatur exuritur; alibi nequaquam vel si speculo proximius id fiat. [...] At eo redeamus unde digressi sumus, et iuxta dictam lucis reflexionem conspissationemque calorem fieri declaremus, ne declarandum quidem amplius; ubique enim perpetuo manifeste maior fieri calor videtur, ubi lux magis

To understand Telesio's idea better it is important to note that he conceives light not so much as a linear ray but rather as a three-dimensional cylinder which reaches objects on a small surface. When the cylinder hits a surface perpendicularly, it is reflected onto itself, and so its heating power is enhanced, just as happens in a burning mirror. When light hits a surface obliquely, however, the reflected cylinder only partially overlaps with the incident one, and therefore its heating power is smaller. Telesio states that because of this effect northern countries are colder then southern ones, and summer is warmer than winter.

This theory is especially worth noting for a number of reasons. First of all, the connection between the optical and the burning properties of mirrors, although it may appear straightforward today, only became generally established around the middle of the sixteenth century, shortly before Telesio started publishing his works.[56] Burning mirrors made out of polished metal had of course been known and studied since Antiquity, but the optical properties of a curved surface, such as image projection and inversion, only started being appreciated in the Renaissance when mirrors and lenses made out of transparent white glass (the so-called "crystal" glass) became available in increasing quantities for artisans and interested scholars. The earliest evidence that a connection was being made between the focus of a burning mirror and the "point of inversion" of an optical one was being made is found in the manuscript versions of the *Theorica* by Ettore Ausonio and in the first edition of Della Porta's *Magia naturalis* (Natural Magic, 1558).[57] As noted by Luigi De Franco in his discussion of Telesio's theory of light, we possess no information on Telesio's sources on optics, but there can be little doubt that he was aware of and reacting to the new optical knowledge available in his time.[58] In the second edition of *On the Nature of Things* (1570) Telesio only discusses the issue of light reflection and heat generation qualitatively, while in the third edition (1586) he expands more on the subject, adding a geometrical 'demonstration' that light always reflects at right angles.[59] These passages have been analyzed by various historians

---

in se ipsam reflectitur magisque colligitur. Propterea scilicet maior in Meridionalibus quam in Borealibus Terris, et ab aestivo Sole et meridiano quam ab hyberno, et exoriente occidenteve fit calor, quod quo puncto verticis proximior fit Sol, eo magis in se ipsam reflectitur lux, magisque sibi ipsi unitur."

56 Dupré, "Ausonio's Mirrors and Galileo's Lenses," Smith, *From Sight to Light*, 333–349. On the complexity of the notions of focus and point of inversion of a spherical mirror in the Renaissance see Zik and Hon, "Giambattista Della Porta."

57 Duprè, "Ausonio".

58 De Franco, "Luce", 58–62.

59 De Franco, "Luce", Maierù, "Telesio". Of course Telesio was well aware of the law of reflection which was known since Antiquity, but he tried to show how that law, which was

and philosophers, but so far no one has highlighted how Telesio was building upon very new experiences which had become possible thanks to technological developments. For example, Mulsow suggests that the writings of Gaetano da Thiene might have provided the inspiration for Telesio's connection of heat and light, but he does not consider the fact that that connection was becoming increasingly evident during the second half of the sixteenth century.[60]

The fact that Telesio referred to new optical developments was not the only innovative feature of the text: even more original was his use of reflection to explain the correlation between solar heating power and the inclination of solar light. The (correct) assumption that the higher the Sun is above the horizon the more intense the solar heat will be had so far not been stated with such clarity in the Western meteorological tradition, although it had been noted by Arabic-Islamic scholars.[61] To my knowledge, the idea of linking this correlation to the law of reflection appears here for the first time, at least as far as the Graeco-Latin tradition is concerned. This very original answer to a meteorological question seems to have prompted Telesio to engage further with the issue of light reflection and heat generation, since, as we saw, he considerably expanded on the theme in the final edition of his main philosophical work. In conclusion, Telesio's meteorological booklet can be regarded as an example of both how his natural philosophy could explain observation and of how an interest in understanding specific phenomena could prompt further theoretical reflection. In this context, the rarefaction and condensation of matter offered a heuristic tool to connect philosophical reflection and natural investigation. As we shall see, it was an approach that Della Porta would appreciate.

4 Giambattista Della Porta's *On the Transmutations of Air*

4.1 *Della Porta's Approach to Natural Philosophy and Meteorology*

Giambattista Della Porta (ca. 1535–1615) was an outstanding representative of the vitality and variety of southern Italian culture in the late Renaissance, but at the same time his work displayed features resonating with the many approaches to the experimental philosophy of nature being developed all over

empirically valid, could be understood as light reflected by right angles. For details of the demonstration see Maierù, "Telesio."

60 Mulsow, *Selbsterhaltung*, 127–130.

61 Lettinck, *Aristotle's "Meteorology" and its Reception*, 194–487, Sezgin, *Geschichte des arabischen Schrifttums*, 302–304.

Europe.[62] Della Porta had already achieved fame in Europe in the early years of his long life with his *Magiae naturalis libri quatuor* (Four Books on Natural Magic), published in 1558 and expanded into a twenty book edition in 1589.[63] The *Natural Magic* was a collection of experiments on a very broad range of subjects presented in a form appealing to a general audience. Later on, Della Porta took up and expanded upon many of these areas of knowledge in thematic monographs directed at a more learned public—and he also wrote a series of quite successful theatre plays. Della Porta's natural philosophical publications covered topics which today we regard as part of the natural sciences and technology, like optics, pneumatics, mathematics, or cryptography, and also subjects which today are classified as 'occult', such as astrology, alchemy, or physiognomy. Della Porta lived only a few decades later than Telesio, but at a time when the political situation in southern Italy was rapidly changing. While Telesio could work and publish relatively undisturbed until his death, Della Porta had a close encounter with the Inquisition and, especially in the final years of his life, faced increasing difficulties in obtaining permission to publish his work. In fact, his planned magnum opus on marvels (*Taumatologia*) remained unpublished and is only partly extant as a manuscript, which has recently been edited.[64] The last book that Della Porta saw in print was his treatise on meteorology, published in 1610 thanks to the financial and political support of Federico Cesi. By that time Telesio's work was on the Index, since his writings around the end of the sixteenth century had come to be associated with southern Italian reform movements regarded as politically and religiously subversive.[65] However, Telesio's natural philosophy remained a most important guideline for those who, like Della Porta, wanted to experimentally explore the secrets of nature in Renaissance Naples and, as historian Sean Cocco wrote, "their type of experimentation fulfilled the promise of Telesian natural philosophy".[66]

Della Porta's meteorological treatise is longer than Telesio's booklet and comprises four books in which the various meteorological phenomena are discussed following the traditional subdivision corresponding to the four elements: Book I is devoted to air and wind, Book II to the watery "meteors" like rain, hail, and snow, Book III to the fiery ones like thunder and lightning, and

62 On Della Porta's life and cultural context, and for further references, see Eamon, *Science and the Secrets of Nature*, id., *Professors of Secrets* and, most recently, id., "Theatre of Experiments," as well as Borrelli, "Giovan Battista Della Porta's Neapolitan Magic."

63 Della Porta, *Magiae naturalis libri IIII*, Della Porta, *Magiae naturalis libri XX*.

64 Della Porta, *Taumatologia e criptologia*.

65 Agrimi, "Telesio", Borrelli, "Neapolitan Magic", Cocco, "Locating the Natural Sciences," Eamon, *Books of Secrets*.

66 Cocco, "Locating," 457.

Book IV to phenomena which we might characterize as geophysical, such as the sea and its saltiness, rivers and springs, and earthquakes.[67] Della Porta has never been known as a particularly refined theoretical philosopher, and his last published work confirms this view, although compared with his earlier books it definitely contains more extensive philosophical discussions and learned quotes from ancient and modern authors. In my earlier discussion of Della Porta's meteorology, I characterized it as 'humanistic', not so much in the traditional sense of Renaissance humanism but because Della Porta regards weather and climate not primarily as abstract subjects of theoretical reflections but as fundamental components of human life and culture. Meteorological phenomena are for him research objects that should always be studied by looking at them through human senses and emotions. Knowledge of their origin, traditions, and utility is not a theoretical aim but is essential both for practical purposes and for freeing man from superstition and unfounded fears. Because of this epistemic attitude, Della Porta discusses all meteorological phenomena according to the same scheme. The first items to be listed for each "meteor" are its "utilities" (*utilitates*), i.e. what it positively represented for everyday human life, so that destructive events like hail or earthquakes have no section on utility. After this, Della Porta goes on to expound the opinions of various ancient and contemporary scholars on how that specific weather phenomenon comes to be, and finally states "his own opinion" (*opinio propria*).

Unlike Telesio, Della Porta is not interested in embedding the explanations in one coherent philosophical world view but rather seeks to find an empirically satisfactory view of how individual phenomena such as wind, rain, or lightning originate. Very often his explanations are based on the description of experiences which analogically show how the weather phenomenon in question comes to be. For example, winds are explained by describing how an amount of air trapped in a glass vessel expands and contracts with heat, and rain is discussed by describing the procedure of water distillation and condensation.[68] After the explanation of a meteorological phenomenon comes the description of its observed features, such as the names and direction of the winds, or the different forms of lightning seen in the sky. The concluding paragraphs for each "meteor" deal with another extremely important side of Renaissance weather: how to predict certain phenomena and, if possible, to try

67 Here I can only offer a brief overview of Della Porta's meteorological treatise. For a more detailed treatment with specific references to the sources and literature see Borrelli, "Neapolitan Magic" as well as the latest edition of the work Della Porta, *Aeris*.

68 Borrelli, "Neapolitan Magic".

and prevent them from happening. Here Della Porta quotes weather wisdoms found in the ancient authors but also criticizes some popular remedies against weather disasters. Given the difference in scope and methodology between the two authors, a general comparison between Della Porta's treatise and Telesio's booklet would make little sense. I will only set Della Porta's explanatory sections and Telesio's statements side by side, and I will mainly analyze passages from Book I where air, heat, and wind are discussed, offering only a brief survey of Books II–IV.

4.2 *Air in Della Porta's Meteorology*

The broad framework of Della Porta's treatise is expressed in the title: *On the Transmutations of Air*. Its general assumption is that all weather phenomena are due to transmutations of the element "air", which for Della Porta is both a substance and an active principle of life. By often quoting Seneca's *Natural Questions* when expounding his views, Della Porta connects his writing with the tradition of Stoic philosophy, yet he also cleverly inserts passages from Christian authors into his text, chosen and at times adapted to support his views. After a short introductory chapter on the utility of meteorology, Chapter 2 of Book I deals with air, and in the beginning states:

> [Air] makes winds by flowing, when it is excited more strongly and made rarer [it makes] lightning and thunder, when contracted clouds, when condensed rains, when frozen snow, when frozen in a more turbulent way hail, and when it is relaxed [it gives rise] to serene weather.[69]

These words almost, but not quite, coincide with a passage from the *Etymologies* of the revered authority Isidore of Seville:

> [Air] makes winds when moved, when it is excited more strongly lightning and thunder, when contracted clouds, when condensed rains, and frozen clouds make snow, and denser clouds freezing in a more turbulent way [make] hail, and when [air] is relaxed it gives rise to serene weather.[70]

69 Della Porta, *Aeris*, 14: "Nenpe fluctuans ventos facit; vehementius concitatus et attenuatus ignes et tonitrua; contractus nubila; conspissatus pluviam; congelatus nivem; turbulentius gelatus addensatusque grandinem; distentus serenum facit."

70 Isidore, *Etymologiae* XIII, 7, 1: "Nam commotus ventos facit; vehementius concitatus ignes et tonitrua; contractus nubila; conspissatus pluviam; congelantibus nubilis nivem; turbulentius congelantibus densioribus nubilis grandinem; distentus serenum efficit."

The two passages are almost identical, but Della Porta's modifications are worth noting: he lets air flow (*fluctuans*) instead of being moved (*commotus*), adds that it is "made rarer" when it turns into lightning, and explicitly states that air (and not air which had previously turned into clouds) freezes into snow and hail. In short, in Della Porta's version, the transmutations of air takes center stage and have the form of rarefaction and condensation, just as was the case for Telesio's vapors. As noted in section 2 above, employing what we would call pneumatic and thermodynamic observations to explain the behavior of air in general and weather-air or weather-vapors in particular became increasingly common in Renaissance meteorology, but it is nonetheless interesting how Della Porta, like Telesio, elevates rarefaction and condensation to the status of fundamental principles of nature. In the beginning of the treatise, Della Porta presents a *Synopsis aeris transmutationum* (Synopsis of the Transmutations of Air)[71] in which all meteorological phenomena are listed and presented as air in different states of rarefaction or condensation, from the rarest state, in which air becomes fire, to the densest one, when it forms composites with earth and gives rise to earthquakes.

In line with the Stoic tradition, as well as with the alchemical one, Della Porta regards air as something more than an element: it is also a principle of life, a spirit. He quotes Seneca to state that "air links the Earth to the sky [..] passing upwards whatever it receives from Earth, and transferring to earthly things the celestial strength from above."[72] Air, Della Porta explains, is the reason why all animals live and why "all sound arrives to the ears, images of things to the eyes, smell to the nose and all other sensations to the other senses".[73] In conclusion, there are certainly parallels between Della Porta's air and Telesio's vapors and spirit, although they can be ascribed to the common background they shared with other Renaissance scholars. Moreover, Della Porta's main aim in his treatise is not to closely connect the explanations of individual meteorological phenomena with the general principles stated in the beginning but to find in each separate instance a causal account of events which fits known empirical results. However, processes of rarefaction and condensation, and of heating and cooling, constantly turn up in his explanations, and I will argue that it is at that level that deeper similarities between Della Porta and Telesio come to light.

---

71 Della Porta, *Aeris*, 11.

72 Ibid., 14: "Hic est enim quid coelum terramque connectit [...] Supra se dat quicquid accepit a terris, sursum vim syderum in terrena transfundit."

73 Ibid., 15: "[aer] cunctos sonos ad aures, rerum imagines ad oculos, odores ad nasum refert et reliqua ad reliquos sensus."

Chapters 3 and 4 of Book I expand on Della Porta's notion of air, criticizing Aristotle's views according to which air is hot and humid and is a purely passive component of meteorology. Della Porta rejects these ideas, claiming that air has only one property, coldness, and actively uses it to oppose solar heat in an interplay generating all meteors. Della Porta supports this view by referring to Stoic writings:

> The Stoics stated that air is cold and misty, so as to oppose the brightness and heat of fire. [...] I, too, believe a thing to be cold if through its virtue it can cool down and freeze other things; but the strength and power of air can cool down snow, hail, and ice: therefore it will itself be of extreme coldness.[74]

In Della Porta's meteorology, heat and cold are in turn linked with rarefaction and condensation of air. Air, when heated or moved, can rarefy to the point of becoming fire, while cold makes it denser, and these two principles constantly act on it in opposite directions: "Air is always in the middle of these two opposite principles, heat and cold [...] and is constantly agitated by the intense battle raging between these two opposites".[75] Della Porta's depiction of heat and cold as fighting within air is once again reminiscent of Telesian views. Although Della Porta does not delve into the role of heat and cold as active principles, in the following pages they appear again and again as key factors in the production of meteorological phenomena by making air rarer or denser and causing it to transmute. In this sense, apart from the fact that the fluid in question is air and not vapor, some of Della Porta's explanations appear as sharpened versions of Telesio's arguments. For example, Chapter 5 of Book I explains how air becomes lighter or heavier. In his meteorology, Telesio states that there is a correlation between the density of vapors and the height to which they can rise, so that winds cannot move upwards beyond a certain height unless they become as rare as air. Della Porta makes a similar statement, links it to the weight of air—a notion which was slowly emerging in the Renaissance—and supports it by extending Archimedes' principle to air-like fluids.[76] Thus, Della Porta's treatment appears as a quantified, sharpened version of Telesio's reflections.

---

74 Ibid., 16–17: "Stoici rigidum et caliginosum dixerunt, ut ignis splandori et caliditati opponentur. [...] et mihi illud frigidum videtur, cuius virture catera rigent frigeriscuntque sed aeris vi et potestate nives, grandines, glacies rigent: ipse igitur frigidissimus erit."

75 Ibid., 19: "aer semper in medio duorum contrariorum est, caloris et frigoris, [...] et ab his duobus contrariis praeclara pugna semper exagitatur."

76 Ibid., 21–23.

4.3 *Solar Heat and Its Action on Air*

Chapter 7 of Book I of Della Porta's *On the Transmutations of Air* is devoted to "How the Sun Heats" (*Quomodo sol calefaciat*).[77] The aim of the section is to explain how the sun heats earth and air, and thus makes sense of the seasonal differences in temperature. Della Porta starts by criticizing Aristotle's notion that air is heated by the rotation of the celestial spheres, and then goes on to offer his own explanation, which turns out to be a more detailed version of the one Telesio had offered:

> When solar rays multiply in their own innate nature they shine and heat, even if they are in themselves neither hot nor shiny. [...] We see how solar rays impinging on a concave spherical mirror, and even more strongly on a parabolic one, are concentrated in a single point, and are so multiplied, that they not only can generate a great light, but even ignite fire. Therefore we have to assume that solar rays, when they come from the sky to the Earth, are reflected when they hit the ground. Since the convex surface of the ground is not polished like that of a mirror, however, the rays do not burn, but only produce a mild heat, and the more vertical they are, the more intensely they heat.[78]

Della Porta provides a geometrical drawing showing how the overlap between incident and reflected rays is larger or smaller depending on how high above the horizon the sun is. The fact that the same explanation is found in Della Porta's and Telesio's meteorology is in this case striking since, as already mentioned in the previous section, no other version of this argument in Western literature is known so far. In an earlier discussion of Della Porta's meteorology, I was not yet aware of Telesio's text and put forward the hypothesis that Della Porta might have taken the idea from some Arabic manuscript.[79] However, it now seems more probable that Della Porta's source is either Telesio's text or possibly that both authors relied on the same material, which is possibly not extant. Finally, there is also the possibility that the idea originally came from Della Porta through non-written channels, since in 1565, when the first edition

77 Ibid., 25–29.

78 Ibid., 27: "Solares radii suapte ingenio cum multiplicantur nitent et incalescunt, etsi ex se neque calidi neque lucidi sint. [...] Videmus solis radios ad concavum spheralem speculum pervenientes, et valentius ad parabolicum et recurrentes ad punctum unum, ita multiplicari ut non solum maximum fulgorem, sed ignem excitare valeant. Putandum est solis radios a coelis in terram incumbentes obiectu terrae repercussos replicari, cuius facies convexa quia non perpolita ut speculum, non comburuntur, sed tepent mitiusque agunt et quo magis verticales fuerint intentius urunt."

79 Borrelli, "Neapolitan Magic".

of *On the Nature of Things* appeared, Della Porta had already published the first edition of *Natural Magic*, where spherical mirrors are discussed.[80] Renaissance sources on the question of the origin of seasons still await a systematic analysis, and it is possible that further searches in published and unpublished material will provide additional evidence on this issue.

Book I continues with a series of chapters reporting and criticizing the opinions of previous authors on the origin of wind, until in Chapter 16 Della Porta finally expounds "his own opinion" (*opinio propria*) on the matter.[81] For Della Porta, the cause of winds lies in the fact that air is rarefied and made mobile by heat, but as soon as the action of heat ceases, air "reverts and remains in its own consistency and takes pleasure in its own consistency".[82] Interestingly, Della Porta states here that air "takes pleasure" (*gaudet*) in its consistency, an expression that recalls Telesio's description of the vapor's active reactions to attempts at expanding or compressing them and in general to his views of the sensibility of matter. Moreover, the notion that air somehow strives to conserve its consistency might be linked to Telesian notions of "self-conservation", as was also recently suggested by Oreste Trabucco.[83] In the meteorological treatise, Della Porta does not speak of conservation, but in his writings on pneumatics he attributes the suction power of siphons to the air's desire to "conserve its own essence" (*suae essentiae conservatio*). When suddenly expanded, air reacts by holding together and resisting rarefaction, and therefore water comes to fill the space available. The reason why air strives to remain as unified in itself as possible is that "conservation obtains in unity", and so air, by resisting rarefaction, is trying to conserve itself.[84]

While the desire of air to keep its consistency only exhibits a vague similarity to Telesio's natural philosophy, a clear correspondence to Telesio's meteorology can be found in the process of thermal rarefaction and condensation which, as we will presently see, was at the core of Della Porta's theory of winds. To explain his views, Della Porta described an experience involving an inverted glass vessel in which the expansion and contraction of air by heat and cold can be visualized, an experience which at that time was already connected to discussions of the origin of winds.[85] For our present concern it is interesting to note how Della Porta then uses the experiment to explain not only the mechanism of wind generation but also the variety of local winds. As he had already done

80 Della Porta, *Magia naturalis*.

81 Della Porta, *Aeris*, 43–45.

82 Ibid., 43: "in suam consistentiam redit conseditque gaudet enim sua consitentia."

83 Mulsow, *Selbsterhaltung*, Trabucco, "Nel cantiere della *Magia*," 229.

84 Della Porta, *Pneumaticorum libri III* [1601] (2008), 15: "conservatio fit in unitatem."

85 For a detailed analysis of the inverted-glass experiment and its connection to the question of the origin of winds see Borrelli, "Weatherglass," Borrelli, "Pneumatics."

in his pneumatic writings, he states that, by measuring the volume occupied by the air in the glass vessel before and after rarefaction, it is possible to quantify "in how many parts of rarer air one ounce of air in its consistency [*consistenza*] can dissolve".[86] However, the results of the measurement are not constant but depend on the intensity of heat and on the original consistency of air, i.e. its rareness or denseness, which in turn depends on local circumstances:

> [I]t is worth noticing that this experience never obtains in the same way, since sometimes the air contained in the vase is denser, for example when it has been in a cold, humid place, and sometimes it is finer, if it stayed in dry places. And one part of denser air will dissolve in a larger quantity of air.[87]

This passage expresses ideas very similar to those Telesio had presented in his meteorological booklet. At the end of the chapter, Della Porta summarizes his views in a description of wind behavior very similar to that offered by Telesio:

> We have to know that the Sun, as it circles the Earth and hits it with its rays, gives rise through reflection to a large amount of heat, so that the air heats up and becomes rarer, takes up a larger space and moves upwards, to the right and to the left, and all around. And having achieved a monstrous dimension and not being able to contain itself any more, it approaches the air nearby and pushes it away and the air pushed away fights with the other air near itself, and the vanquished is compressed and gives way, and occupies space near the winner, and goes where a weaker opposition invites it, and where it finds some vacuum, as long as the air which has become larger pushes it, and there the air is violently pushed around until it has become wider.[88]

---

86 Della Porta, *I tre libri de' spirituali* [1606] (2008), 144–145: "un'oncia d'aria nella sua consistenza in quante parti d'aria più sottile si può dissolvere." Please note that *consistenza* seems to be a property related to what we would now call density, although it should not be called as such in order to avoid an anachronism.

87 Della Porta, *Aeris*, 44–45: "Hoc animadversione dignum est experimentum nunquam eodem modo succedere, nam aliquando aer vase contentus crassior erit, utpote si vas locis frigidis, humidisque steterit, aliquando aridior, si siccis, et pars una crassi aeris in maiorem aeris soluti quantitatem se solverit."

88 Ibid., 45 "Sciendum quod Sol dum terram circumibit, et verticalibus suis radiis eam verberat calorem ingentem ex reflexione gignit, unde is aerem excalefaciendum attenuat, hic in vastum locum se explicans superna petit, dextra sinistrave et circumquaque, sic immani mole exauctus nex sui capax proximum sibi aerem facessit, disploditque hic

Of course, as has already been noted, for Della Porta the material of wind is air and not vapors. Yet we have to remember that neither Telesio nor Della Porta saw a sharp difference between different air-like substances, since Telesio states that wind-vapors can in principle become air and Della Porta allows air to become water or mix with earthly matter to generate fiery meteors.

4.4 *Watery and Fiery Meteors in Della Porta's Meteorology*

With the statement of his theory of the origin of winds, Della Porta concludes the explanatory part of Book I and goes on to discuss the names and places of winds and how they are linked to weather forecasting. When discussing "watery" meteors in Book II of *On the Transmutations of Air*, he offers an account in terms of the condensation of air with cold, supporting his views by presenting a series of distillation experiments and thus implicitly rejecting Telesio's idea that rain and hail do not need cold to be generated. Della Porta's connection of rain, hail, and snow to alchemical experiments on the thermodynamic transformations of water largely corresponds to the later modern views, but at the time it represented an innovative step for Western scholars.[89] Here we see how Della Porta exploits the conceptual scheme of the interplay of heat and cold and dense and rare as a heuristic tool for conceptualizing weather in terms of simpler phenomena constructed in the laboratory.[90]

Books III and IV of *On the Transmutations of Air* are devoted respectively to 'fiery' meteors like thunder and lightning and to the phenomena taking place on or under the earth, such as the properties of seas and rivers or earthquakes. On these subjects there is not much material for a comparison with Telesio's meteorological booklet, in which only one chapter is devoted respectively to lightning and thunder (Chapter 10) and to earthquakes (Chapter 11). Telesio explains thunder and lightning along Aristotelian lines: rare vapors trapped in the cooling clouds seek to escape and in the end throw themselves towards the ground. Earthquakes, instead, are seen as analogous to gunpowder explosions, an idea which, as we saw, might have been influenced by Paracelsian meteorology but may also have independently occurred to Telesio. By comparison, the Paracelsian tradition is almost certainly behind Della Porta's explanation of thunder and lighting, even though he for obvious reasons never referred

displosus cum altero sibi propinquo colluctatur, hic victus comprimitur, ceditque locum proximus victoris occupat, et ubi debilior exitus invitat, et aliquid reperit vacui, se recipit, et id usque donec amplior factus aer vehementius impellit."

89 Borrelli, "Neapolitan Magic".

90 On Della Porta's use of standardized descriptions of experimental procedures to conceptualize experience see Borrelli, "Thinking with Optical Objects," Borrelli, "The Recipe as a Heuristic Tool."

explicitly to it. Thunder, lightning, and all other 'fiery' meteors, Della Porta explains, are due to a rare, fat matter which the Sun extracts and raises up from trees and bituminous lands and which, once in the highest and hottest air region, catches fire with more or less violence:

> The Sun absorbs a rare, fat [exhalation] from fat trees producing resin and from bituminous lands and other places, and raises it upwards to the burning hot region of the sky, and there suddenly it catches fire and disappears in various different visual effects.[91]

Della Porta's explanation follows the tradition of Paracelsian aerial nitre and adds to it by stating that both thunder and the booming noise of gunpowder explosions are due to the sudden expansion of heated air, which leads to the compression and displacement of successive layers of air.[92] This explanation is a good example of Della Porta's syncretism: he employs both the Aristotelian idea of exhalation, which he had previously rejected, and the Telesian notion of rarefaction and condensation as heuristic tools to explain in detail a specific meteorological phenomenon. It is not by chance that this explanation appears correct in modern terms: Della Porta's approach to making sense of weather—and of nature in general—by prioritizing the explanation of the sensually perceivable features of individual phenomena over the coherence of the theoretical framework allows him great flexibility in exploiting a broad range of older and newer notions, transforming and combining them to construct what might in modern terminology be referred to as a "phenomenological model".[93] As a final note I wish to add that, in his unpublished treatise *De fulmine* (On Lightning), Telesio discussed in detail the earthly origin of the exhalations causing it, so that his treatment was nearer to Della Porta's views than the statements he had made in *On What Takes Place in the Air*.[94]

---

91 Della Porta, *Aeris*, 123: "Sol enim ex aboribus resiniferis et pinguibus, locis bituminosis, et alliis, tenue pingue absorbet atque in sublime evehit ad flagrantem et torridam coeli plagam, ibique illico accenduntur et in varias diversasque abeunt facies."

92 Debus, "Aerial Niter".

93 Borrelli, "Neapolitan Magic", Borrelli, "Recipe".

94 Telesio, *De fulmine*, in Delcorno, "Il commentario 'De fulmine'".

5 Conclusions

Although Della Porta never mentioned Telesio in his works, it would be extremely implausible to assume that he was not at least to a certain extent familiar with Telesian philosophy, which had contributed a great deal to shaping Neapolitan culture in the late sixteenth century. In my paper I have suggested that Della Porta not only generically knew Telesio's work but had actually read at least his meteorological booklet, exploiting some of its key principles as a heuristic tool to conceptualize and explain certain weather phenomena in terms of simple experiences such as those he had presented in his *Natural Magic*. Particularly significant as evidence of a direct reading of Telesio's meteorology by Della Porta is the latter's discussion of the connection between solar heat, the inclination of the Sun above the horizon and the law of reflection. Yet I believe that also the general approach to explaining weather and climate by means of the thermal rarefaction and condensation of air might owe much to Telesio's template.

Telesio set heat and cold as the two opposite principles governing nature in general and weather and climate in particular and the rarefaction and condensation of air-like vapors as the primary kind of transformation responsible for the variety of meteorological phenomena. In the hands of Della Porta, when combined with ideas of other authors and with his own reflections, these principles turn out to be heuristically fruitful, leading to explanations for winds and their seasonal and local varieties, as well as for the formation of rain and the occurrence of thunder and lightning. I am however not claiming in any way that Della Porta's meteorology should be regarded as Telesian. My claim is that Della Porta liberally chose, employed and adapted some of Telesio's notions in the same way he did with those of authors whom he explicitly quoted, such as Aristotle, Seneca or Girolamo Cardano, appropriating and adapting what he could use in a specific situation and rejecting the rest. In the end, unlike Telesio, Della Porta was not primarily interested in constructing a coherent theoretical framework to explain nature top-down on the basis of a few principles. What he sought was to build new ways of conceptualizing specific experiences from the bottom-up, leaving open whether they might or might not in the end be mutually connected. In many instances, such as optics, pneumatics or meteorology, this approach was quite fruitful.

CHAPTER 5

Telesian Controversies on the Winds and Meteorology

Oreste Trabucco

In 1592, the reputable Aristotelian philosopher Federico Bonaventura published his work on winds *De causa ventorum motus peripatetica disceptatio* (*Peripatetic Discussion on the Cause of the Motion of Winds*) in Urbino. About ten years later this book was reprinted in Venice by the distinguished printer Francesco de' Franceschi. Bonaventura dedicated his work to Alessandro Giorgi, who was one of Federico Commandino's pupils and the Italian translator of the widely read *Pneumatics* by Hero. In Bonaventura's *De causa ventorum*, one can find some pages, so far neglected, in which the author criticized theses stated by Bernardino Telesio in his booklet *De iis quae in aere fiunt* (On Those Things Occurring in the Air). It was reprinted in 1590 in *Varii de naturalibus rebus libelli* (*Various Booklets on Natural Subjects*), the miscellanea published in Venice by Antonio Persio, one of Telesio's most important pupils. Bonaventura's pages, written in opposition to Telesio, testify to the wide reception of the latter's views, their involvement in Aristotelian controversies, and his impact on sixteenth century natural philosophy.

In order to stress the impact of Telesio's anti-Aristotelian meteorology, it is expedient to begin with the seventeenth-century assessment by a critical intellectual such as Alessandro Tassoni. His unprejudiced intellect was not afraid of being irreverent, so much so that he was amongst the first to ignite the long-running *querelle des anciens et des modernes*.[1] In the tenth book of his *Pensieri diversi* (*Various Thoughts*), when discussing *De gl'ingegni antichi e moderni* (On Ancient and Modern Minds), he affirmed that,

> Even Aristotle, in order to present difficult arguments clearly, made use, in his works on logic, of examples taken from mathematics, which, instead, appear to be very difficult for the youth of today because they have little mathematical knowledge. Notwithstanding this, original theories were elaborated even in modern times, thanks to Tartaglia, Commandino, Tycho, Clavius, Copernicus, Galileo, Gaurico, Magini, Regiomontanus,

1 See Fumaroli, "Les abeilles et les araignées."

 | DOI:10.1163/9789004352643_007

> and other famous scholars, who extended the knowledge introduced by Archimedes, Euclid, Eudoxus, Proclus, and other mathematicians of antiquity. Modern mathematicians have revealed many things that were unknown to the ancients in terms of the centre of the Earth, the number of planets and their paths, the distance of the comets, and the dimensions and nature of the seas.[2]

With deliberate capricious swagger, the *Pensieri* aimed at embracing the entirety of knowledge. It included some books dedicated to questions of *naturalis philosophia*, where, in line with this intellectual capacity, Tassoni allowed himself complete freedom to confute Aristotle and the peripatetic school. In the fourth book, on the question of *Perché i venti feriscano di traverso* (*Why the Winds Hit Obliquely*), he observed:

> In the fourth chapter of *De iis quae in aere fiunt* (*On Those Things Occurring in the Air*) Telesio affirms, as a supporter of a new doctrine, that the vapors, out of which winds are born, split the air obliquely because, when they begin to shrink and condense to avoid becoming air, they are nourished on new matter and grow so much that it becomes necessary to expand laterally, if there is available space, so satisfying their very nature. They do this violently if the air is saturated with other vapors and these continue to expand due to rarefaction brought on by the power of the sun, since, given that they are constricted from all angles, they have to make space for themselves forcefully in order to find an outlet. This, though, does not answer the question of why winds move in one direction rather than another.[3]

2 Tassoni, *Pensieri e scritti preparatori*, 561: "[…] Aristotile anch'egli ne' suoi libri disputativi, per dichiarar cose difficili con argomenti piani ricorre ad esempi di matematica, che all'incontro i giovani nostri, che non hanno di quella dottrina alcuna precognizione, paiono intricatissimi e strani. Con tutto ciò anche nella teorica l'età nostra ha veduto il Tartaglia, il Comendino, il Ticone, il Clavio, il Copernico, il Galileo, il Gaurico, il Magino, il Regiomontano e altri famosi, i quali all'invenzioni d'Archimede, d'Euclide, d'Eudosso, di Proclo e degli altri antichi hanno trovato che aggiugnere […] Del centro della Terra, del numero de' pianeti, del corso de' cieli, della distanze delle comete e dell'ampiezza e qualità del mare hanno trovate e insegnate molte cose i moderni che non conobbero gli antichi."

3 Ibid., 128: "Il Telesio nel IV capo *De his quae in aere fiunt*, inventore e maestro di nuova dottrina, disse che i vapori onde si formano i venti fendono l'aria per traverso, perché mentre si restringono e si condensano per non esser convertiti in aria, sopravvenendo tuttavia loro nuova materia, crescono a tanta quantità che necessitati dalla forza del luogo è forza ch'esalino ne' lati piacevolmente, se trovano il campo libero e senza impedimento da potersi diffondere; ma con impeto, se trovano l'aria ingombrata d'altri vapori e che di continuo sopravvenga loro

Without embracing Telesio's opinion, Tassoni thus summarized some of the central theses of *De iis quae in aere fiunt*, one of the Telesian *libelli*, or booklets. The latter appeared in 1570 to support the second edition of *De rerum natura iuxta propria principia* (*On the Nature of Things according to their Own Principles*) and to contribute to the general subversion of Aristotle's authority, in this case regarding the argument of the *Meteorologica*. Nor was this the only occasion on which Telesio was mentioned in Tassoni's *Pensieri*. As a heterodox author, adverse to the Scholastic tradition, Telesio was well suited to the anti-dogmatic flamboyant strategy which Tassoni—taken on by the Accademia degli Umoristi in Rome under the name of *Bisquadro*—was so fond of pursuing. He constantly displayed reluctance to *iurare in verba magistri*, that is, 'to swear on the magisters words', as can be seen in this extract from a letter of 1613 addressed to Camillo Baldi,[4] a prestigious lecturer of Philosophy at the University of Bologna:

> It is incredible how you Aristotelians distort the words of your prophet when they are wrong, bending them to your aims [...] but you are right to do this because, if you did not turn to superstition to dazzle the minds of the young, we would return to philosophising with the freedom of old and you would lose the salaries that the university gives you to defend Aristotle's doctrine and chimeras sophistically.[5]

Hence Tassoni could, with sharp realism, go further and say in the *Pensieri* that Telesio had become the emblem of the 'moderns', together with Pierre de la Ramée and Girolamo Cardano.[6]

---

aiuto di nuova materia e 'l sole li rarefaccia col suo calore, sì che trovandosi angustiati da ogni banda e premuti, bisogni che per aver luogo con violenza s'aprano il passo. Ma né questa scioglie le dette già difficultà, perché il vento si muova più ad una parte che all'altra."

4 In Tassoni's *Secchia rapita* (*The Stolen Bucket*) (II, 13), Baldi is called "principal scholar [...] / highly stroked / and housed at public expense".

5 Tassoni, *Lettere*, vol. 1 and 2, 165–166: "Ma è certo bellissima cosa di voi altri Aristoteleschi che, quando il Profeta vostro non dice bene, subito cominciate a negare il senso, ch'è chiaro e piano, e vogliate adattare alle sue parole quello che a voi torna bene [...] Ma voi altri avete ragione, che se non vi servite di questa superstizione ad offuscar gl'intelletti della gioventù, si tornerebbe a filosofare con l'antica libertà; e voi correreste rischio di perdere i salari, che vi dà il Pubblico, perché con sofisticherie difendiate le dottrine d'Aristotile e tutte le sue chimere." Reeves, *Evening News*, 121.

6 Tassoni, *Pensieri*, 297 (own translation): "I know that the objection can be given that Pierre de la Ramée, Girolamo Cardano, and Bernardino Telesio, those who wanted to contradict Aristotle, made themselves not only objects of mockery, but caused their works to be

Meteorology was well suited to Tassoni's *Pensieri*, as it was a variation of the very wide-spread genre of the *Problemata*.[7] Meteorological issues had enjoyed widespread success, as much in the high literature of Aristotelian comments as in that for wider consumption, given that meteorological phenomena are closely connected with everyday life and have military and civil applications.[8] As for the question of the origin, nature, and movement of the winds, debated *a latere* of *De rerum natura*, it reached into the depths of Telesio's philosophy, as he intended the *libelli* to apply the fundamental principles of his major work to the different areas of the *physiologia*. This was the core of Aristotelian anemology in the *Meteorologica*:

> We recognize two kinds of exhalation, one moist, the other dry. The former is called vapor: for the other there is no general name but we must call it a sort of smoke, applying to the whole of it a word that is proper to one of its forms. The moist cannot exist without the dry nor the dry without the moist: whenever we speak of either we mean that it predominates. Now when the Sun in its circular course approaches, it draws up by its heat the moist evaporation: when it recedes the cold makes the vapor that had been raised condense back into water which falls and is distributed over the earth. (This explains why there is more rain in winter and more by night than by day: though the fact is not recognized because rain by night is more apt to escape observation than by day). But there is a great quantity of fire and heat in the earth, and the sun not only draws up the moisture that lies on the surface of it, but warms and dries the earth itself. Consequently, since there are two kinds of exhalation, as we have said, one like vapor, the other like smoke, both of them are necessarily generated. That in which moisture predominates is the source of rain, as we explained before, while the dry one is the source and substance of all winds [...].

---

proscribed. I will answer this by saying that the works of the first two were not proscribed because they contradicted Aristotle's text, which is not the undeniable gospel, but because, in religious matters, they contained a number of heresies. Telesio's [works], on the other hand, were not proscribed, but just suspended because that sharp wit, in the desire to deny what Aristotle had said, also denied certain propositions which are theological principles [...] However, when the heretical contents were removed, the three authors, who had previously been ridiculed by the Aristotelians, were by then well known by those who knew how to comprehend them [...] and Telesio was forming a new school and the Telesians were mentioned in public places of teaching, particularly among their fellow Calabrians."

7 Blair, "The *Problemata*," Cherchi, "Il quotidiano, i *Problemata* e la meraviglia."

8 Martin, *Renaissance Meteorology*, 1–20.

> The course of winds is oblique; for though the exhalation rises straight up from the earth, they blow round it because all the surrounding air follows the motion of the heavens. Hence the question might be asked whether winds originate from above or from below. The motion comes from above: before we feel the wind blowing the air betrays its presence even if there are clouds or a mist; for they show that the wind has begun to blow before it has actually reached us; and this implies that the source of winds is above. But since wind is a quantity of dry exhalation from the earth moving round the earth, it is clear that while the origin of the motion is from above, the matter and the generation of wind come from below. For the direction of flow of the rising exhalation is caused from above; for the motion of the heavens determines the processes that are at a distance from the earth, and the motion from below is vertical and every cause is more active where it is nearest to the effect; but in its generation and origin wind plainly derives from the earth.[9]

Telesio had argued against Aristotle regarding the origins and nature of the winds:

> [...] erroneously, it is also believed that rain is caused by vapors which come from more humid lands and winds from vapors in drier lands; indeed, Aristotle did not verify his beliefs, considering the two emissions in terms of their natures, and, therefore, it seemed to him that the earth emits a dry emission and water a humid emission. Aristotle should not, though, have considered the matter of the emissions, but rather their nature and peculiar substance, such that, on ascertaining that the earth emissions are very faint, he should have considered them very humid, and so, on observing that the emissions from water are much less delicate, he should, in the same way, have considered them less humid; and, he considers those things which are faint and soft more humid than those things that moisten.[10]

9 Aristotle, *Meteorologica* (1984), 359b–361b.

10 Telesio, *De iis quae in aere fiunt et de terraemotibus*," in Telesio, *Varii de naturalibus rebus libelli* (1981), XIII: "At nec recte nec proprias intuito positiones exhalatio utraque veluti rei, e qua fit, servans naturam, quae e terra educitur sicca, quae vero ex aqua humida videtur; neque enim materia, e qua educuntur, sed propria utriusque natura propriaque substantia spectanda Aristoteli erat; et longe tenuissima, quae e terra educitur conspecta, longe humidissima; quae vero ex aquis longe quam illa minus tenuis, minus humida videri debuit; cui nimirum [...] quae tenuia sunt molliaque, non quae humefaciunt, humida videntur."

Therefore, regarding their movement:

> He [Theophrastus] does not even agree with Aristotle about the causes of the oblique movement of vapors [...] Perhaps in the light of this, Theophrastus, leaving aside Aristotle, does not attribute the causes of the oblique movement of the wind to the circular rotation of air, but to the dual nature of the emission, affirming: "Since the two emissions are mixed together, so that the dry one tends to rise while the humid one pulls it down, they move obliquely;" but this is neither righteously said nor does it respect the positions of the two emissions, since everything, according to the Peripatetics, is moved according to the nature of whatever dominates and is superior. Although they are mixed, the two emissions cannot flow with great impetus, and certainly not downwards to the ground, since even if the humid were not as light as the dry, it would be drawn upwards anyway and would not drop back down, if not condensed into water.[11]

These objections rested on a cardinal principle of Telesian philosophy,[12] the "greatest desire for conservation and [the] greatest hate for one's own destruction",[13] bestowed upon all entities by the two acting natures of heat and cold. Indeed Telesio affirmed in *De iis quae in aere fiunt*:

---

11 Ibid., XVIII: "At neque obliqui vaporum motus causam edocens placere Aristoteles potest [...] Haec forte animadvertens Theophrastus, Aristotele posthabito, obliquae ventorum motionis causam non circulari aeris vertigini, sed duplici exhalationis attribuit naturae: 'Commixtae—inquit—inter se et sicca sursum tendente, humida vero deorsum illam trahente, in obliquum feruntur;' non recte, ut videtur, nec iuxta proprias positiones: composita enim omnia iuxta praedominantis naturam Peripateticis feruntur. At neque vel sibi ipsis commixtae impetu tanto nec deorsum omnino ad terram usque delabuntur; ut enim humida siccae aeque levis non sit, ad superiora tamen ferri et ipsa videtur; et nisi in aquas coacta, delabi nunquam."

12 Badaloni, "Sulla costruzione e sulla conservazione della vita," 11–49; Lerner, "Le 'parménidisme' de Telesio," Mulsow, *Frühneuzeitliche Selbsterhaltung*.

13 Telesio, *De rerum natura iuxta propria principia* [1570] (2010), I 34, Telesio, *De rerum natura iuxta propria principia libri IX* (1586), III 25: "Quoniam vero entium operatio a propria eorum substantia a proprioque manat ingenio nihiloque fere ab ea minus quam a propria natura servari; et ea dempta nihilo fere minus prompte quam propria dempta natura corrumpi videntur entia; quae operatione eadem oblectantur ac servantur, et eadem privata corrumpuntur, ab eadem prorsus natura constituta sunt;" IV 29: "Mundum non veluti casu quodam, sed ab opifice et longe eo sapientissimo longeque potentissimo et longe optimo ipso nimirum a Deo constructum fuisse, eumque non coelum modo terramque, e quibus mundus constare videtur, sed entia prorsus omnia, maria, aquas reliquas, lapides, metalla, sulfurea, bituminosa, plantarum animaliumque genera omnia, et humanum in primis, qualia fiunt, fieri conservarique voluisse; nulli quidem, qui caeli terraeque vires

> Those vapors which become winds, are perhaps moved by their own spontaneously oblique motion, since, given their mobile nature and the fact that they still have not achieved the tenuity of the air, they cannot be carried to the higher place, nor, in truth, do they desire to be taken there. However, although they do not tend to move, it may anyway seem that they are moved and pushed by themselves, since, being tenuous by nature, they greatly hate and abhor to thicken and condense and, therefore, turn completely into an alien substance.[14]

While Tassoni, when discussing the Telesian doctrines, combined his own idiosyncrasy for Aristotle's authority with themes suited to a popular encyclopaedia, Campanella, when quoting fully from Telesio in *Epilogo magno* (*Great Epilogue*), was moved by very different reasons:

> [...] Divine wisdom says that the winds were created, constituted of vapors that are not as thick as those of water or fog, nor so thin that they could rise without due cause. Therefore, they are, neither in winter nor in summer, created in excess, but of mediocre consistency, so that, while remaining above the face of the earth, they serve for the aforementioned purposes; when the vapors come together in great quantity, they feel restricted, which is against the nature of tenuity and brings about its death, as the thickness dies from its expansion. And, therefore, the vapors escape to find a space which can accept their ample tenuity until, colliding and reduced by the motion, they rise towards greater space. And the winds are more vehement as the nature of the place in which they are born provides them with more vapors, to the point that they most feel constrained and risk extinguishing; so the vapors attempt to escape to a different region, as happens to powder in a cannon, which, reduced and in need of more space, violently escapes and pulls everything it meets along with it.
>
> [*Advice.*

---

magnitudinemque et illius constructionem motusque paulo diligentius attendat, dubium obscurumve esse potest."

14 Telesio, *De iis quae in aere fiunt*, IV: "Qui vero in ventos diffunduntur proprio forte motu sponteque sua in obliquum feruntur; bene enim propria natura mobiles nec dum aeris nacti tenuitatem nec ad superiorem efferri possunt locum nec efferri omnino appetunt. At vel nihil moveri appetentes, a se ipsis moveri tamen impellique videri possunt, quoniam enim tenues sui natura in se ipsos spissari densarique, in alienam omnino agi substantiam summe odio habent summeque aversantur."

> *a*. Aristotle makes a great error when he says that the winds are drawn from the circulating air, since the air has such weak motion that it can hardly move the treetops. How then can it give such motion to the wind that it can blow down towers and trees? Therefore, the motion of the wind is spontaneous, in the search for an appropriate space, *ut supra*].[15]

This is the young Campanella, the forceful supporter of Telesio, who, in the near-contemporaneous *Philosophy as Demonstrated by the Senses* …,[16] which was written in response to *Pugnaculum Aristotelis adversus principia Bernardini Telesii* (*Aristotelian Fortress against Bernardino Telesio's Principles*) (1587) by Giacomo Antonio Marta,[17] says that the natural propensity of vapors to rise is an example of the principle of entities' self-conservation.[18]

---

15 Campanella, *Epilogo magno* (*Fisiologia italiana*), 279: "[…] commandò il primo Senno che si facessero i venti, di vapori non tanto grossi quanto quelli dell'acqua e della nebbia, né tanto sottili che possano andar in alto senza pro'. Onde né di verno né di state troppo se ne fanno, ma di mediocre consistenza, acciò restando sopra la faccia della terra servano alli predetti usi: dove a copia ragunandosi si sentono stringere, il che è contra la natura della sottigliezza e sua morte, come è morte della grossezza lo slargarsi; et però sfuggono in giro per trovar spatio capace della loro ampia sottigliezza, finché insieme azzuffati o assottigliati dal moto vanno in alto a più spatio. E tanto più sono vehementi, quanto che la minera del paese onde elli nascono somministra a loro più copia di vapori, sì che si sentono più stringere et morire: onde fanno ogni forza per uscir da quella regione in un'altra, come la polvere dentro lo scoppio conversa in sottigliezza cercando ampio spatio esce con impeto all'aria et porta seco ciò che s'incontra. [*Avertimento*. *a*. Grand'errore d'Aristotele, dicente che i venti son agitati dall'aere in giro: poiché l'aere ha moto sì debole che non può quasi muovere zima de arbori, come può dunque dar moto al vento che gitta torri et arbori? Dunque si muove da sé per trovar luogo, *ut supra*]."

16 The full Latin title is: *Philosophia sensibus demonstrata … ubi errores Aristotelis et asseclarum ex propriis dictis, et naturae decretis convincuntur … cum vera defensione Bernardini Telesii Consentini, Philosophorum maximi*. See Firpo, *Bibliografia degli scritti di Tommaso Campanella*, 65–67.

17 Firpo, *Bibliografia degli scritti di Tommaso Campanella*, 31.

18 Campanella, *Philosophia sensibus demonstrata* (1591), 496 (own translation): "But to give satisfaction to the Aristotelians, to whom we believe that [Telesio] gave full satisfaction, regarding the reasons for which fire and the light bodies tend without delay to rise, water and earth to descend, neither adding perfection in them nor subtracting perfection from them, there is no need to say that a body changes place, since the place from which it distances itself is not as congenial as that which it comes to occupy. However, rather because the place it leaves holds something which it finds intolerable and harmful, as it finds it convenient to occupy another in order to escape. This is shown by the flames and vapors, which rise and do not descend, because they feel constricted and endure suffering due to the density or cold of the contiguous bodies and, so, wish to escape from that torment. Each body possesses the sense and inclination to save itself and never abandons them."

This principle which, referring back to *De sensu rerum* (*On the Sense of Things*),[19] animates the Campanellan conception according to which "all things multiply, generate, and spread to the place of the enemy, crushing all others and craving the security to save themselves, live forever and make themselves gods if possible":

> If then the space is such a divine creature, one may conjecture that things are pleasantly attracted to it and that, in order to occupy the space which is the basis of being, things are driven to fill it, almost voluntarily acquiring a new place and existence; and that there is not only the reciprocal contact that holds the world together, since the air adheres to its contrary element in order to avoid the void, but that there is the pleasure of filling the void and that the air does not expand to prevent the void, but to occupy it and dominate the space, since we ascertain that the love of expanding, multiplying, and conquering more life with a vaster existence is common to all beings, which multiply, generate, and spread in the place belonging to the enemy being, crushing all others and craving to impose themselves so as to have a guarantee of surviving forever and, if they could, becoming gods; since all beings imitate eternal God, and aspire to become similar to Him and His cause.[20]

Thus Campanella clearly recognized which general principles Telesian anemology had to return to and this was generally also well recognised by those who were active in other opposing philosophical groups. In 1592, Federico Bonaventura published a *De causa ventorum motus peripatetica disceptatio* (*Peripatetic Discussion on the Cause of the Motion of Winds*) of his own in Urbino. Bonaventura was a philosopher who was tied, with unshakeable loyalty and pride, to Aristotle's authority and enjoyed great prestige at Francesco Maria II Della Rovere's Urbino court; Bonaventura was the 'philosopher to

---

19 Firpo, *Bibliografia degli scritti di Tommaso Campanella*, 58.

20 Campanella, *Del senso delle cose e della magia* (2007), 27: "Ora se lo spazio è sì divina creatura, si può conietturare che le cose sien tirate con voluptà a lui, e che per occupar lo spazio, ch'è base dell'essere, le cose corrono ad empirlo, quasi acquistando volentieri nuovo regno et esistenza; e che non solo ci sia lo scambievole contatto che il mondo unisce, poiché va l'aria a toccarsi col suo contrario per proibire il vacuo, ma che ci sia il gaudio dell'empire il vacuo, e che non corra per proibire il vacuo, ma per regnare e dilatarsi in lui, poiché l'amor di dilatarsi e moltiplicarsi e viver vite assai in spaziosa esistenza proviamo in tutte le cose trovarsi, che si moltiplicano, generano e diffondono nel luogo del nemico, scacciando ogni altro, e sole esser bramando per sicurtà di conservarsi et eternarsi e deificarsi se potessero; poiché tutte imitan Dio eterno, e a lui, come a sua causa, simili farsi bramano."

the Prince', with whose authority he wrote his most celebrated work, the tract *Della ragion di stato* (*On the Reason of State*).[21] Honoured with this authority, Bonaventura was the highest voice of official philosophy in the duchy, which was recognised by Guidobaldo del Monte. Even though del Monte contemporaneously supported mathematics and mechanics which were far removed from Aristotle and also obtained a remarkable reputation at court, he only ascribed to himself the role of mechanic, and was thus subordinate to the philosophy personified by Bonaventura.[22] Bonaventura had also established relations with the school of Commandino, of whom Guidobaldo was the best student and heir; indeed the *De causa ventorum motus* was dedicated to Alessandro Giorgi,[23] another of Commandino's students who continued his master's work. Also, Giorgi printed an Italian version of Hero of Alexandria's *Pneumatics* in 1592, which was based on Commandino's Latin edition of 1575, the text which originated the *Hero-Renaissance* at the end of the sixteenth century.[24] Giorgi, in the introduction to his translation of the *Pneumatics*, took care to connect his work with Bonaventura's erudition.[25] The fact that Bonaventura was an

21 Firpo, "Bonaventura, Federico," Panichi, "Premesse teoriche della filosofia politica di Federico Bonaventura," 7–58.

22 Bertoloni Meli, "Guidobaldo dal Monte and the Archimedean Revival," Frank and Napolitani, "Il giovane Galileo e Guidobaldo dal Monte," Omodeo, "Riflessioni sul moto terrestre nel Rinascimento."

23 Bonaventura, *De causa ventorum motus*, 4: "Iam vero dum mecum ipse haec animo versabar, ecce mihi literae tuae allatae sunt, in quibus te in eandem incidisse quaestionem scribis atque in summa propter tantorum hominum dissidia esse difficultate, rogans, ut meam tibi de hac quaestione sententiam aperirem. Magna quidem postulatio, mi Alexander; si enim te (cuius tamen eruditio ingeniique acumen nobis satis notum est) e tantis angustiis educere non potes, ut id ego praestare queam, non est ut expectes; sed et oportuna rursus postulatio, quae non omnino imparatum me offendit, ut pote eadem haec cogitantem; quin potius pro iure summae amicitiae nostrae calcar admovit, ut dies noctesque postea cogitare non cessaverim, quo pacto tibi morem gerere possem."

24 Trabucco, "*L'opere stupende dell'arti più ingegnose*".

25 Hero Alexandrinus, *Spiritali*, 4*v*–5*r*: "Spirito quello che sia, e da diversi diversamente considerato, è diffinito, e li Medici dissero che era quella facoltà divisa in tre parti, cioè animale, vitale e naturale, onde l'anima fa le sue operationi. Aristotile volse che fosse aura o vento cagionato da l'essalationi calde e secche, quali ascendendo alla seconda regione dell'aria e quivi perché ella move in giro, prendendo il movimento loro laterale, divenissero venti. Ma i Latini furono di diversa opinione, cioè che per esser l'aria di sua natura fredda, et in conseguenza contraria di qualità a l'essalationi che sono calde e secche, le si opponesse e contrastando le ricacciasse in giù per forza, onde in quel contrasto si cagionasse gran movimento d'aria e successivamente il vento; intorno a che si potrebbono addurre di molte cose che non occorre in luogo riferire. Ma chi desidera di haverne compita notitia, ricorra a l'*Anemologia* del nostro Molto Illustre Signor Federigo Bonaventura, che pur hora se ne viene in luce, e vi trovarà tutta la materia de' venti sottilissimamente essaminata e con molta dottrina pienamente rissoluta. A noi basti dire che spirito secondo

obstinate, pugnacious custodian of Aristotle's teaching did not lower his esteem within the *respublica literaria*, which went well beyond the function he performed in Urbino; this esteem was gained through his profound knowledge of Aristotle's original texts and the Aristotelian school. In 1593, Bonaventura placed the *De causa ventorum motus* in a massive anthology of works concerning the arguments of anemology, of which a large part was made up of translations of and commentaries on Theophrastus' meteorological texts. Here he recalled the friendship he cultivated with his fellow countryman Alessandro Giorgi and with Gianvincenzo Pinelli.[26] Indeed he had been able to make use of Pinelli's rich Paduan library,[27] fully participating in the entourage which revolved around it, as well as collaborating with Luigi Lollino,[28] the eminent Cretan scholar of Greek language and culture who had settled in Venice and was soon to become the bishop of Belluno. The rapport Pinelli and his associated *hommes de lettres* had with the contemporary exponents of Urbino culture was very strong: it is sufficient to think of his support of the Italian translation of Guidobaldo del Monte's *Mechanicorum liber* (*Book of Mechanics*).[29] Moreover, Pinelli was held in high consideration by Duke Francesco Maria II Della Rovere himself.[30]

---

il nostro Herone è propriamente l'aria commossa nelle machine e nelli vasi spiritali, mediante il contrasto che fanno alcuni elementi uno con l'altro."

26 Bonaventura, *Anemologiae pars prior*, 62: "cum exemplarium veterum nulla nobis esset copia, etsi ipsa per omnes illustres Italiae bibliothecas diligenter conquisiverimus, emendationes nonnullae Adriani Turnebi, quas ex bibliotheca literatissimi ac integerrimi viri Vincentii Pinelli habuimus, et sectio vigesimasexta Problematum Aristotelis, unum fuere nobis in tanta re subsidium; ex quo factum etiam est, ut dum libellum hunc diligenter cum Aristotelica oratione conferebamus, ipse quoque Theophrastus non semel iacenti praeceptori manus admoverit; nobisque propterea plura loca, et ea quod difficillima, intelligendi restituendique dederit facultatem [...] Illud etiam sciant lectores velim, nos in multis doctissimorum virorum opera et iudicio fuisse usos; videlicet ex nostris Petri Pauli Florii et Alexandri Georgii; item Aloisii Lollini Patritii Veneti, viri omnibus artibus et disciplinis ornati, et graeci sermonis scientissimi."; Schmitt, "Theophrastus," 287–288.

27 Grendler, "Greek Collection in Padua," Nuovo, "Creation and Dispersal."

28 Canart, "Alvise Lollino et ses amis grecs."

29 Keller, "Mathematics, Mechanics and Experimental Machines", Micheli, *Le origini del concetto di macchina*, 163–167.

30 Gualdo, *Vita Ioannis Vincentii Pinelli*, 40: "Franc. Maria Urbini Dux, quem veterum Regum studia aemulantem admiratur aetas nostra, Pinellum consulere haud gravabatur. Idemque cum acta rerum in orbe nostro quotidie gestarum, quae Imperiorum olim instrumenta qui rerum potiebantur in arcanis habebant, colligenda sibi decrevisset, ad unum Ioh. Vincentium confugit, qui eius votis egregie respondit." Continisio, "Scritture politiche urbinati."

In *De causa ventorum motus* Bonaventura placed Bernardino Telesio at the forefront of a gallery of examined ancient and modern authors (up to his contemporaries). He presented Telesio's theses in the field of anemology which contrasted with Aristotle's (*Bernardini Telesii opinio in Aristotelem animadversio*, i.e., *Bernardino Telesio's Rejection of Aristotle*)[31] and then he subjected them to detailed confutation (*Bernardini Telesii opinio reiicitur*, i.e., *Refutation of Telesio's Position*).[32] The core of Bonaventura's criticism of Telesio was expressed as follows:

> [...] the cause attributed to the motion of these vapors [...] is surprisingly inadequate; according to Telesio it is nothing more than compression: when they are compressed, vapors begin to move to avoid the compression. However, how could there be such compression without thickening? If it is true that vapors condense and, thus, thicken when compressed, the thickening, as was clear to Telesio himself in chapter IV [of *De iis quae in aere fiunt* / *On Those Things Occurring in the Air*], turns the vapors into water and, therefore, it would seem appropriate that there was rain rather than wind [...].[33]

Bonaventura's objection was rigorous and had an impact on the profound structure of Telesian physics, far beyond the case of the nature and motion of the winds. It developed from reasons that had already touched the debate on the principal theories in *De rerum natura iuxta propria principia* and that

31 Bonaventura, *De causa*, 40: "Bernardinum Telesium virum aetate nostra non obscurum Aristotelem, in quem unum inveheretur, delegisse neminem latet; itaque nobis haec tractantibus, quae ipse in opusculo *De his quae in aëre fiunt* ventorum motus causam edocens, de hac quaestione statuit atque ipsius ibi in Aristotelem argumenta dissimulanda silentio non fuerunt."

32 Ibid., 75: "Quoniam Bernardinus Telesius non solum in magistri verba non iuravit, sed tanquam acerrimus peripateticae philosophiae hostis principia illius penitus negavit, idcirco in eo refellendo non Aristotelis auctoritate, sed vel ipsiusmet dictis, vel ratione tantum ipsa ac sensu utendum nobis erit."

33 Ibid., 77: "[...] causa motus [...] horum vaporum mirifice claudicat; nulla enim alia est secundum Telesium quam compressio; cum enim comprimuntur, id ut vitent, moveri incipiunt; atqui compressio haec quo modo sine conspissatione esse potest? Siquidem cum comprimuntur, densantur necessario conspissanturque vicissim; at conspissatio Telesio, ut capite 4 patet, vapores in aqua cogit; itaque imbres potius quam ventos oriri oportebit [...]."

were to last as long as its reception.[34] Telesio, as we have seen, attributed the genesis and motion of the winds to the inclination of humid vapors to conserve their level of tenuity. Within a general rejection of the Aristotelian physics of elements and qualities, Telesio had disagreed with Aristotle that the winds originated in dry vapors. He appealed to the principle that the tenuity of these vapors implied humidity. As his pupil, Sertorio Quattromani, synthesized his view, "humid does not mean something that makes wet, but something tenuous and subtle."[35] And, as Quattromani also noted, the property of tenuity played a central role in Telesio's physics:

> Heat and cold alone are the primary principles of everything that is seen in the world. Humidity and dryness are just dispositions of matter, that is humidity is matter that is rendered thinner by the heat and dryness is matter that is thickened by the cold; and they are both produced by these two acting natures.[36]

---

34 Bacon, "De principiis atque originibus secundum fabulas Cupidinis et Coeli. Sive Parmenidis et Telesii et precipue Democriti philosophia tractata in fabula de Cupidine," in Bacon, *Works*, vol. 5, 325, 329–330: "Videndum igitur deinceps, qualia sint ea quae a Telesio dicuntur circa dispositionem materiae, in quam calor agat; cuius ea est vis, ut actionem ipsam caloris promoveat, impediat, immutet. Eius ratio quadruplex. Prima differentia sumitur ex calore praeinexistente aut non praeinexistente. Secunda, ex copia aut paucitate materiae. Tertia, ex gradibus subactionis. Quarta, ex clausura vel apertura corporis subacti [...] Sed interim satagit et aestuat Telesius, et miris modis implicatur, ut expediat modum divortii et separationis qualitatum suarum primarum connaturalium, caloris, lucis, tenuitatis, et mobilitatis, ac quaternionis oppositae, prout corporibus accidunt: cum corpora alia inveniantur calida, aut ad calorem optime praeparata, sed eadem inveniantur quoque densa, quieta, nigra; alia tenuia, mobilia, lucida sive alba, sed tamen frigida; et similiter de caeteris; una quapiam qualitate in rebus existente, reliquis non competentibus; alia vero duabus ex istis naturis participent, duabus contra priventur, varia admodum permutatione et consortio. Qua in parte Telesius non admodum feliciter perfungitur, sed more adversariorum suorum se gerit; qui cum prius opinantur quam experiuntur, ubi ad res particulares ventum est, ingenio et rebus abutuntur, atque tam ingenium quam res misere lacerant et torquent; et tamen alacres et (si ipsis credas) victores suo sensu utcunque abundant. Concludit autem rem per desperationem et votum, illud significans, licet et caloris vis et copia, et materiae dispositio, crasso modo et secundum summas distingui et terminari possint; tamen exactas et accuratas eorum rationes, et distinctos et tanquam mensuratos modos, extra inquisitionis humanae aditus sepositos esse; ita tamen, ut (quo modo inter impossibilia) diversitas dispositionis materiae, melius quam caloris vires et gradus, perspici possit; atque nihilominus in his ipsis (si qua fata sinant) humanae et scientiae et potentiae fastigium et culmen esse." See Iovine, "Henry Savile lettore di Bernardino Telesio."

35 Quattromani, *La philosophia di Berardino Telesio*, [12].

36 Cf. Telesio, *La natura secondo i suoi principi*, ed. Bondì (2009), 57, 59; Telesio, *De rerum natura* 1570, I 18.

As Telesio stated,

> The consistence and the tenuity are [...] properties of mass and matter and are, without doubt, a disposition of matter, albeit they differ from one another profoundly. Matter itself cannot in itself differ with respect to itself, if not simply in the fact that it is more or less reunited, condensed, and shrunk; it can therefore be considered that tenuity only differs from consistency in this way [...]. consistency is [...] the maximum condensation of matter [...] tenuity [...] appears to be the maximum dispersion, extension, and amplification of matter [...]. It is not possible to observe or explain the countless, tiny steps through which consistency passes to reach tenuity [...]. Tenuity occupies the fifth and final stage and does not appear to oppose the things it touches, but, rendered incorporeal, avoids being touched or seen [...]. The vapors [...] so to speak intermediaries between fluidity and tenuity, certainly remain hidden from touch, but not also from sight [...] and when they condense a little, they become water or even hail.[37]

The last quotation comes from the 1570 edition of Telesio's *De rerum natura*. I refer to this edition because this is the text that Francesco Patrizi took into consideration in 1572 at the invitation of the Telesian physician Antonio Persio. With reference to the chapters in the first book where tenuity and density were dealt with, Patrizi made an objection which is very similar to Bonaventura's.[38] As Patrizi wrote:

37 Telesio, *La natura*, 65, 67; Telesio, *De rerum natura* 1570, I 20: "Crassities tenuitasque vel omnium consensu molis est materiaeve et materiae omnino dispositio, at penitus ab altera differens altera. Materia porro ipsa per se non alio a seipsa differre potest, at eo tantum quod sibiipsi vel magis unita existit magisve in seipsam conspissata et in angustum acta vel minus. Hoc igitur tantum a crassistie tenuitas differre videri potest [...] crassities [...] summa materiae conspissatio existit [...] summa [...] materiae explicatio extensioque atque amplificatio tenuitas videtur [...] At [...] gradus, quibus a crassitie ad tenuitatem itur, innumeros et longe illos minutissimos intueri explicarive impotentibus [...] Quintum atque extremum spatium tenuitas occupat, quae nimirum non modo nihil contingenti obniti videtur, sed veluti incorporea facta tactum fugit visumque [...] Nam vapores, veluti fluoris tenuitatisque medii, tactum quidem, at non et visum etiam latent et lucem, si non reiiciunt, at imminunt tamen foedantque et modicum quid in seipsos spissati, aqua fiunt aut etiam grando."

38 Bonaventura, *De causa*, 76: "Sed nec firmiora mihi videntur quae ab eodem [Telesio] scribuntur de loco, in quo vapores istos moveri voluit; facit enim eos inter aërem et aquam medios, ut vel in aquas cogi possint vel tenuiores facti aëris naturam obtinere subindeque in superiorem locum ferri; antequam vero in aërem mutentur, in loco inferiore acquiescere."

> It is a beautiful consideration from all points of view, such as to embrace the entirety of your philosophy; but more metaphysical than physical, for being an, as it is said, abstract treatise; because, if it is physics, I believe that it should have had examples in order to be more in line with the sense that you say guides your doctrine [...].
>
> The degrees of density and rarity are laid out in a continuous series; I would like you at least to rectify this by also attributing to the vapors that you placed between fluidity and density an intermediate position and assigning a fifth level, and a sixth to tenuity, unless of course you have a stronger motive; the whole question needs examples to be added to chapters XXI and XXII.[39]

As is well known, both Telesio and Persio replied to Patrizi. Telesio accepted—or rather, he partially accepted, as we will soon see—Patrizi's solicitation: "Sense shows that the vapors, as you justly observe, are placed at the fifth level of intermediation, between fluidity and tenuity [...]."[40]

Persio, for his part, reiterated the master's idea, corroborating it with a mention of the tenuity-humidity dyad, which also brought *De rerum natura* and *De iis quae in aere fiunt* closer to each other:

> Telesio did not confer a specific level on the vapors, as they did not completely possess one of their own, appearing closer to tenuity than fluidity [...].
>
> That earth changes in water is shown as much by the senses as reason; following the senses, it can be seen that vapors, extracted from the earth and condensed into clouds, pour water back onto the earth. However, you will reply to Aristotle and say that vapor comes from water, not earth, and that earthly emissions are dry. We, instead, will say: an earthly emission is very tenuous and not even Aristotle can deny this; but the emission which is very tenuous is also very humid, as is proven in chapter XXV

39 Patrizi, "Obiectiones," in Telesio, *Varii*, 470: "Contemplatio omnium pulcherrima, quaeque universam tuam philosophiam comprehendit; sed magis metaphysica videatur quam physica; ita veluti in abstracto, ut dicitur, est a te pertractata, ut physica videatur, puto eam exemplis indigere, ut sensui, duci tuo, familiarior fiat [...] Perpetuo gradus densitatis raritatisque dispositi sunt; id modo velim corrigi vapores, quos medios inter fluorem et densitatem facis, medios quoque, quinto scilicet gradu statuas, tenuitas sextum teneat, nisi maior tibi subsit ratio, exemplisque tota res illustranda; quod et sequentibus vigesimo primo, vigesimo secundo est faciendum."

40 Telesio, "Solutiones obiectionum Francisci Patritii," in Telesio, *Varii*, 461: "Vapores, quod recte admones, medios inter fluores tenuitatemque quinto gradu positos sensus ostendit [...]."

> of the second book [of *De rerum natura*] and in *De iis quae in aere fiunt*; if indeed the emission of the earth were dry, according to Aristotle, it never might be drawn upwards and that which is dry is also dense.[41]

It would seem that the dispute led to Telesio's verifiable reworking of the final part of Chapter XX of the first book, indeed persisting with the premise according to which "it is not possible to observe or explain the uncountable tiny steps which one goes through from consistency to tenuity." As one reads in the 1586 edition of *De rerum natura*:

> Occupying the fifth position are the vapors generated by the more rarefied fluids such that they will turn to fluids if condensed; from what we perceive, these do not offer any resistance to compression, accept light much more quickly than fluids do and obfuscate it much less. In the sixth position, there is tenuity [...].[42]

However, the examples that Patrizi wanted to support Telesio's thesis did not come in the following chapters. Telesio did not correct the reported incongruence because he could not correct it. In replying to Patrizi, Telesio invoked the senses, saying that they could be the basis upon which the level of rarefaction

41 Persio, "Responsiones ad obiectiones Francisci Patritii contra Telesium," in Telesio, *Varii de naturalibus rebus libelli*, 489: "Vapores ideo forte a gradu exemit, quia non integrum facerent, et magis tenuitati proximi videntur esse quam fluoribus [...] Terras in aquas agi et sensus et ratio ostendit; sensus, quia ex vaporibus, qui e terra educi videntur et conspissati ipsi veluti in nubes aquas emittunt in terras. At dices cum Aristotele vaporem non e terra, sed ex aqua educi; e terra vero siccam tamen exhalationem. At dicimus nos: quae e terra educitur exhalatio tenuissima est, quod nec Aristoteles negare potest; ac quae tenuissima est, ita humidissima, ut probatum est libro secundo, capite vigesimo quinto et in *De iis quae in aere fiunt*. Sicca vero si esset, nunquam vel Aristoteli sursum trahi posset; et siccum quod est crassum etiam."

42 Telesio, *De rerum natura* 1586, I 20: "Ac limitum terminorumque, quibus crassities ad tenuitatem proficiscens dividi potest, is primus poni potest, quo lentor flexibilitasque contineri videtur, bene ampla quidem et ipsa et quae et ipsa terminis multis dividi possit. At, ut dictum est, gradus, quibus a crassitie ad tenuitatem itur, innumeros et longe illos minutissimos intueri explicareque impotentibus, lentor omnis omnisque flexibilitas, vel valde a se ipsa differens, una poni potest [...] Quintum spatium vapores occupant, qui e fluoribus fiunt amplius attenuatis et qui in se ipsos conspissati in fluores coguntur nihilque prorsus, quod sentiri possit, comprimenti renitantur et lucem multo quam fluores admittunt promptius multoque minus faedant. Sextum vero atque extremum tenuitas, quae scilicet non tactum modo, sed quantumvis in se ipsam coacta, visum etiam—quod vapores non faciunt—penitus lateat et quantavis facta, lucem nihil imminuit faedatve usquam, ut a vaporibus seiungenda ideo sit et coelo ea universo inesse videtur."

to assign to the vapors could be established. This, though, was an inadequate argument. Patrizi himself had objected to Telesio:

> You continue to say that mass [...] is that which the ancient philosophers called matter; as it happens, I think that this matter is the greatest of fictions. It is certain that not one of the senses, which you declared you would make use of at the beginning of your work, has ever shown matter to anyone.[43]

In the opening chapter to *De rerum natura*, Telesio affirmed:

> And if it is seen that my pages do not contain anything divine or worthy of admiration, at least they will never be contradictory or irreconcilable about things, given that I have done nothing other than follow my senses and nature, which, always in complete agreement with itself, always does the same things in the same way and equally.[44]

Now, in appealing to the senses over the presumed "fifth level" of vapors' tenuity which they should have revealed, he slipped into evident contradiction as soon as he proposed the premise that "it is not possible to observe or explain the countless, tiny levels which one goes through from consistency to tenuity." This premise, as has been mentioned, would remain the same for the 1586 edition. Patrizi noted this contradiction when he observed that such a treatment pertains "more [to] metaphysics than physics." In light of these considerations, Bonaventura's Peripatetic criticism of Telesio looks reasonable: "how can there be [...] compression [of vapor] without there being thickening?" Once elements and quality are abolished and Aristotle's theory of forms is rejected, the questions stemming from the problem of the intensification and attenuation of qualitative forms (*intensio et remissio formarum*) are also

43 Patrizi, "Obiectiones", 467: "Declaras molem [...] materiam illam veterum philosophorum scilicet intelligere te; forte, ut mea fert opinio, figmentorum omnium figmentum maximum materia haec est. Certe sensus nullus, quo solo duce te in omnibus usurum initio es professus, materiam ulli unquam indicavit." Bondì, *Introduzione a Telesio*, 54.

44 Telesio, *La natura*, 5; Telesio, *De rerum natura* 1570, I 1: "[...] si nihil divinum, nihil admiratione dignum, nihil etiam valde acutum nostris inesse visum fuerit, at nihil ea tamen vel rebus vel sibiipsis repugnent unquam, sensum videlicet nos et naturam aliud praeterea nihil sequuti sumus, quae summe sibiipsi concors idem semper et eodem agit modo atque idem semper operatur."

rejected. Telesio's physics, partially or ambiguously imposing the acting natures upon the forms,[45] is subsequently not able to provide a solid explanation for the phases of material transformations, and thus the processes of condensation and rarefaction.

The pages Bonaventura dedicated to Telesio's *De iis quae in aere fiunt* are important for many reasons. First of all, they are testaments to a little known and totally unexplored episode in Telesian polemics. Second, they are also important in consideration of the date of their printing. They appeared two years after the Venetian edition of the *Varii de naturalibus rebus libelli* edited by Antonio Persio in 1590.[46] Among the *libelli*, there was *Quod animal universum ab unica animae substantia gubernatur* (*All Animals Are Governed by a Unique Substance of the Soul*), which was to enter Pope Clement's *Index* together with *De rerum natura* and the other booklet *De somno* (*On Sleep*).[47] Persio dedicated his edition of Telesio's *Quod animal universum* to Gianvincenzo Pinelli, who he said was a reader and admirer of Telesio, in spite of the latter's forceful anti-Aristotelianism; nor did Persio forget to mention his own conversations with Pinelli regarding Telesian philosophy.[48] During the years of his stay in Venice, Persio built up a solid network of intellectual relationships: he became part of the cultural society that was stimulated by the presence

45 Quattromani, *La philosophia*, [11]: "Forma è quella sostanza che dà l'essere alle cose e che fa ogni azione ed ogni operazione che vediamo fare alle cose. E perciò si dice che il caldo è forma del fuoco e il freddo è forma della terra." Kessler, "Metaphysics or Empirical Science?"

46 Firpo, "Appunti campanelliani III: La perduta *Apologia pro Telesio*," Garin, "Nota telesiana."

47 Firpo, "La proibizione di Telesio."

48 *Io. Vincentio Pinello Antonius Persius*, in Telesio, *Varii de naturalibus rebus libelli* (1590), [1]: "Nullus est non in hac urbe solum, sed ne in tota Europa quidem locus, quo maiores Doctorum atque Insignium in qualibet liberali arte virorum concursus, ac frequentiores fiant quam ad aedes tuas, Ioannes Vincenti Pinelle nostrae Deus aetatis atque ornamentum. Confluunt enim ad te quotidie ex diversis orbis regionibus, qui te aut officii causa invisant aut de gravi aliqua disputatione consulant aut ignotam sibi antea faciem tuam contemplentur. Ita sit, ut cum istic plures eodem tempore convenerint, nullus sit dies, quo non de quam dignissimis scitu rebus sermones habeantur. Multique quorum hic sedes est ac domicilium, limina istaec tua inprimis terunt. Sic enim illi, ac recte quidem, et mecum sentiunt nullum esse ludum, Academiam nullam, unde quis doctiorem se ac prudentiorem abiise gloriari possit. Experior id ego in me ipse quotidie, qui tamdiu frequento aedes tuas, neque aliud est, quo malim hic esse quam diutissime. Qui cum enim honestius atque eruditioribus colloquis diem traducam, ne fingi quidem potest. Collocuti autem praeter caetera saepe sumus de Telesiana philosophia, quam etsi longissime a peripatetica abhorrentem, sic tamen laudas, ut admirandum esse Auctorem eius ingenue fatearis."

and untiring activity of Pinelli[49] to such an extent that Paolo Gualdo, Pinelli's biographer, refers to him as one of the most assiduous and appreciated in Pinelli's circle.[50] This was the same environment which Bonaventura was in dialogue with from his residence in Urbino. There seems to be good reason to believe that Bonaventura's polemical intervention, which hinged upon the examination of *De iis quae in aere fiunt*, was provoked by the appearance, two years before, of Persio's edition of Telesio's *libelli*. This is a relevant fact as it contributes to certifying the impact that that edition had within one of the most fervid areas of Italian culture at the end of the sixteenth century, a space traversed and fed by many voices, such as the very diverse and distant voices of Persio and Bonaventura. The circulation of Telesio's ideas in the transregional context of culture which flowered around Pinelli's library over some significant years is well documented; these were the years immediately prior to the censorship of Telesio's works, when the Aristotelians could discuss Telesio in a very different manner compared to the theologians and philosophers of Padua. This is a macroscopic fact which does not need a detailed comparison between Bonaventura's arguments and the text of the *Censura in opus Bernardini Telesii Consentini, quod inscribitur De rerum natura iuxta propria principia* (*Censure of Bernardino Telesio's Work Entitled On the Nature of Things according to their Own Principles*). The Paduan theologian Girolamo Pallantieri drafted this work at the end of November 1600 and it was supported by Cesare Cremonini around the time he succeeded Francesco Piccolomini as lecturer *primae sedis* of philosophy and when he was destined to become the emblematic representative of the Aristotelian position as professed in Padua;[51] different reasons, different tones.

---

49 Maccagni and Derenzini, "Libri Apollonii qui ... desiderantur," Artese, "Una lettera di Antonio Persio al Pinelli."

50 Gualdo, *Vita*, 47–48: "Immo et cum amici adessent, nunquam non illi [Pinelli] ad manus fuit calamaria theca, ut summa rerum quae placuissent capita, ne exciderent, annotaret. In hoc commentarios si quis forte studiosorum inciderit, is tum denique leonem ex unguibus, ut dicitur, aestimabit. Testatus sane mihi est Antonius Persius, selecta doctrina non minus quam Pinelli antiqua familiaritate illustris, consuevisse Ioh. Vincentium adnotare ad omnes quotquot nancisceretur alicuius pretii auctores, uberiores et politiores notas, quas acerrimo studio ab aliis non conquirebat modo, sed excogitabat ipse docte et eleganter."

51 Firpo, "La proibizione", 40: "Quantam utilitatem ex doctrina Aristotelis omnis schola, omnis Academia et universa respublica literaria semper acceperit, accipiat et acceptura sit, et quam libenter gravissimi viri, nedum gentiles, sed etiam fideles, docti, pii et sancti, circa illam comparandam, interpretandam, defendendam insudaverint, lippis et tonsoribus (ut proverbio dicitur) est manifestum. Quapropter, si ii superstites essent inter nos et pravam Bernardini Telesii eius abolendi intentionem persentirent, nemini dubium quin ad tantam temeritatem omnes uno ore essent exclamaturi. Si enim de natura, de moribus,

In conclusion, the content of Bonaventura's criticism of Telesio further clarifies which weapons the better equipped sixteenth-century Aristotelianism had at its disposal when faced with Telesian philosophy. By referring to the case of condensation and rarefaction, Bonaventura's polemic against Telesio's doctrine indicates how it inevitably fell into contradiction, because it was tied to a theory of the structure of matter framed in qualitative physics, which was a theory that was difficult to oppose to that of an Aristotelian mould.[52]

---

de virtutibus et vitiis occurrat disputatio, quo confugient principes scholae et humanae et divinae philosophiae, si ipsis unus Aristoteles adimatur? Certe, cum tot et tanti viri in huius philosophi doctrinam, quamtumvis loco et tempore et moribus disiuncti, tanquam veram consenserint, divinitus videtur velut magister orbis pro rebus lumine naturali cognoscendis constitutus. Nam omnis sapientia a Domino Deo est et veritas, a quocumque dicatur, a Spiritu Sancto est."

52 Valleriani, "From *Condensation* to *Compression*."

CHAPTER 6

Telesio and the Renaissance Debates on Sea Tides

Pietro Daniel Omodeo

In the concluding sections of the naturalistic treatise *De mari* (On the Sea), Bernardino Telesio outlines an explanation of the tides, consistent with the general plan of his natural philosophy *iuxta propria principia*. In the same pages he distances himself from well-established explanations based on the remote action of the Sun and Moon. His criticism is part of a historical-scientific trend toward a physical explanation of the phenomenon that excludes common astrological accounts. In this chapter, I will provide the essentials for an understanding of Telesio's speculations within a process of empirical-rational emancipation that ideally links Pico's criticism of the belief in astral influences with the Copernican-mechanical explanation of the tides by Galileo in his *Dialogo sopra i massimi sistemi del mondo*.

1 Telesio's Explanation of the Tides

Telesio first deals with sea tides, the topic of the so-called *fluxus et refluxus maris*, in Chapter X of *De mari*, "Motum mari necessarium fuisse et a quibus intumescit moveturque" (The necessity of the motion of the sea, and by what causes it is moved and raised). According to Telesio, the general cause of the continuous motion of the sea is its simmering, produced by solar heat and the formation of vapors "which attempt to come out but are hindered in this attempt by the sea above; as a consequence they raise and swell it just like the spirit that fire generates in the water."[1] Hence the waves are analogous to the boiling of water in a pot.

> [...] and so one sees the alternating ebb and flow of the sea, mostly in equal time intervals of six hours. The reason is that during this time interval so many vapors are generated in the sea capable of raising and

1 Telesio, *De mari* in *De iis que in aere fiunt et de terremotibus* (1990), 113: "li quali sforzandosi di uscire et essendo prohibiti di farlo dal mare sopraposto, a guisa di spirito generato nell'acqua dal foco, lo innalzono e lo fanno gonfiare."

 | DOI:10.1163/9789004352643_008

> pushing it for that time. However, the rapidity of the sea seems marvellous in narrow and hollow places. In fact, in an open space nothing forces it [the sea water] to accelerate its motion if it is pushed by the sea behind. If it [the sea water] goes through narrow places it cannot pass all at once but only in part; as [this portion of the flowing sea water] hinders the flow of the [water] behind and hence is pushed and spurred, it accelerates its motion by necessity [...].[2]

The tides are thus a marine simmering, which must be related to the flow of currents in the straits. Telesio advances the idea of a circular motion of the waters toward the west observed not only in the Mediterranean but also in regard to ships crossing the oceans. The Spanish navigators employ 24 days to reach the New World but a full three months to return, "as they are hindered by the opposite waters, flowing in the opposite direction."[3] Similarly, the Portuguese navigators are hindered by the flow of water in their journeys to the East Indies beyond the Cape of Good Hope but are aided during the return voyage for the same reason.

In Chapters XI and XII of *De mari*, Telesio discusses and refutes the theories of those who believe that the phenomenon of the tides results from the remote action of the sun or moon. Some conceive such action as an influence (*influxus quidam a Luna in mare missus*), others as a magnetic action (*quomodo ferrum ad magnetem*). The main difficulty of these explanations is in assuming the possibility of long-distance actions without intermediate causes, which we could call 'arcane' or 'occult'. Telesio puts the reader on guard against those who assume that there is an "incorporeal faculty emitted by the moon" (*incorporea facultas a Luna emissa*).

> One should not listen to them just as [one should not listen to] those who assert that the Moon emanates an incorporeal faculty (something that the human intellect finds impossible to understand) which raises and

2 Ibid., 115: "[...] e così si vede, come per vicenda, il flusso e reflusso del mare, et il più delle volte in tempo eguale, e l'uno e l'altro in spatio di sei hore; e questo per che in quello spatio di tempo si generono nel mare tanti vappori, che per quel tempo lo possono inalzare e spignere; ma ne' luoghi stretti e ne' cavati apparisce maravigliosa la celerità del mare, per che scorrendo egli per loco aperto e non essendo dal mare anteriore sospinto, non ha cosa che lo stimoli a concitare il moto; ma passando per lochi stretti non può correre tutto insieme, ma solamente una piccola parte, e quel che rimane dando impedimento al corso di quello che li vien dreto, e per ciò essendo sospinto e stimolato, è necessario che egli acceleri il moto [...]."

3 Ibid., 117: "repugnando loro le acque contrarie, in contraria parte correnti."

> moves the whole sea with such great and many motions without moving any other being, even if there is earth between it and the Moon.[4]

In Chapter XIII, Telesio provides an explanation of the tides, particularly of the variability of their amplitude at different locations. The variations depend on the different water depths and meteorological, climatic, and seasonal variations, which result in unequal amounts of solar heat. The moon in opposition would produce a greater simmering of the seas. In other words, Telesio does not deny that the sun and moon should be taken into account in tidal theory, but he 'rationalizes' their action in terms of a heating by the rays of the sun and their propagation by the moon's reflection of sunlight. The basic periodicity of six hours is also not questioned and it is not linked to the diurnal cycle and the position of the moon. Rather, it is considered a natural cycle of the waters, in which the flow is followed by an ebb aimed at restoring the original condition.

It seems that Telesio was not completely satisfied with his own theory, so much so that he deleted the last three chapters of *De mari* from the Neapolitan edition of the naturalistic *opuscula* of 1570. These sections only appeared in the posthumous Venetian edition edited by his pupil Antonio Persio in 1590.[5]

Let us summarize the main points of Telesio's tidal theory. First the phenomenon is inserted in the more general topic of the motion of the waters. This fits within the Aristotelian-Scholastic discussion of the natural motions of the elements, especially connected with the reception and comment of *De coelo* and *Meteorologica*. Instead of a 'natural motion' (*motus naturalis*), Telesio considers the motions of the waters to be processes of boiling dependent on the action of the Sun. He also believes that the overall motion of the seas is a circular flow toward the west, a hypothesis supported by the ocean voyages of the Spanish and Portuguese seamen. The tides have a periodicity of six hours although their amplitude varies according to a number of local climatic, meteorological, and geographical conditions. For example, the presence of straits or the conformation of the coasts either facilitates or hinders the marine flows. Hence Telesio's explanation is in contrast to the theories of those who assume that the origin of the tides is a remote action without intermediate causes. In this sense, Telesio's theoretical attempt is part of a process of 'mechanization'

---

4 Ibid., 123: "[Costoro] non dovrebbero essere ascoltati come quelli che vogliono che la Luna emetta una facoltà incorporea (cosa che non è possibile comprendere con l'intelletto umano) la quale senza muovere le altre acque ed alcun altro ente [...], innalzerebbe e muoverebbe con così grandi e numerosi moti tutto il mare, persino quello al quale la terra stessa s'inframmezza tra esso e la Luna."

5 Cf. De Franco, "Nota introduttiva."

or at least of anti-astrological physical explanations since it excludes factors operating at a distance. All of these aspects must be considered in detail with reference to the wide-ranging Renaissance debates on sea tides and maritime flows.

2 The Tides as an Astrological Subject: Albumasar's Legacy

It should be stressed that in the late Middle Ages and Renaissance the subject of the tides had a clear astrological connotation. This explains Telesio's concern about explanations of sea tides assuming a distant action of the Moon reminiscent of astrological influences. Indeed a reference source on this topic was one of the most widespread texts devoted to astrology, Albumasar's *Introductorium in astronomiam* (Introduction to Astronomy): "One can say that it is from this book—wrote Pierre Duhem—that all the [scholars of the] Latin Middle Ages learned the laws of the ebb and flow of the sea."[6] For the famous ninth-century Persian astrologer, the phenomenon of the tides is clear and incontrovertible proof of the influence of the stars on terrestrial events. The action of the moon on the waters is second in magnitude and visibility only to seasonal effects determined by the sun, by its annual journey along the zodiac and by its variations in declination. In other words, in the *Introductorium* the consideration of the tides is functional to an astrological interpretation of sublunar events.[7]

Albumasar carefully examines the subject in the third book of the treatise from the fourth chapter to the eighth. The lunar influence on the seas is introduced as a special relationship of the celestial body of the moon with the watery element. The sun has a particular influence on two of the four natural elements of Aristotelian tradition: fire and air; the moon affects the remaining two: earth and water. The explanation of the tides is based on a triad of causes: the conformation of the locality (depth, length, and breadth), the particular condition (*habitudo*) of the water (agitated or not because it is brought by rivers or is near sources, the density, salinity, and mixture of vapors) and lunar motion.[8]

---

6 Duhem, *Le Système du Monde*, vol. 2, 369: "C'est dans ce livre, peut-on dire que tout le Moyen Age latin a appris les lois du flux et du reflux de la mer. La doctrine d'Albumasar mérite donc que nous y arrêtions avec quelque complaisance."

7 Albumasar, *Introductorium in astronomiam*, a3r–v.

8 Ibid., c1v: "Dicimus igitur quod nunquam huiusmodi accessus et recessus nisi trium rerum conventu gignitur, loci videlicet natura, aquarum habitudinem [sic!], [et] motu <Lunae>. [1.] Loci natura est ut aquarum locus profundus, longus et latus vix temporis impendio transfretandis montuosus asper et durus quale vi quolibet motu acriter repercusse multe unde

The moon's influence is the basis of the fundamental regularity of the tides, whose variability is linked to several factors, not least the latitude of the seas subject to the lunar action. The moon attracts the waters through a particular force, which acts by affinity or sympathy. In Latin this is called a *cognata virtus* (force of affinity), a type of astrological consonance acting without intermediate causes. The docility of the waters, induced to follow the lunar motion, is indicated as a "spontaneous inclination to be pulled" (*spontaneitas ad tractionem*). The rise of the waters, as Albumasar explains, is primary and depends directly on this astrological traction. The ebb is instead secondary, being a movement only aimed at restoring the original condition prior to the external action of the moon.[9] Moreover, the former tendency is a kind of boiling and thus is warmer than the latter: "The flow [*accessus*] is a warming of the water and the ebb [*recessus*] a cooling down. In fact, during the flow it boils from the profoundest abyss; during the ebb the [water that] overflowed becomes colder."[10] Note that this idea is also used by Telesio in his explanation of the waves but in a completely different natural-philosophical framework. One could say that he does not reject the observation and description of the phenomenon but seeks a physical cause different than the astrological one.

In the seventh chapter of the third book of the *Introductorium*, Albumasar also addresses the opinions of critics of the moon's remote action as the cause of the tides. In other words, he deals with those who deny one of his main arguments in favor of the astral influence and of astrology. According to them, the tides are caused by tendencies intrinsic to the waves, and thus the tides are reduced to a wave phenomenon. Albumasar's refutation is based on several considerations. Firstly, if the tides were determined by the *nature* of the waters, i.e. by an internal tendency, there would not be variations in timing and in amplitude nor a parallelism with respect to the lunar motion. Secondly, it is not credible that the waters of basins overflow from their usual place by means of spontaneous motion. Moreover, waters have an inherent tendency to move downward, as Albumasar argues, following Aristotle's physics. Hence

tumidos fluctus concipiant. [2.] Aquarum habitudo est tantas in huiusmodi loco longo ex temperie aquas esse confusas ut nec in fluxu fluminum nec ex collatione fontium augerive minui sentiatur. Que quanto ipse condensate salebre calefacte densos vapores agitent qui terre vaporibus permixti agitandis undis aspirent. [3.] Motus autem <Lunae> desuper orientis atque occidentis sepius repetitus cognata virtute eiusmodi aquas trahit: quem tractum sponte sequens quousque illa accedit: accedunt usque adeo quoad diffusius eferventes loco suo minus contempte extremis inundent littoribus."

9 Ibid., c3v.

10 Ibid., c2r: "Est autem accessus quidem aque calidior et recessus frigidior. In accessu namque ex imis abyssis ebulliunt in recessu forinsecus expanse infrigidant."

it is difficult to comprehend by what natural tendency they can be induced to rise. Albumasar concludes that the cause must be extrinsic and the only such ascertainable cause is the moon.

3 Pico's Criticism of the Lunar Explanation of the Tides

The centrality of the lunar theory of the tides, as well as the breadth of discussion in Albumasar's *Introductorium*, meant that the critics of astrology could not ignore the topic. Before Telesio's refusal of astrological tides, Giovanni Pico della Mirandola dedicated several pages to this problem in his famous argument against astrology, *Disputationes adversus astrologiam divinatricem* (Disputations against Divinatory Astrology, published posthumously in 1496). Chapter xv of the third book is entitled "The Sea Tides Can Be Explained through a Cause Different than the Moon; Even if This Were the Explanation, It Would Not Support Astrology" (*Aestus maris in aliam causam quam in Lunam referri posse, in quam et si referatur nihil inde iuvari astrologiam*). Pico observes that the question is complex because *everyone* is of the opinion (*cum omnibus videatur*) that the moon is the origin of the phenomenon. It is worthwhile considering his critique since it constituted an important point of reference for subsequent anti-astrological explanations of marine motions.

Pico begins with a series of considerations taken from Adelard of Bath who explained the tides as a motion caused by the tendency of waters to reunite when they are separated by land, according to the principle that "the elements, as parts of a whole, have a natural tendency to restore the [original] integrity" (*naturali propensione feruntur elementi cuiusque partes ad suam integritatem*).[11] This tendency did not coincide with the natural motion of the element but rather a law regulating the behavior of the parts of a whole.[12] As one reads in Pico, the phenomenon can be explained without recourse to celestial causes simply by considering the tides as a process of boiling followed by an expansion of the element (i.e. *accessus*) and then by cooling and contraction (*recessus*):

> Therefore, some will consider the following cause of the sea tides as clear enough and sufficient: that vapors and [...] winds rise from the earth and water so that wind and turmoil is to be found in the water and,

---

11 Pico della Mirandola, *Disputationes adversus astrologiam divinatricem* III 15 (2005), 488.

12 Ibid.: "Hunc accedendi recedendique motum naturalem esse aquae non quatenus aqua est, sed quatenus partes habet elementi principalis, objectu molis terrenae diremptas et separatas."

> in particular, that this motion and impulse, and the mixing of vapors, warms [the water]. As a consequence, it needs a wider space to expand. By contrast, if that force extinguishes because the vapors are dissolved, [the waters] go down and flatten. They contract in narrow places and descend from the places they occupied.[13]

Such thermal theory anticipates the more articulated one by Telesio. Pico stresses that this explanation is based on nothing other than the *nature* of the sea. He concludes by referring to natural reason and experience: "If [this explanation] is in agreement with natural reason and if it is in agreement with experience, why should we regard them as less likely? Moreover, why should we add the lunar motion to these causes [...]?"[14]

Pico then refutes the "Saracens," supporters of a celestial theory of the tides, which violates rational-empirical evidence. He names two of them: Albumasar (*Aboasar*) and Alpetragius. The former is mentioned in regard to lunar causation. The latter is attributed with a general theory of celestial causation of the sublunar elements.[15]

Alpetragius (al-Bitruji, XI century), a contemporary and fellow countryman of Averroes (Ibn Rushd), dealt with the motion of the elements (*motus elementorum*) in the fourth chapter of his physical-astronomical work *De motibus celorum* (On Celestial Motions), which strongly influenced Girolamo Fracastoro and other homocentric astronomers of the Italian Renaissance.[16] As one reads in Alpetragius' work, all of the sublunar elements are affected by the circular motion of the stars but in a progressively reduced manner the farther they are from the celestial sphere. Fire is moved at almost the same velocity as the celestial bodies, as evidenced by comets or similar 'meteorological' phenomena (*quod apparet de similibus stellarum que videntur in quibusdam*

13 Ibid., 488–489: "Poterit autem cuipiam hinc apparere satis aperta et sufficientis causa, marinae reciprocationis, siquidem de tali terra et aqua, vapores [...] ventique suscitantur, unde in aqua sit ventus, tumultus, praesertim quod ex motu impulsuque isto, et vaporum admixtione, calescit, quare locum quaerit ampliorem quo se diffundat. [...]. Rursus ubi vis illa dissolutis vaporibus conflamescit, subsidunt atque sternuntur, et in angustiis se contrahentes, ab occupatis locis abscedunt."

14 Ibid., 489: "Quae si rationi consonant naturali, si consonant experimentis, cur parum probabilia iudicari debent? Aut cur necessarium praeter has causas addere Lunae motum [...]?"

15 Ibid.: "Alpetragius attulit huius motus coelestem causam, nec a Luna, verum a diurno motu quo movent omnia. Sed inferiora minus. Ignis enim sphaera rotatur in orbem. Inordinatus motus in aere fit, quae in aqua definit in accessum atque recessum."

16 Al-Bitruji [Alpetragius], *De motibus celorum*, 80–82. Cf. Di Bono, *Le sfere omocentriche di Giovan Battista Amico*.

horis incensis in aere). Air is slower; whereas the motion communicated by the heavens to water is that of the tides, attributed by many (mistakenly) to the moon:

> And the motion of the water is less rapid than the air. For this reason it is believed that its motion follows that of the moon, in consideration of the closeness of their motions (of the moon and the sea); and hence, it was also believed that it [the sea] follows it [the moon] and is attracted by the latter.[17]

Alpetragius' explanation of the tides is based on the action of the heavens, the *ponderositas* (weight) of the waters and their *multitudo* (which probably means volume). The movement of the tides would go unnoticed if it could not be seen on the seashores (*ubi non invenitur nisi una ripa propter sui magnitudinem et profunditatem*). The east to west flow of the waters, of celestial origin, strikes the coast. The result is an oscillating motion of the tides due to a triple impulse: the westward tendency (communicated by the stars), a resistance in the opposite direction (because of the "weight") and a downward resistance, called *multitudo* or quantity (which is the natural tendency of waters):

> The motion of the water from the east is a motion that imitates that [motion] which is above it. Its motion backwards is due to its weight [*ponderositas*] and its downward thrust is due to its quantity [*multitudo*].[18]

"But it is evident—Alpetragius concludes—that the earth rests in its entirety"[19] even though its parts can be subjected to local movements. In short, the work of this medieval scholar provides an account of the phenomenon of the tides within an astronomical-cosmological context that comes closer to early-modern mechanic accounts (which I will soon discuss) than to the thermal theories of Pico and Telesio. Alpetragius explains the tides in causal terms and without recourse to 'occult' or hidden forces (*virtutes occultae*). His

17 Ibid., 81: "Et motus aque est minus velox motu aeris; et propter hoc creditur quod ipse sequitur in motu suo motum Lune propter propinquitatem suorum motuum (Lunae scilicet et maris); et propter hoc credebatur quod ipsum sequebatur ipsam et incurtat ab ipsa."

18 Ibid.: "Tunc motus aque qui est a parte orientis est motus quo consequitur ipsum quod est superius ad ipsum, et sua reversio est propter eius ponderositatem, et eius declinatio deorsum propter sui multitudinem."

19 Ibid.: "Sed terra apparet quod quiescit simpliciter in toto."

considerations concern the nature of the waters and the universal circulation of the cosmos toward the west rather than the distant action of the moon.

Let us return to Pico's *Disputationes*. After maintaining the plausibility of a non-astral explanation of the tides, Pico admits—for the sake of argument—but does not concede that the moon might be responsible for them. He stresses, however, that the only possible actions of the heavenly bodies on terrestrial ones occur through motion or light and not through occult influences (*occultis influxibus*). "Whichever of these traditions one accepts, it is clear that we are not forced to ascribe any new power to the moon producing the motion of the sea, except for motion and light."[20] The concomitance of lunar motion and tides can be considered a sort of parallelism rather than a causal relationship: "if we allot that effect to the moon, we will refer such motion to a tacit natural harmony according to which motion is imitated when the occasion is given, as it [the sea] rises when the moon rises and descends when the latter sets."[21] However, despite whatever concession might be made to the astrologers as far as natural philosophy is concerned, Pico does not compromise regarding the ethical implications. It is not permissible to infer from the observation of celestial causes that human actions, be they small or large (*et parva et maxima*), individual or collective, are guided and sustained (*duci et regi*) by stars and planets.[22]

To summarize, Albumasar provided the Latin Middle Ages and the Renaissance with an astrological interpretation of the tides based on the remote action of the moon on the waters of our planet. This influence was an unspecified force of affinity (*virtus cognationis*). Such an explanation approaches the modern post-Newtonian theory based on the law of universal gravitation in different ways, but was also accused of being an undue recourse to occult influences (*occulti influxus*). Pico's refutation, included in the fiercest anti-astrological indictment of the fifteenth century, appears to be particularly important. It opened two paths to those who shared his suspicion of an astrology-based theory applied to the lunar explanation of the tides: they could either reject the lunar influence *in toto*, and thus venture into a search for a new explanation, or accept some account in which the moon was granted nothing more than an action through movement and light. Pico provided an intrinsic explanation of

20 Pico, *Opera*, 492: "Verum quaecunque potius recipiatur harum traditionum, patet nihil nos cogi novam comminisci potestatem in Luna, praeter motum et lucem, quare mare commoveat [...]."

21 Ibid., 491: "[...] nos si pertinet ad Lunam talis effectus, ad eius id motum referamus quem tacito naturae consensu, occasioni motus imitatur, quare ascendit cum ascendente [Luna], descendit cum descendente."

22 Ibid., 490.

the phenomenon (which he saw as rational and empirical) relating the *aestus maris* to a kind of alternating expansion of the waters through boiling and contraction of them through cooling. Telesio's account followed in his footsteps but did not constitute the only alternative to astrology.

4 Giordano Bruno's Vitalistic Approach to the Tides

The denial of the lunar causation of the tides—which was closely related to the criticism of astrology—called for a revision of the explanation of the phenomenon during the sixteenth and seventeenth centuries. Giordano Bruno, who was among the authors who sought an alternative, offered a vitalistic account in the fifth dialogue of *La cena de le Ceneri* (The Ash Wednesday Supper, 1584). He denied that the tides were caused by the distant action of the moon, and accepted that there is a correspondence between the positions of the moon and the movements of the waves, but thought that it depends on a kind of harmony of nature by which the laws regulating one process can correspond to those governing a parallel one without there being a causal link. This means that the positions of the moon can be considered *signs* of the rise and fall of the sea level without being considered a *cause*.

The starting point for Bruno is a type of astrobiology according to which the planets' bodies "possess the principle of intrinsic motion [through] their own natures, their own souls, their own intelligence."[23] Since the principle of motion is inherent to the moving object, there is no need to invoke some "tractive or impulsive force or something similar, which cannot be done without the contact of at least two bodies."[24] Bruno maintains that each body that moves without *appreciable contact*, as if affected by another *driving or attracting* body, should be explained on the basis of a spontaneous internal principle (*appulso*). Everything that moves with respect to something else without contact and as if propelled by some deprivation or desire does so by spontaneous motion: it is the iron object that moves the magnet, not the latter that forces the movement; similarly the straw moves spontaneously toward the amber, the feather toward the jet, the sunflower toward the sun. The motion of the tides must also be discussed and explained in this perspective:

---

23 Bruno, *The Ash Wednesday Supper* (1995), 206. Cf. *La cena de le ceneri*, in *Opere italiane*, 427–589, 547–548: "Questi corridori hanno il principio di moti intrinseco la propria natura, la propria anima, la propria intelligenza."

24 Ibid., 206. Cf. Bruno, *La cena*, 548: "virtù trattiva, o impulsiva, et altre simili, che non si fanno senza contatto di dui corpi almeno."

> Upon the consideration that nothing moves in space on account of an extrinsic principle, without a contact more forceful than the resistance of the medium, depends the further consideration that it is solemn foolishness and an impossible thing to persuade an orderly mind that the moon moves the waters of the sea [causing tides], [...] since for all these things it is properly a sign and not a cause. It is a sign and indication, I say, because the observation of these things [together] with certain dispositions of the moon [...] proceeds from the order and correspondence of things, and from the laws of one mutation which are in conformity and correspondence with the laws of another.[25]

The idea that motion is caused from within the moving body infringed against an established Aristotelian principle that "nothing moves by itself." By contrast, Bruno sought an inner cause of motion, which is well in accordance with his vitalistic philosophy of nature. Living beings act following their inner tendencies. A similar idea was proposed by Francesco Patrizi, who wrote that the motion of the seas originated in an intrinsic impulse, and that this impulse was similar to the one that moves the animal-like planets in ethereal space.

Bruno concludes his remarks on the tides with a criticism of *so many strange philosophies* that confuse *signs* and *causes*. The reference is to astrology, which mistakenly believes that the movements of the stars are causes and not *signs* of earthly affairs.

Similarly, geometricians often confuse *signs* and *natural causes*, for example when they state that the perpendicular rays of the sun cause more heat, when the cause of the heat can only be material, i.e. the greater or lesser persistence of the sun on the earth. "It is one thing to play with geometry and another to verify with nature. It is not lines and angles which make the heat of fire more or less, but distance and nearness, long and short duration."[26] Such preference accorded to natural causation over mathematical modelling makes Bruno's path

25 Ibid., 207. Cf. Bruno, *La cena*, 548–549: "Da questo considerar che nulla cosa si muove localmente da principio estrinseco senza contatto più vigoroso della resistenza del mobile, depende il considerare quanto sii solenne goffaria, e cosa impossibile a persuadere ad un regolato sentimento, che la luna muove l'acqui del mare, caggionando il flusso in quello [...]: atteso che quella [...] è propriamente segno, e non causa [...], perché il vedere queste cose con certe disposizioni della luna [...] procede da l'ordine e corispondenza delle cose, e le leggi di una mutazione, che son conformi o corrispondenti alle leggi de l'altra."

26 Ibid., 208. Cf. Bruno, *La cena*, 549–550: "Altro è giocare con la geometria, altro è verificare con la natura. Non son le linee e gli angoli che fanno scaldar più o meno il fuoco; ma le vicine e distanti situazioni, lunghe e brieve dimore."

to science closer to Telesio than to the physico-mathematical path that, in the Italian Renaissance, would culminate in Galileo's work.

5 Mechanical Explanations of the Tides: Galileo's Teacher Cesalpino

In spite of the evident methodological differences, Giovanni Aquilecchia highlighted a possible link between Bruno's rejection of the remote action of the moon on the tides and the much better known and articulated discourse on the tides by Galileo in the concluding part of *Dialogo sopra i massimi sistemi del mondo*. It should be stressed that Telesio also contributed to the calling into question of the lunar action. More specifically, for Galileo the tides are not ascribable to an action of the moon but constitute tangible proof of the motion of the earth.[27] His explanation, on closer inspection, is rather mechanical and not based on a vitalistic and teleological concept of the inherent impulse of bodies to motion or to some tendency to self-preservation. If there is a common foundation of the criticism of the lunar theory in Telesio, Bruno, and Galileo, it is the criticism of astrology and the distant influence of the stars, not the explanations they provided, which are rather different indeed.

The fundamental thesis of "Day Four" of the *Dialogue Concerning the Two Chief World Systems: Ptolemaic and Copernican* (1632), a section entirely dedicated to the tides, is "that if the terrestrial globe were immovable, the ebb and flow of the oceans could not occur naturally; and that when we confer upon the globe the movements just assigned to it [by Copernicus], the seas are necessarily subjected to an ebb and flow agreeing in all respects with what is to be observed in them."[28] In the tides Galileo sought incontrovertible proof of the earth's motion. Renouncing both the action of the moon and proto-gravitational considerations, he explained the phenomenon in what one can call inertial terms. The rise and fall of the waters depend on the combined action of daily axial rotation and annual revolution.[29]

Galileo claimed that his explanation of the tides was absolutely original. Certainly he took a step forward with respect to his predecessors in combining tidal theory, Copernican hypotheses, and mathematical physics. Nevertheless,

27 Aquilecchia, "I *Massimi Sistemi* di Galileo e *La Cena* di Bruno."

28 Galilei, *Dialogue Concerning the Two Chief World Systems*, 417. Cf. Galilei, *Le opere*, vol. 7, 443: "che quando il globo terrestre sia immobile, non si possa naturalmente fare il flusso e reflusso del mare; e che quando al medesimo globo si conferiscano i movimenti già assegnatili [da Copernico], è necessario che il mare soggiaccia al flusso e reflusso."

29 For a recent examination of Galileo's tidal theory, see Clutton-Brock and Topper, "The Plausibility of Galileo's Tidal Theory."

in light of the broad Renaissance discussion on terrestrial motion (*motus terrae*), elemental motion (*motus elementorum*) and tides, the words of the Copernican Salviati in Galileo's *Dialogue* sound ironic: "[…] what I am about to say, I propose merely as a key to open portals to a road never before trodden by anyone […]."[30] Indeed, the portal (*la porta*) and the road (*la strada*) which led to an explanation of the tides based on the Earth's motion rather than on "occult qualities […] and similar idle phantasies" (*qualità occulte e […] simili vane immaginazioni*) had already been opened by one of Galileo's Pisan teachers, Andrea Cesalpino. It is worthwhile to dwell briefly on his theory.

In Chapter III 5 of *Peripateticae quaestiones* (Peripatetic Questions, 1571), Cesalpino, the famous professor of the University of Pisa, demonstrates the thesis that "the ebb and flow of the sea is produced by the motion of the earth and not of the moon" (*maris fluxum et refluxum ex motu Terrae non Lunae fieri*).[31] He discusses not only the motion of the waters but also that of the earth. Firstly, Cesalpino refutes the theory of the moon's influence on the movements of the sea. The observation of a correspondence between lunar motion and tides is the basis for the belief that there is a causal relationship between the former and the latter. Cesalpino argues, however, that if this were true the seas would always flow in the same direction, accompanying the moon, instead of having an alternate motion.

Secondly, the remote action is inexplicable. By what mystery would the moon act on water and not the intermediate elements fire and air (ordered according to the peripatetic doctrine of natural places)? "In fact [the moon] cannot move [the water] by itself, because there is no contact.[32] Or is it necessary to assume that there is a hidden virtue (*virtus quaedam occulta*) such as the one the magnet exerts on iron? But even in this case the *virtus* must communicate the motion through an intermediate motion.

Thirdly, Cesalpino deals with the theory that the moon acts on the waters through its light, which would cause heating (*calefactio*), expansion of the waters (*tumor*), and a consequent flow (*exundatio*). As we have seen, this was the theory considered most plausible by Pico. However, Cesalpino disagrees. If it were true, the same "sympathetic" action (*huiusmodi sympathia in aqua*) would have a greater effect on concentrations of water smaller than seas and oceans,

30 Galilei, *Dialogue*, 418. Cf. Galilei, *Le opere*, vol. 7, 444: "E quello che io sono per dire, lo propongo solamente come una chiave che apra la porta di una strada non mai più calpestata da altri."

31 I discuss Cesalpino's theory in "Riflessioni sul moto terrestre nel Rinascimento."

32 Cesalpinus, *Peripateticarum quaestionum libri quinque*, 60r: "Nam [Luna] se ipsa [aquam] movere non potest, quia non tangit."

as in the case of lakes and ponds: "in fact, that which is smaller is moved by the same force more easily."[33]

After clearing away the arguments based on the Moon's action, Cesalpino moves on to consider Aristotle's *Meteorologica* II 2, in which the origin of the tides is sought in the combination of two causes. The first is the massive entry of water from rivers into the sea, especially in the eastern Mediterranean Basin. The waters of the Mediterranean flow from east to west, from the Black Sea and the Aegean to the Tyrrhenian Sea. The second cause is an oscillatory balancing (*libratio*) of the waters: "another one is the measured oscillation/balancing of the entire sea which in fact often oscillates [*libratur*]."[34] The parallel between the oscillatory motion of waters and the behavior of a balance, implicit in the concept of *libratio*, does not convince Cesalpino. If in fact the weight were greater in one part of the balance, it follows that there would not be a rebalancing of the distribution of the waters but rather a flow in a single direction:

> Aristotle assumes that the same balancing that occurs to a steelyard can be ascribed to the sea. If they receive an initial motion, they alternatively incline towards one side and the other, owing to the equality of the weights. Actually, if the weight on the one side would be greater, the whole would incline and would not be lifted back again.[35]

Cesalpino additionally observes that if the element of water encloses that of earth everywhere, there would be no explanation why an alternate motion like the one in question originates.[36] It could certainly not be a 'violent motion', according to the Aristotelian distinction between natural and violent motions. Indeed "nothing produced with violence is perpetual" (*nullum violentum sit perpetuum*).[37]

---

33 Ibid.: "quod enim minus est, ab eadem virtute facilius movetur."

34 Ibid.: "alteram autem esse modicam quandam totius maris librationem: huc enim illuc libratur saepe."

35 Ibid.: "Quod igitur stateris accidit aequilibris, mari vult contingere Aristoteles. Accepto enim principio motus inclinant modo in unam partem, modo in alteram saepe, propter aequalitatem ponderis. Nam si in altera parte pondus superaret, in eam totum vergeret, nec in alteram resurgeret."

36 This is an old argument, used in antiquity by Strabo (basing himself on Archimedes) against Eratosthenes, supporter of a hypothesis similar to that of Aristotle mentioned here.

37 Cesalpinus, *Peripateticae quaestiones*, 60r.

A similar criticism of the relevant passage of the *Meteorologica* is found in Telesio's *De mari* (vi–vii). Aristotle's *libratio* also does not seem plausible to him, "because the Earth is spherical, therefore it is impossible that its northern part, nor any other of its parts, can be higher or lower."[38] Hence the parallel of the balance scale is wrong, but even if one accepts such an absurdity, Telesio writes, one must consider the phenomenon of balancing and draw consequences that are different (both from Aristotle and from Cesalpino):

> If it has been inclined and no external force pushes nor moves it, it will remain forever at rest. Aristotle should strongly agree on this, as he upholds that all the elements [...] are at rest in their natural place and benefit from immobility.[39]

Let us return to *Peripateticae quaestiones*. According to Cesalpino's reasoning, there only remains one option: that the tides are an "incidental motion" dependent on the container (*continens*), which refers to the basins containing the seas. The extrinsic violent motion has already been excluded by the observation that no violent action can last uninterrupted. However, he excludes the possibility that the seas tides could be "natural motions". According to Aristotle, each element has a unique natural motion and, in this case, the four elements would have a single downward or upward vertical tendency (or, better said, they would move either *toward the center* and *away from the center* of the elements) to return to their natural place. If the tides were a natural motion of the water it follows that this element has more than one natural motion. Thus having eliminated the explanation based on what Cesalpino calls a *per se* cause (the natural motion of the waters), Cesalpino investigates the incidental one, caused by the motion of the "container". The tides would result from the action of one of the two elements contiguous to the sphere of water, i.e. the air or earth. Since the only motion of air is the disordered one of the winds, which at best can ruffle the waters of the seas, Cesalpino infers that the motion of the tides depends on that of the earth. Q.E.D.: *maris fluxum et*

---

38 Telesio, *De mari* vii, *ed. cit.*, 101: "Per che la terra è sferica, né può parere che la parte boreale, o qual si voglia altra sia più alta o più bassa."

39 Ibid.: "Ma dove gli sia una volta declinato, quivi non essendo da alcuna forza esterna sospinto o mosso, rimarrà perpetuamente immobile; e questo debbe parere particolarmente ad Aristotile, al quale pare che tutti li primi corpi [...] sieno nel proprio loco immobili, e che della immobilità godino." Telesio's conclusion is remarkable in the context of the Renaissance *equilibrium controversy* over the behavior of balances that are displaced from their equilibrium as reconstructed in Renn and Damerow, *Equilibrium Controversy*.

refluxum ex motu terrae fieri (the ebb and flow of the sea is produced by the motion of the earth).

Cesalpino explains that the tides should be conceived in a manner similar to the behavior of water in a low, wide container in motion. At first the liquid in the container resists the motion, then follows it and quivers as if seeking its equilibrium:

> As one can see in a small vessel, which is more wide than deep, if it is moved the water first resists in the part opposite [to the direction of the motion] and often oscillates [*libratur*] here and there searching for its equilibrium. Thus, after the earth has moved a bit, the water, which has first remained behind and is out of balance, flows in the other direction, but surpassing the point of equilibrium as a consequence of the imparted motion. For the same reason, it returns back, in the opposite direction, and continues to do that in the search for an equilibrium, in which it can rest naturally.[40]

The next step is to identify the nature of the earth's motion, the necessity of which is evident from the discussion on the tides. Cesalpino observes that the movement of the container—meaning the earth—will cause greater agitations where the masses of water are greater, namely in the oceans. Consequently the frequency and amplitude of the tides will not be equal in all the seas: "from this it is evident that the tides do not always follow the moon."[41] Moreover the *circumvolutio* of the earth must be minimal (*parva tamen*), otherwise the marine fluctuations would be much greater than they appear. Such motion will have consequences on celestial phenomena, a *variatio stellarum fixarum*, which Cesalpino identifies in the millennial motions. According to him, the *theorica planetarum* (planetary theory) can do without the spheres placed by astronomers beyond the eighth one, or rather those assigned, according

---

40 Cesalpinus, *Peripateticarum quaestionum libri*, 60v: "[...] ut videre licet in parvo vase, cuius amplior sit latitudo quam profunditas: si enim dimoveatur, resistit aqua a tergo priusquam in opposita partem, et saepe huc atque illuc libratur quaerens aequilibrium. Cum igitur terra modice praetergressa fuerit, aqua autem posterius derelicta, extra suum aequilibrium existens, in alteram partem ruit, sed ultra aequilibrium ob acceptum motus principium. Inde iterum ob eandem causam in oppositam partem vergit, et saepe id facit, quaerens aequilibrium, in quo naturaliter quiescat."

41 Ibid., 61r: "ex quibus etiam patet non ubique aestus insequi Lunae cursum."

to Peurbach, to the precession of the equinoxes and its irregularities (the so-called *titubatio* or *trepidatio*):[42]

> If the sea perpetually oscillates forth and back, it is necessary that the earth moves. If this is correct, it is necessary that the position [*aspectum*] of the fixed stars changes accordingly. It is the motion of trepidation, discovered by the astronomers in the eighth sphere, that most likely depends on the motion of the earth rather than on its own motion [of the eighth sphere]. Moreover, if this oblique and discontinuous motion of the earth is sufficient to account for the observed change of position [*aspectus*], one does not have to posit any other spheres above the eighth sphere.[43]

This is a moderate Copernican position. Of the three terrestrial motions postulated by Nicolaus Copernicus in *De revolutionibus orbium coelestium* (On the Revolutions of the Celestial Spheres, 1543), Cesalpino retains only the one that seems least plausible to the modern reader: neither the diurnal rotation nor the annual revolution but the third motion called *motus declinationis* which would account for the precession of the equinoxes, for the variability of the earth's axis as well as the presumed irregularities of the precession.

Thus Cesalpino's doctrine is a historical precedent of the Galilean theory of the tides. Galileo's basic thesis is very close to that of his teacher in Pisa, beginning with the experiment (whether mind experiment or not) of the 'vessel': "But if, by simply setting the vessel in motion, I can represent for you without any artifice at all precisely those changes which are perceived in the waters of the sea, why should you reject this cause and take refuge in miracles?"[44] The same applies to the conclusion: "[…] [Y]ou have explained very persuasively why it would be impossible for the observed movements to take place in the ordinary course of nature if the basins containing the waters of the seas

42 I dealt with this topic in relation to Peurbach, Copernicus, and Bruno in Omodeo, "Giordano Bruno and Nicolaus Copernicus," 49–51.

43 Cesalpino, *Peripateticarum quaestionum libri*, 61r: "Si enim libratur huc illuc mare perpetuo, necesse est terram moveri. Si vero hanc, necesse est aspectum stellarum fixarum variari. Motus ergo trepidationis ab astrologis inventus in octava sphaera, ex motu terrae rationabilibus contingit, quam ex motu illius. Si igitur parvus hic atque obliquus et inaequalis terrae motus sufficit ad mutationem aspectus, quae notata est, non sunt ponendi alii orbes supra octavam sphaeram."

44 Galilei, *Dialogue*, 421. Cf. Galilei, *Le opere*, vol. 7, 447: "Ma se co'l far movere il vaso, senza artifizio nessuno, anzi semplicissimamente, io vi posso rappresentar puntualmente tutte quelle mutazioni che si osservano nell'acque marine, perché volete ricusar questa cagione e ricorrere al miracolo?"

were standing still [...]."[45] Galileo and Cesalpino were both driven by a radical rejection of the lunar option, which introduced occult qualities and remote influences into physics. Galileo's anti-astrological rancor can be seen in the following passage concerning the monthly and annual periodicity of the tides, to be considered along with the daily periodicity:

> Now two other periods occur, the monthly and the annual. These do not introduce new and different events beyond those already considered under the diurnal period, but they act upon the latter by making them greater or less at different parts of the lunar month and at different seasons of the solar year—almost as though the moon and the sun were taking part in the production of such effects. But that concept is completely repugnant to my mind; for seeing how this movement of the oceans is a local and sensible one, made in an immense bulk of water, I cannot bring myself to give credence to such causes as lights, warm temperatures, predominances of occult qualities, and similar idle imaginings. These are so far from being actual or possible causes of the tides that the very contrary is true. The tides are the cause of them; that is, make them occur to mentalities better equipped for loquacity and ostentation than for reflections upon and investigations into the most hidden works of nature. Rather than be reduced to offering those wise, clever, and modest words, "I do not know," they hasten to wag their tongues and even their pens in the wildest absurdities.[46]

45 Ibid., 461. Cf. Galilei, *Le opere*, vol. 7, 486: "Molto concludentemente si dichiara, che stando fermi i vasi contenenti le acque marine, impossibil sarebbe, secondo il comun corso di natura, che in esse seguissero quei movimenti che seguir veggiamo."

46 Ibid., 445. Cf. Galilei, *Le opere*, vol. 7, 470: "Seguono ora gli altri due periodi, mestruo e annuo, li quali non arrecano accidenti nuovi e diversi, oltre a i già considerati nel periodo diurno, ma operano ne i medesimi con rendergli maggiori e minori in diverse parti del mese lunare ed in diversi tempi dell'anno solare, quasi che e la Luna e il Sole entrino in parte dell'opera e nella produzion di tali effetti: cosa che totalmente repugna al mio intelletto, il quale, vedendo come questo de i mari è un movimento locale e sensato, fatto in una mole immensa d'acqua, non può arrecarsi a sottoscrivere a lumi, a caldi temperati, a predominii per qualità occulte ed a simili vane immaginazioni, le quali *tantum abest* che siano o possano esser cause del flusso, che per l'opposito il flusso è causa di quelle, cioè di farle venire ne i cervelli atti più alla loquacità ed ostentazione, che alla specolazione ed investigazione dell'opere più segrete di natura; li quali, prima che ridursi a profferir sulla savia ingenua e modesta parola *Non lo so*, scorrono a lasciarsi uscir di bocca, ed anco della penna, qual si voglia grande esorbitanza."

6 Pandolfo Sfondrati's Middle Way: the Sun's Heat as the Cause of the Tides

Alongside those who wrestled with vitalistic, peripatetic, or mechanical theories aimed at refuting the idea of a remote action of the moon, there were those who took the other path opened by Pico aimed at limiting the action of the heavenly bodies to motion, light, and heat. For example, a theory based on the action of the sun's heat was proposed by an eclectic atomist from Cremona, Pandolfo Sfondrati, in a work entitled *Causa aestus maris* (The Cause of Sea Tides). The first edition, now lost, must have appeared in Turin around 1582; a second was printed in Ferrara by the typographer Mammarello in 1590 with the *imprimatur* of the local Inquisition and the indication of approval of the preceding edition by the Inquisitor of Turin.[47] Hence the book appeared in the period between the first (1570) and second edition (1590) of Telesio's *opuscula*, during which time Cesalpino's *Peripateticae quaestiones* (1571), Bruno's *La cena de le Ceneri* (1584) and the third edition of Telesio's *magnum opus* (1586) were also published.

Sfondrati, linked to the Savoy court, belonged to a distinguished Cremonese family which, in the person of the Milanese senator Paolo Sfondrati, represented the Habsburg interests, that is those of Milan and of Philip II in Turin. Paolo's brother was Nicolò Sfondrati who, taking the name Gregory XIV, occupied the papal throne between 1590 and 1591. Pandolfo dedicated the second edition of *Causa aestus maris* to him.

In this book Sfondrati begins with eclectic positions in philosophy. He argues for the concordance between Plato and Epicurus in natural philosophy: "I found so much solidity in the teachings of the academics and the Epicureans as far as the natural causes are concerned that I would not move away from their schools [*gremium*]."[48] Using this background Sfondrati proposes an original heliothermal explanation of the tides. The centrality of the action of the sun's heat is emphasized from the first lines of the book, with recourse to a pseudo-epistemological comment on the term *aestum*: "They derive the word 'aestum', tides, from 'aer', air, and assume that it properly means warmth. From

47 Sfondrati, *Causa aestus maris*, 44v: "Frater Vincentius Vaschinus de Calvisano Vicarius generalis Sanctiss. Inquisitionis Status Sereniss. Ducis Ferrariae, visa subscriptione Reverendi Patris Inquisitoris Taurini 1582, a quo probatum fuit opus in exemplari veteri, imprimatur."

48 Ibid., 31v: "Me tantam naturalium causarum soliditatem in dogmatibus Academicorum et Epicureorum reperisse, ut ab eorum gremio discedere nequeam, nec nisi ad satietatem eorum aquis etiam, atque etiam ablutum, curare ad alios me conferre [...]."

it derives the word 'aestas', summer."[49] Like Telesio, Sfondrati introduces the 'heliothermal' theory of the tides with the metaphor of a boiling pot in which the vapors caused by heat tend to move upward, producing a rising effect.[50]

Sfondrati hypothesizes that the tides are generated by the action of the sun on the water particles and thus the phenomenon should be considered in terms of changes in the mutual relations of the sun and the earth. This would result in a perpetual flow of the seas and oceans around the earth, a circular motion similar to that of the celestial revolutions: "The sea flows eternally, running through the entire terrestrial globe, with the same order of all celestial bodies."[51]

The contrary motion of the ebbing of the tides would be linked to the collision of the primary flow against barriers, particularly in the vicinity of straits. These would hinder the impetus of the waters and partly push them back, producing recoils strong enough to explain the eastward tidal ebb.[52]

7 Patrizi's Appraisal of the Debate about the Tides

The major philosophical work by Francesco Patrizi, *Nova de universis philosophia* (A New Philosophy of the Universe, 1591), was published less than a year after the second edition of Sfondrati's *Causa aestus maris* by the same Ferrarese typographer; it was dedicated to the same patron, Pope Gregory XIV, and contained a broad discussion of the tides. Six chapters of *Pancosmia* (XXIV–XXIX) and the fourth book of *Nova de universis philosophia* (which followed *Panaugia*, *Panarchia*, and *Pansychia*) were devoted to the

49 Ibid., 3r: "Aestum ab aere deductum volunt, et proprie calorem significare, unde etiam aestatem derivatam esse." Cf. ibid., 4r: "aestum non ab aere simpliciter, sed ab aere usto derivatum esse [...] quasi quaedam ebullitionem, unde aestatem, et per methaphoram aestum maris nominata esse videmus [...]."

50 Ibid., 3r.

51 Ibid., 8r: "Mare transiit universum terrarum globum perenniter currendo in orbem circulariter, eodem ordine quo sydera omnia." Cf. ibid., 28v: "[...] et ideo mare suo cursu, cursum stellarum ab aeterno imitatur."

52 Ibid., cap. 3, *Causa vera aestus marini*, 8r–v: "Cum via illa a Natura fuerit constituta, per quam Mare transiens universum terrarum globum perenniter currendo in orbem circuiret, eodem ordine quo sydera omnia, licet partim velocius, partim tardius indesinenter volvuntur, per illasque [Magellanicas] fauces concitato cursu ad rapidi flumini instar ferantur, neque omnes uno impetu tantorum Marium undae, per angustas huiusmodi fauces partransire possint, coguntur ex fuga contrarii contra oppositas ex adverso aquas sequaces regurgitare, et cum unda palpitatione quam vocant, undam proximam impellat, de necessitate ad oppositas partes quantumvis remotas, aestus concitatur, maior aut minor iuxta oppositionis distantiam, et aquarum multitudinem."

subject of waters and seas.[53] The specific topic of the tides was dealt with in Chapters XXVIII and XXIX.

Chapter XXVIII of *Pancosmia*, entitled "De maris affluxus et refluxus varietate" (Various [Opinions] on the Ebb and Flow of the Sea), is an overview of the positions expressed in the intense sixteenth-century debate. Patrizi reviews the extensive literature of his time starting with *De fluxu et refluxu maris* (1588) by the physician and natural philosopher Federicus Chrysogonus of Zadar.[54] Chrysogonus wrote about the variable periodicity of the tides, which he attributed to the combination of solar and lunar cycles and calculated starting from the conjunction of the two heavenly bodies. Patrizi considers Chrysogonus the first of a host of Aristotelians who followed the Greek philosopher more or less slavishly. They include the Paduan professor of mathematics Federico Delfino, the celebrated Julius Caesar Scaliger, the natural philosopher Girolamo Borri, and the physician and astrologer Annibale Raimondo.[55]

Patrizi dedicates a separate discussion to Niccolò Sagri of Ragusa (Dubrovnik) and underscores his diligence and originality. This little-known Dalmatian had written a curious dialogue, *Ragionamenti sopra le varietà de' flussi del mare oceano occidentale* (Reasoning on the Variety of the Tides of the Western Ocean) (Venice, 1574), in which he tried to reconcile the doctrine of lunar traction, based on the analogy between the moon acting on the seas and the magnet attracting iron, and the mechanical doctrine based on the analogy between the tides and the oscillations of a balance:

> However, if one concedes that the opposite part does not have enough force as [to counterbalance] the moon, I affirm that the small amount of force that you concede will be sufficient to move the waters, if not much then at least a little bit [...]. This [imparted motion] added to the past motion of the moon acts like a magnet on a compass. As one observes, when the [compass] is moved, [it] also [keeps moving] after [the magnet] has been removed from its sight [and] would never stop moving, if the stone was shown to it from time to time in the appropriate

53 xxiv De aqua et mari; xxv De aquae rotunditate; xxvi An aqua et terra unum efficiunt globum; xxvii De maris universi motibus; xxviii De maris afluxus, et refluxus varietate; xxix De causis affluxus et refluxus maris.

54 Chrysogonus, *De modo collegiandi, prognosticandi et curandi febres*.

55 Cf. Delphinus, *De fluxu et refluxu aquae maris*; Borri, *Del flusso e reflusso del mare* and Raimondo, *Trattato ... del flusso e riflusso del mare*. For an essential review of the fourteenth-century debate on the tides, see Ventrice, *La discussione sulle maree tra astronomia, meccanica e filosofia nella cultura veneto-padovana del Cinquecento* (1989).

> manner [...]. The same occurs if one touches a balance with equal weights: it needs some time to stop, alternately rising on the one side and on the other.[56]

Patrizi also considers Sfondrati's text in his review of his predecessors' works but he dismisses it as unfounded. The interpretation of the tides as a phenomenon resulting from the contrasted impetus of the waters, which would then be redirected eastward, seems to him implausible, indeed ridiculous.[57] Patrizi also rejects the atomistic-mechanical approach to the heliothermal theory of the tides. Indeed Sfondrati had proposed that the sun's heat has an impact on the water particles. Patrizi's rejection of Sfondrati's theory of the action of the sun's rays on the waters does not imply a rejection of Telesio's heliothermal theory. It only criticizes Sfondrati's mechanistic approach. As I will explain shortly, Patrizi reformulates Telesio's theory in vitalistic terms. In fact his opinion on Telesio is completely different from that of the other authors who had discussed the tides: Telesio is presented as the one who came closest to the solution of the problem.

Patrizi admired Telesio although he was also critical of various aspects of his philosophy.[58] Antonio Persio dedicated the opusculum *De mari* in the Venetian edition of the *Opuscula* (1590) to Telesio. The dedication began by recalling the common philosophical discussions:

---

56 Sagri, *Ragionamenti sopra le varietà de i flussi et riflussi del mare oceano occidentale*, 90: "Tuttavia qualora si volesse concedere che la parte opposita non habbia tanta forza, quanto la Luna, almeno dico, con quella poca forza che mi concedete che lei habbia, bastaria far movere l'acque, se non tanto almeno poco manco [...] e questo sarebbe con l'aiuto del passato moto della Luna a guisa come fa la calamita nella bussola, che quando viene ad essere mossa, avanti che si fermi, come si vede, ancor che sia levata la pietra dalla sua vista, e se da tempo in tempo convenevole li fosse rimostrata, non si fermerebbe mai [...] siccome viene a uno trabucco, o bilanza, che sia toccata, e datali causa che tra pesi equalmente, prima che si fermi tarda assai, hora alzandosi d'una parte hora dall'altra [...]."

57 Patrizi, *Nova de universis philosophia*, 139v(b): "Paucos ante menses editus est liber, titulo *Cause aestus maris*, magno sane apparatu, sed cause redditu ut videtur et exitu ridiculo. Ait, omnes aquas ad quaslibet fauces naturali cursu properare. Oceanum, a Laboratoris terra, ad Magellanicas fauces decurrere. Tum etiam ab Oriente easdem ad fauces accurrere. Per quas cum transire nequeat omnis, inde retro regurgitat, et aestum, in Africa, atque Hispaniae littoribus excitat: aqua, aquam proximam impellente; et ea palpitatione in opposita parte intumescente. Sed quot nam horis, aut diebus, aut hebdomadibus, aut etiam mensibus ea palpitatio retrocedat? Cur item in proxima, faucibus illis Brasiliae ora tam parvum facit, in longiquissimis, Lusitano et aversus etiam Gallico, Britannico, ac Belgico? [...]."

58 On the "friendly polemic" between Patrizi and Telesio and the involvement of Persio, see Puliafito, "Introduzione," xxxiii—xlv.

> Very erudite Patrizi, you remember that, when we sojourned together in Venice, I often recommended to you Telesio's new philosophy and his approach to philosophy; I urged you to carefully read his natural books [...]. I was then glad to explain to you any passage that might be obscure to you and I solved your doubts and criticism whenever I could.[59]

Thus, Persio's dedication attested to an intellectual affinity and mutual respect between Patrizi and Telesio. It continues,

> When I prepared the new edition of his booklet on the sea—which he had first published and was now augmented with the addition of some writings of his pertaining to the same subject—I judged that no better father and patron than you, Patrizi, could be found. Therefore, I decided to entrust it to you.[60]

Therefore, it is not surprising that Patrizi discusses Telesio's tidal theory in *Pancosmia*. He looks favorably on *De mari*, appreciating more the theoretical and natural profundity underlying the explanation of the tides than the special solution which he partly rejects.

> Telesio, that excellent man who dared to mint a new philosophy with the force of his ingenuity (and for this reason we admire him deeply) is the one who dealt with this issue in the most fitting manner. He affirms that [1] the sea is naturally warm and inclined to move whereby it is preserved and pleased; and [2] thereby it flees from the action of the sun in order to avoid excessive evaporation. The first affirmation is perfectly true. However, I reject the second.[61]

---

59 Telesio, *Varii de naturalibus rebus libelli* [1590] (2012), 2r: "Meministi eruditissime Patriti, cum Venetiis commoraremur, me tibi novam Telesii Philosophian, ac Philosophandi rationem saepius commendare, et te hortari, ut libros eius de natura legere diligenter. [...] Ego igitur libenter, et obscura quaecunque tibi essent interpretabar, et obijcientium sese dubitationum scrupulos eximebam, quando poteram."

60 Ibid., 2v: "Cum igitur libellum eius de mari ab ipso primum editum, atque aliquibus ex eiusdem scriptis ad eandem rem pertinentibus auctum, denuo imprimendum curarem, patrem ipsi, ac patronum nullo Patricio aptiorem invenire me posse existimavi, tuaeque idcirco ipsum fidei comendare decrevi."

61 Patrizi, *Nova de universis philosophia, Pancosmia*, 140r(b)-v(a): "Telesius vir ingens, qui proprii viribus ingenii novam cudere est ausus philosophiam, quem ea de re, nos maxime admiramur, quaestionem etiam hanc, omnium optime videtur perfecturus. Mare inquit [1.] sui natura calidum, pronum est in motum, quo et servetur, et oblectetur. Et [2.] quo solis actionem fugiat, ne ab eo usto plus solvatur in vapores. Pars prior verissima est [1.]. Secunda haec non placet [2.]."

Patrizi accepts the basic thesis according to which the sea is warm by nature and as such is naturally led to undergo those motions that ensure its conservation. He does not accept, however, the explanation of the tides attributed solely to the action of the sun. Telesio's heliothermal theory is not able to explain why the sun does not act on all waters in the same way nor the difference in the behavior of salt and fresh waters in response to its radiation. It also cannot account for the fact that similar tides are found at different latitudes, which seems to be at odds with the variations in intensity of the solar rays.[62] Finally Patrizi criticizes Telesio's hypothesis because it assumes seasonal variations which do not exist.[63]

In conclusion, Patrizi rejects the theory that the tides are solely an effect of solar heat. He emphasizes above all the difficulty of matching the implications of the theory with the empirical evidence. He would not be the only one to make this criticism. For example, we can recall the much more corrosive polemic against the heliothermal theory of the tides advanced by Galileo in the *Dialogue*:

> As for those who make the temperate heat of the moon able to swell the water, you may tell them to put a fire under a kettle of water, hold their right hands in it until the heat raises the water a single inch, and then take them out to write about the swelling of the seas.[64]

---

62 Ibid., 140v(a): "Cur enim omnia maria, ea fuga non cientur? Cur aquae dulces nullae? Cum et tenuiores sint, et solutu faciliores? Sed et causa haec communis motibus maris omnibus est. Fluxus vero et refluxus propriam dicit esse, quia sol in mari ingeneret vapores, qui egressum molientes, a mari superposito prohibiti, ipsum attollunt, et agitant. Idque vere et Autumno maxime, quia medius sol, plurimos crassioresque educit vapores. Sed causam reddat, cur in Aremoricis, et Belgicis, quae a medio sole longe distant, par aestus fit, ac in Taprobana, quae aequinoctiali et medio soli est subiecta? [...] Aestate inquit, minor sit, quia sol tenuissimos vapores gignit qui facile elabuntur, et ipsum non attollunt. At et aestate, aestus hic aeque attollitur, atque alias. Hieme item, inquit minus, quia sol languidissimus per paucos ingeneret, qui sint mare attollere impotentes. At et hoc salsum est, hieme aequalem aliis temporibus, aestum non fieri."

63 Ibid.: "In Pleniluniis, inquit, maior, quia multa a luna resiliens lux, multos educit vapores. At quae nam lunae lux resilit, in nostra maria, cum luna est apud antipodas? In noviluniis, ait, quia refrigerato aere, internus maris calor, se se colligens, valentior factus, plures facit vapores et emittit. Sed si a superposito mari prohibiti egressu ipsorum attollunt, quo modo eos emittit? Et si emittit, quo modo egressu prohibentur, et attollunt? In lunae quadratis, addit, non multa a luna resiliente luce, nec proprio maris calore in se collecto, minime attollitur. At cur non saltem dimidio attollitur, ut et lux ei est dimidiata a plenilunio? Et calore dimidiate in se collecto? Hae fluxus ei viro causae funi. Refluxus vero hae aliae."

64 Galilei, *Dialogues*, 420. Cf. Galilei, *Le opere*, vol. 7, 446: "A quelli del calor temperato, potente a far rigonfiar l'acqua, dite che pongano il fuoco sotto di una caldaia piena d'acqua, e che vi tengan dentro la man destra sin che l'acqua per il caldo si sollevi un sol dito, e poi la cavino, e scrivano del rigonfiamento del mare."

8 Patrizi's Vitalistic Theory of the Tides

After discussing the hypotheses of his predecessors and his immediate interlocutors and having discussed Telesio's doctrine, Patrizi advances his own explanation in Chapter XXIX of *Pancosmia*, "De causis affluxus et refluxus maris" (On the Causes of the Ebb and Flow of the Sea).

Firstly, he pronounces against lunar causality alone.[65] The moon, he maintains, is not alone in presiding over the tides *in universale*. The sun is the life-giving principle that communicates warmth and life and renders earthly things able to move. The moon instead has a deep affinity with the earth, which explains the parallelism of its celestial motions and numerous terrestrial cycles. Nevertheless, Patrizi believes that the celestial bodies are universal causes (*causae universales*) and thus unsuitable to account for precise phenomena such as marine motions, for which it is necessary to identify the particular causes (*causae propriae*).

What then is the cause of the motion of the seas? Patrizi identifies it as an internal and vital impulse. In the same way in which the stars move about the ethereal heavens thanks to an autonomous impetus, like the birds in air and the fishes in water, the waters of our globe are moved by an intrinsic principle of life and movement.

> Why should we not allot this to the inner nature of the sea? In fact, just as we have taught that the stars are carried through the ether by their intellect, soul, and spirit, and that the planets, the sun and the moon, as well as the air below them, are carried by the same causes, in the same manner, why should it be a miracle that the sea is carried by its own nature in various directions not differently than the planets? Among those motions are the ebb and flow, [produced] by its own intellect, soul, and spirit.[66]

---

65 Patrizi, *Nova de universis philosophia*, 141r(b): "Si Luna, uti aiunt, dux aquarum esset omnes aquas aeque duceret, non aliter ac igni, omnia comburitur ustilia. Sol omnes discutit tenebras, Luna ipsa omnia maria, omnes lacus, stagna omnia, amnes omnes, quando lucet, suo collustrat lumine, at non omnes ducit aquas. Non est erto aquarum omnium dux, non tractrix omnium, non avectrix."

66 Ibid., 142*r*(b): "Sed quid vetat maris propriae naturae hoc tribuere? Nam sicuti stellas propria natura, ab intellectu, ab animo, a spiritu, in aethere ferti docuimus, planetas quoque eisdem causis ferti, solemque lunamque, et sub eis aerem, quid miraculi est, mare, quoque pluribus natura sua, non aliter, ac planetae motibus cieri? Inter quos et affluxus sit, et refluxus? Ab intellectu nimirum, ab animo, a spirito suo."

According to Patrizi, the moon and sun "impress" a motion on the waters but this relationship is not causal. Rather it is an approximation: the sea or the ocean mimics the celestial motions "but in its own way" (*sed suo modo*). The vital motion of the waters is precisely the *mimesis* and *variatio* of those of the sun and the moon.[67] The impulse to life implies the search for self-preservation. The tide is a kind of breathing of the living sea that is nourished by an alternating generative exchange with the shores:

> If we only consider the ebb, which we have attentively observed on so many shores, the issue does not seem to be deprived of reason. In fact, in calm sea, [the water] shows a motion to and from the plane shore, at regular intervals, producing a continuous motion, which we have called a sort of respiration. While part [of the waters] moves back and flows down into lower places, another part arises, merges, passes over it and covers the shore. The first [wave], as if it grew shy and sought its own safety, hides itself in the belly [of the sea], impregnates itself and grows. Once it has grown it flows quicker to the shore. Thus, at regular intervals, a flowing forth and back is produced.[68]

The origin of this phenomenon is a process of rarefaction and boiling similar to the process of thermal expansion advanced by Telesio. The heating of the waters results from the combination of the heat proper to them and the life-giving action of the sun, moon, and stars.[69] The *causa propriissima* (the most

67 Ibid., 142v(b): "A Luna ergo, et a sole in mare astrorum motus veluti imprimuntur, tum eorum quae perpetuo, uniformique circumeunt mundum motu, tum eorum, quae variis multiplicibusque feruntur; qualibus, et maria feruntur, et Oceanus. Sed suo modo. Nam illos quidem non assequitur, sed aemulatur. […] Inesse autem plures salsedini spiritus, multa docent experimenta."

68 Ibid., 143v(b): "Nam si modus refluxus consideretur, quem nos in multis littoribus studiose spectavimus, non videbitur quaestio carere ratione. Namque tranquillo mari, moto eo, quem perpetuam quasi eius respirationem appellavimus, continue, et fluere, et refluere ad plana littora, tempore eodem conspicitur. Parti enim eius recurrenti, et ad humiliora refluenti, pars alia altior supervenit, et priorem illam obruit, eique superequitat, et super eam in littus currit. Illa, quasi timida, salutem sibi quaerens, in ventrem se obruentis, conditur; et se ipsa gravidam eam reddit, et altiorem facit. Haec altior facta, citatior ad littora affluit. Eodem igitur tempore, affluxus fit et refluxus."

69 Ibid., 144r(a): "Dum vero bulliunt, et attolluntur, necessario rarescunt. Per ergo rarefactionem, quae attolluntur aquae, altiores se ipsis fiunt. Rarefactio autem, non nisi vacui atomis, quae omni (uti ostensum antea est) insunt aquae dilatatis, et maioribus redditis. A calore nimirum in vapores partes aquae soluta. Eodem hoc modo (nullum enim alium experientia ostendit ulla) necesse est mare intumescere, ob poros eius rarefactos,

direct cause), however, is an intimate impulse generated by the spirit inherent in the waters and concentrated in the salt. "Many experiences show that there are several spirits in that which is salty."[70] Hence the salt is the direct cause of the sea's motions and the phenomenon of the tides in particular:

> Saltiness—a nature that [the sea] does not share with any other [element]—is the most direct cause [*propriissima causa*] accounting for the variety of motions of the sea. In fact, no sweet water or water with another taste is moved in so many ways. Aside from the salty [water] none has an ebb and flow [...]. Saltiness is therefore the closest, internal, and most direct cause of the marine motions.[71]

On this basis, Patrizi is able to indicate a cause intrinsic to the waters which accounts for the phenomenon of waves and tides. At the same time he does not reject the importance of celestial causes acting *in universale* rather than *in particulare*. Indeed he proposes a vitalistic and thermal theory able to hold together and go beyond the astrological theories and the heliothermal ones. Concerning the link between marine phenomena and astronomy, Patrizi's perspective allows for the abandonment of astrological causality without losing sight of the cosmological framework:

> Through these motions the sea, which is like a terrestrial ether, imitates the motion of the ethereal stars.[72]

7 Concluding Note

Telesio's tidal theory in *De mari* X–XIII is part of an extremely varied Renaissance discussion of the phenomenon. On the one hand, geographical explorations, colonial enterprises, and the needs of navigation expanded and diversified knowledge of the seas and oceans, ocean currents, straits, and tides.

---

et in vapores aquae parte acta, ab insito maris calore, a Sole interdiu, a Luna etiam noctu, sideribusque concalefacto."

70 Ibid., 142v(b): "Inesse autem plures salsedini spiritus, multa docent experimenta."

71 Ibid., 142v(a): "Salsedo igitur propriissima est causa, nulli alii communis natura, cur mare tot motibus agitetur, nulla enim dulcis, nulla alterius saporis aqua, tot agitatur; nulla affluxum, et refluxum [...] patitur, praeter unam salsam. Salsitudo ergo motum marinorum proxima, et interna, et propriissima est causa."

72 Ibid.: "Per hos motus, mare, quasi terrenus quidam aether, aethereos stellarum imitatur motus."

On the other hand, attempts at rational-empirical emancipation from astrology, which was increasingly seen as an occult, superstitious, and dubious doctrine from both the theoretical and ethical point of view, coincided with the search for new explanations of the tides, whose theory appeared to be refuted by some of the commonest astrological explanations. Particularly influential was the discussion by Albumasar, whose *Introductorium in astronomiam* contained pages and pages on the influence of the moon on the waters. The astrological treatise, a standard reference on the phenomenon in question, also dwelt on the exact relationship between the tidal cycle and lunar phases and on the variations of tides in relation to the changing positions of the sun and moon. After Newton, it would be understood that the basis of these correspondences between celestial motions and the tidal cycle was to be found in the law of universal gravitation, but for Albumasar's successors it was an astral influence. Indeed, the celestial origin of the sea's motions was an indisputable empirical proof of the action of the stars on earthly events. In fact the origin of the explanation of the tides through the remote action of the sun and moon was astrological. By contrast, the philosophical, religious, and ethical criticism of astrology also included the rejection of the lunar or solar-lunar tidal theory and was faced with the challenge of indicating an alternative explanation of the tides. A clear testament to the link between criticism of the lunar theory and the rejection of astrology is Pico's *Disputationes*, which had a broad influence on the scientific rationalism of the Renaissance.

Although the theory of the remote attraction of the moon and sun continued to be followed throughout the sixteenth and seventeenth centuries, especially in university circles linked to Aristotelianism, there was a growing number of those subjecting this concept to severe criticism for the reasons just mentioned. Many scholars rejected the possibility of a remote action by means of a *cognata virtus* (as the interpreters of Albumasar called it), viewed as a *qualitas occulta* (in the terminology of the detractors). Very diverse natural philosophers and mathematicians, such as Giordano Bruno, Andrea Cesalpino, and Galileo Galilei, completely denied the possibility of a lunar influence. For them, in the best of cases one could speak of a parallelism whose reasons were to be sought in a common root of the phenomena and not in a direct causality of one with respect to the other. The complete renunciation of celestial causes led either to vitalistic explanations by which the seas are moved by an intimate vital impulse or to a mechanical explanation.

Other Renaissance authors tried to take a middle way. They accepted Pico's invitation to limit the action of the heavenly bodies to motion, light, and heat. For them, heat, not so much the lunar heat as the solar one, would be the cause (or a contributing cause) of waves and tides. This was the path taken not only by Bernardino Telesio but also by lesser-known thinkers such as the eclectic atomist Pandolfo Sfondrati.

In summary, four explanatory models of the tides merit our consideration as the ones dominating the Renaissance debate on this phenomenon:

1. The astrological approach in which the lunar, or lunar-solar, tidal theory makes use of the analogy of the magnet. The moon is a kind of movable pole that attracts water to itself. This idea remained firm in the Scholastic Aristotelian tradition, as shown by its longevity among scholars at the University of Padua.
2. Vitalistic explanations such as those by Bruno and Patrizi. If we wish to indicate a metaphor for this option, it would be that of the living organism whose movements are conceived as teleological impulses aimed at self-preservation. In this perspective, Patrizi considers the tides as a motion of breathing by the living sea whose movement mimics "in its own way" that of the heavenly bodies, which are also free within the heavens. It is curious to note that although Patrizi rejects Telesio's heliothermal tidal theory he retains a thermal explanation of the tides and ultimately bases his explanation on Telesian premises, in particular the assumption that the sea is intrinsically warm, thus apt to move, and that its motions respond to a principle of universal animation.
3. The mechanical explanations of those who consider the motion of the tides analogous to the oscillation of a balance scale or to the behavior of a vessel in movement which communicates its motion to the liquid it contains. This is the theory Galileo developed from premises set by his Scholastic teacher Cesalpino.
4. The heliothermal explanation seeks to rationalize the celestial influence by reducing it to a heating action. The metaphor is that of boiling liquid in a pot. Telesio worked along this line of thought.

Within the pluralism of positions expressed in the Renaissance debate on the tides and in the variety of explanatory models (sometimes intertwined and not always clearly distinguishable from one another), the explanation with an astrological foundation is closest to the modern one. However, it has several limitations with respect to Newton's conclusions. For example, the remote action of the moon is not conceived in terms of gravitational attraction. It is irreducibly qualitative. Indeed the *cognatio* of moon and waters is qualitative; hence the *virtus tractrix* (tractive force) of the heavenly body is ascribed to an obscure essential affinity. Similarly, the combined influence of the moon and the sun is expressed in astrological terms such as conjunction, quadrature, and opposition. As mentioned above, the Renaissance detractors of this approach were not only worried about the elusive nature of the recourse to occult powers and remote actions but also about the ethical implications inherent in the assumption that the heavenly bodies do not only act on the elements but also on human events and choices.

Johannes Kepler, who did not disdain astrology at all, was harshly rebuked by Galileo for his acceptance of the lunar theory of the tides:

> But among all the great men who have philosophized about this remarkable effect, I am more astonished at Kepler than at any other. Despite his open and acute mind, and though he has the motions attributed to the earth at his fingertips, he has nevertheless lent his ear and his assent to the moon's dominion over the waters, to occult properties, and to such puerilities.[73]

In truth, Kepler opened the way to modern celestial physics by laying the foundation for Newton's explanation. With all due respect to Galileo, it is no coincidence that a supporter of the Copernican system with astrological interests had to be the intermediary between the lunar theory of the tides and the modern gravitational explanation. In fact the heliocentric theory demolished the idea of the uniqueness of the center of gravity in the universe of Aristotelian physics and forced scholars to admit that there are many centers of gravity in the universe. This is an indispensable assumption for a theory of universal gravitation.

The Renaissance debates on the tides remind us that the historical developments of empirical and rational science travelled along roads that were anything but straight. The plurality of opinions and theories about the tides presents an inextricable tangle of philosophical, ethical, and methodological views in which the link between scientific theories, rationality, and experience is extremely complex. Telesio's discussion in *De mari* fits precisely in the core of a debate in which both the precursors of modern scientific rationality and the discredited heirs of doctrines destined to marginalization and decline, such as astrology, contributed to the discovery of the fundamental laws of nature.

---

73 Galilei, *Dialogues*, 462. Cf. Galilei, *Le opere*, vol. 7, 486: "Ma tra tutti gli uomini grandi che sopra tal mirabile effetto di natura hanno filosofato, più mi meraviglio del Keplero che di altri, il quale, d'ingegno libero ed acuto, e che aveva in mano i moti attribuiti alla Terra, abbia poi dato orecchio ed assenso a predominii della Luna sopra l'acqua, ed a proprietà occulte e simili fanciullezze."

CHAPTER 7

In Search of the True Nature of the Rainbow: Renewal of the Aristotelian Tradition in the Renaissance and the *De Iride*[1]

Elio Nenci

From the point of view of the history of science the discussion about how the rainbow is formed is one of the most interesting sections of Aristotle's philosophy of nature (*Meteorologica*, Book 3, Chapter 4). This chapter clearly shows that for Aristotle the explanation of a natural phenomenon cannot be reduced to its mathematical formulation but must consider the whole of the changes that take place during its production. In this case, Aristotle acknowledged the need to resort to mathematics in order to single out the cause of the rainbow. Therefore, he did not hesitate to make use of the results obtained by the contemporaneous science of optics, but he also had to go beyond them since one of the essential aspects of the phenomenon, color, seemed to have been almost ignored in the exact mathematical studies made by the scientists of that time.

With regard to this point, it is relevant to refer to questions extensively dealt with in other Aristotelian works. In the first place, we must refer to the discussion about the so-called "sciences subordinated to mathematics" which takes place in the *Analytica Posteriora* (I.9, 76a9–25 and I.13, 78b36–79a10). These "subordinated sciences", which included optics and harmonics, were devoted to some natural phenomena by assuming principles taken from geometry and arithmetic. These principles explained the cause of a phenomenon by specifying the *reason why* (διότι) it took place, whereas the fact *that* (ὅτι) it took place, i.e. *what* it was, was the object of the subordinated natural science. So the science of harmony studied sounds by expressing them as simple numerical ratios and optics did the same thing for vision by using lines, angles, and triangles.

From this point of view, the case of the rainbow was even more interesting because it was an example of "double subordination". On the one hand, its causes had to be explained by means of optics (*An. Post.*, I.13, 79a10–16);[2] on the other hand, this "science subordinated to geometry" was then based

1 For a classic account on this topic see Boyer, *Rainbow*.

2 Brunschwig, "Aristote et le statut épistémologique de la théorie de l'arc-en-ciel."

 | DOI:10.1163/9789004352643_009

on the theory that the visual rays were expelled from the eye, and this theory was openly in contrast with what Aristotle maintained in the *De anima* (II.7, 418a29–418b14) and in the *De sensu et sensibilibus* (2, 438b2–8): namely, that vision takes place through a change of a *diaphane* (διαφανὲς), i.e. of a transparent substance (such as air, water, etc.), of which light is the activity. By *diaphane* Aristotle meant that which is visible by means of an alien color. Color was considered the "proper sensible" of vision and it was also one of the properties or characteristic qualities of the rainbow. Therefore, the theory of colors developed in Chapter 3 of the *De sensu et sensibilibus* had to play an important role in the explanation of the natural phenomenon of the rainbow.

It is easy to understand why the discussion on the rainbow raised a series of philosophical problems within the Aristotelian tradition. Through the Middle Ages the study of this section of the *Meteorologica* was a source of difficulties for commentators who had to deal with an optical science which was much more advanced than the knowledge of optical phenomena available at the time of Aristotle. Thanks to the progress made by *Perspectiva* it was possible for medieval philosophers to study the optical 'causes' of the rainbow on different foundations, in particular because of the greater importance given to the phenomenon of refraction. We just have to mention Theodoric of Freiberg's (ca 1250–ca 1310) *De iride et de radialibus impressionibus* to give an idea of the impressive progress made in this field.[3]

On the other hand, the important results obtained by Theodoric remained unknown to later generations of philosophers, who largely continued to follow the way in which Aristotle dealt with the problem of the rainbow. This long tradition of comments on the *Meteorologica*, which from the 13th century was part of the *curriculum studiorum* of the main universities in Europe, was disturbed by the reappearance of the *Commentaries* on the *Meteorologica* written by Alexander of Aphrodisias (2nd–3rd century CE)[4] and by Olympiodorus (6th century CE).[5] During the first half of the 16th century, scholars who dealt with still unsolved problems on new foundations, often in contrast with the medieval tradition, mainly referred to these works. Nevertheless, the results of these new studies did not cause the science to progress any further.

3 Theodoric of Freiberg, *De iride*.

4 Alexander of Aphrodisias, *In Aristotelis meteorologicorum libros commentaria*, in *Commentaria in Aristotelem Graeca* [*CAG*], vol. 3,2. Id., *In quatuor libros meteorologicorum Aristotelis commentatio*.

5 Olympiodorus, *In Aristotelis meteora commentaria*, in CAG, vol. 12,2. Id., *In meteora Aristotelis commentarii* (1551).

In this essay I shall select works by Alessandro Piccolomini (1508–1578)[6] and Francesco Vimercati (1512–1571)[7] to illustrate the distinction between mathematics and natural philosophy, which was peculiar to the Aristotelian way of dealing with the rainbow. The discussion of these works will provide a background against which I shall analyze the work of Bernardino Telesio more precisely (1509–1588).[8]

1 Aristotle's Treatment of the Rainbow in the *Meteorologica*

Before going further, I think it will be useful to summarize Chapters 4 and 5 of Book 3 of the *Meteorologica*. In Chapter 2, Aristotle had already dealt with haloes, rainbows, mock suns or parhelia and rods.

> The complete circle of a halo was often visible round the sun and moon and round bright stars, and as frequently by night as by day [...]. The rainbow never formed a complete circle, nor a segmental circle larger than a semicircle. [...] After the autumn equinox it occurred at all hours of the day; but in summer it did not occur round about midday. No more than two rainbows occurred at the same time; of two such simultaneous rainbows each is three colored, the colors being the same in each and equal in number, but dimmer in the outer bow and placed in the reverse order. For in the inner bow it is the first and largest band that is red, in the outer it is the smallest and closest to the red band of the inner. [...] The cause of all the phenomena was the same, for they were all phenomena of reflection [ἀνάκλασις]. They differed in the manner of reflection and in the reflecting surface, and according as the reflection was to the sun or some other bright object (*Meteor.* III.2, 371b22–25, 371b26–27, 371b30–372a5, 372a17–21).[9]

6 Piccolomini, "Tractatus de Iride." On Piccolomini's scientific works see Suter, "Scientific Work of Alessandro Piccolomini." For his participation in scientific disputes see Giacobbe, "Il Commentarium de certitude mathematicarum disciplinarum," Ferraro, "Dimostrazioni matematiche e conoscenza scientifica," Biard, "La certitude des mathématiques."

7 Vimercati, *In quatuor libros Aristotelis Meteorologicorum commentarii*. On Vimercati see *Gilbert, "Francesco* Vimercato of Milan."

8 Telesio, *Liber de iride* (1590). See also id., *De rerum natura* (1586), chapters 10–15.

9 Aristotle, *Meteorologica* (1952), 241–245.

From the last part of this quotation it seems that Aristotle followed the theory of those writers on optics who explained vision by means of visual rays coming out of the eye. According to this theory, the rainbow was produced by a reflection of visual rays towards the sun. Our vision is reflected from substances which have a smooth surface, just like it is from water. In some mirrors, shapes are reflected, and in others only colors. "Colors are only reflected in mirrors that are small and incapable of subdivision by our sense of sight" (*Meteor.* III.2, 372b1–3). For the rainbow, the small mirrors were the little drops of water hanging in some clouds. But how to explain the genesis of colors?

In Chapter 3 of *De sensu et sensibilibus*, Aristotle reviews several hypotheses (1–2) and presents his own solution (3).

(1) Firstly, white and black may be juxtaposed in such a way that by the minuteness of the division of its parts each is invisible while their product is visible; and thus color may be produced. This product can appear neither white nor black, but, since it must have some color and can have neither of the above two, it must be a sort of compound and a fresh kind of tint. In this way, then, we may conceive that numbers of colors over and above black and white may be produced, and that their multiplicity is due to differences in the proportion of their composition; [...] and colors may, indeed, be analogous to harmonies. [...]

(2) This is one of the ways in which colors may be produced; a second is effected by the shining of one color through another. This we may illustrate by the practice sometimes adopted by painters when they give a wash of color over another more vivid tint [...].

According to the theory of juxtaposition, just as we must assume that there are invisible spatial quanta, so must we postulate an imperceptible time to account for the imperceptibility of the diverse stimuli transmitted to the sense organ, which seem to be one because they appear to be simultaneous. But on the other theory there is no such necessity; the surface color causes different motions in the medium when acted on and when not acted on by an underlying tint.

(3) But let us premise that substances are mixed not merely in the way some people think—by a juxtaposition of their ultimate minute parts, which, however, are imperceptible to sense—but that they entirely interpenetrate each other in every part throughout; [...]. On the other hand, things which cannot be resolved into least parts, cannot be mingled in this way; they must entirely interpenetrate each other; and these are the things which most naturally mix. [...] Now, all this being so, it is clear that when substances are mixed their colors too must be commingled, and

> that this is the supreme reason why there is a plurality of colors; neither superposition nor juxtaposition is the cause. In such mixtures the color does not appear single when you are at a distance and diverse when you come near; it is a single tint from all points of view (*De Sensu et Sens.* 3, 439b21–29, 439b32–33, 440a7–10, 440a22–28, 440a34–440b3, 440b10–13, 440b14–19).[10]

From the idea of *mixtio* (μίξις), one could have asked whether the black and white present in any color were related by numerical ratios, or whether one was predominant over the other; then ask to what extent such predominance could be determined more precisely through mathematics. From what Aristotle writes in Chapter 6 of *De sensu et sensibilibus*, it is clear that the infinite divisibility of the *mixtio* implies insurmountable limitations to visual perception, i.e. that which is extremely small could not be perceived unless it is placed within something sufficiently large; only in this case, from being potentially visible it will become actually visible (*De Sensu et Sens.* 6, 446a4–15).

Given these insurmountable limitations, we can formulate our question in a new way: how far can a natural philosopher go in his attempt at mathematicizing reality, when he needs to resort to mathematics in order to explain the causes of some phenomena? If, in the case of the rainbow, Aristotle did not go as far as that, this is probably a consequence of the nature of optical science in his time: it seems that those who dealt with optical problems only discussed colors in connection with other questions, such as the formation of reflected images in small mirrors or the ratio between the increased distance from the seen object and the augmented darkness. Increased distance naturally caused the exact perception of color to become lost.

For Aristotle the "proper sensible" of sight could never become the object of a purely mathematical investigation. This was true even in the case of the rainbow, although essential aspects of it were related to quantitative considerations.

The colors of the rainbow were formed by the reflection of the visual rays coming out of the eye in the little drops hanging in some clouds; these little drops were like small mirrors, and when the cloud, the sun, and the observer were arranged on the same line (with the observer in the middle) the little drops reflected the visual rays towards the sun, so as to present an altered image of the color of this bright body. This alteration was caused by the visual ray meeting the substance of the cloud—water—which is dark by nature. According to the formation of color, which I have previously described, the sunlight operated as the white and the cloud as the black.

10 Aristotle, *De sensu and De memoria*, 57, 59–61.

However, how to explain the formation of the three colors of the rainbow and their order? In addition, why, in the double rainbow, was the external one less bright and why were the colors arranged differently? All these questions must be dealt with in the theory of *mixtio*, which must follow the rules of optical science, according to which the theory of vision was treated in a geometrical manner. Aristotle answered the first of these questions in Book 3, Chapter 4 of *Meteorologica*:

> Bright light shining through a dark medium or reflected in a dark surface (it makes no difference which) looks red. Thus one can see how the flames of a fire made of green wood are red, because the fire-light which is bright and clear is mixed with a great deal of smoke; and the sun looks red when seen through mist or smoke. The reflection which is the rainbow therefore has its outermost circumference of this color, since the reflection is from minute water-drops. [...] We must, as has been said, bear in mind and assume the following principles. (1) White light reflected on a dark surface or passing through a dark colored medium produces red; (2) our vision becomes weaker and less effective with distance; (3) dark color is a kind of negation of vision, the appearance of darkness being due to the failure of our sight; hence objects seen at a distance appear darker because our sight fails to reach them. [...] At any rate, they give the reason why distant objects appear darker and smaller and less irregular, as do also objects seen in mirrors, and why too the clouds appear darker when one looks at their reflection in water than directly at them. This last example is a particularly clear one: for we view them with a vision diminished by the reflection. [...] The reason is clearly that, just as our vision when reflected through an angle and so weakened makes a dark color appear still darker, so also it makes white appear less white and approach nearer to black. When the sight is fairly strong the color changes to red, when it is less strong to green, and when it is weaker still to blue (*Meteor.* III.4, 374a3–10, 374b9–15, 374b17–22, 374b28–33).[11]

As we have seen, the increased darkness of the bright color of the sun, and the formation of the colors of the rainbow which follows from it, is mainly caused by the reflection taking place in the little drops, which act as small mirrors.

Which mathematical aspects of reflection are relevant here? The increased distance of the reflected visual ray compared to direct vision certainly is. However, one must be careful not to regard the phenomenon as solely caused

11 Aristotle, *Meteorologica*, 255–257, 259, 261.

by the different distance and not to strictly apply this explanation to the other examples reported by Aristotle.

Alessandro Piccolomini, in his "Tractatus de iride" published as an appendix to his Latin translation of Alexander of Aphrodisias's *Commentaries* on the *Meteorologica*, thinks that such negligence would be wrong, for if the reflected vision of a cloud in a mirror or on the surface of water is represented by a geometrical figure, one gets two sides of a triangle, whereas the third side corresponds to the direct vision. Now it is true that two sides of a triangle are always bigger than the other side (Euclid's *Elements*, Book 1, prop. 20), but in this case the side corresponding to the visual ray, which goes from the eye to the reflecting surface, is negligibly small compared to that which goes from this surface to the cloud. Hence, one can regard it as minimally affecting the formation of color. According to Vitelo's measurements reported in his *Perspectiva*, clouds can reach a height of between three and five German miles (ca. 5900–7400 meters), whereas the observer's distance is at most four feet (Roman feet ca 30 cm).[12]

In the Aristotelian framework, it was more difficult to establish the cause of the subsequent formation of three colors: red, green, and blue (violet). The slightly weakened view of the original color changed into a view that was increasingly weak. Was the increased distance a sufficient cause for this weakening of the view? And even if this were true, would it have been possible to exactly determine this variation? Aristotle tackled these problems thus:

> In the primary rainbow the outermost band is red. For the vision is reflected most strongly on to the sun from the largest circumference, and the outermost band is the largest: and corresponding remarks apply to the second and the third bands. [...] This, then, is why the rainbow is three-colored and why the rainbow is made up of these three colors only. The same cause accounts for the double rainbow and for the colors in the outer bow being dimmer and in the reverse order. For the effects here are the same as those produced by an increase in the distance of vision on our perception of distant objects. The reflection from the outer rainbow is weaker because it has farther to travel; its impulse is therefore feebler, which makes the colors seem dimmer. The colors are in the reverse order because the impulse reaching the sun is greater from the smaller and inner band; for the reflected that is closer to our sight is the one reflected from the band that is closest to the primary rainbow, that is, the smallest band in the outer rainbow, which will consequently be colored red.

12 Piccolomini, "Tractatus de iride," 121.

> And the second and third bands are to be explained analogously (*Meteor.* III.4, 375a2–4, 375a28–375b9).[13]

In the primary rainbow, Aristotle regards the extension of the bands of colors as the main cause of the weakened vision without considering the variation of distance. However, in the external rainbow he regards the increased distance as the main cause, and seems to put this in relation to the augmented width of the angle of incidence, which according to the optical theories would explain the weakened view through increased departure from the perpendicular.

It seems that these two different explanations could only be reconciled in the case that there was not always a direct relationship between increased distance and weakened view. One could imagine a visual power which kept the same strength up to a certain distance and then quickly weakened. Before that happened, the intensity of the vision would be caused by the small mirror, whereas later the increased distance, or more probably the increased width of the angle of incidence, would be the main cause.

2 Ancient Commentaries on Aristotle's Theory

Aristotle's passage on the rainbow raises a real problem. It seems that he was satisfied with the result that he had obtained, but later commentators did not seem to be equally satisfied. Alexander of Aphrodisias relates that some authors regarded the second rainbow not as a reflection of the visual rays towards the sun but as an image of the internal rainbow reflected in a cloud placed outside the first one.[14] It is likely that this argument was meant to explain the space without color between the two rainbows. Aristotle never pointed out this discontinuity between the two bands of red, but later, as reported by Alexander in his *Commentaries*, some other authors wondered why the empty

13 Aristotle, *Meteorologica*, 261, 265.

14 Alexander of Aphrodisias, *In quatuor libros meteorologicorum*, 81: "Aliqui quidem igitur dicunt, quod secunda iris, non per refractionem ad Solem fieri accidit, sed ad ipsam praeinexistentem iridem. Ita quod usus ipse ad exteriori nube, quae simili quidem modo disposita ad refractionem sit, sicut et prima, in qua prima iris est, ad praeinexistentem iridem refrangatur, et ex tali refractione secunda iris appareat; quapropter et languidiores sunt secundae iridis colores, tanquam ex secunda rursus refractione producti." CAG, vol. 3,2, 159, 9–15: τινὲς μὲν οὖν φασι τὴν δευτέραν ἶριν οὐ κατὰ τὴν πρὸς τὸν ἥλιον ἀνάκλασιν ἔτι γίνεσθαι, ἀλλὰ κατὰ τὴν προϋπάρχουσαν ἶριν, ὡς τῆς ὄψεως ἀπὸ τοῦ ἐξωτέρου νέφους ὁμοίως ἔχοντος, ὡς εἶχε καὶ τὸ πρῶτον, ἐφ' οἷς ἡ ἶρις, ἀνακλωμένης ἐπὶ τὴν προϋπάρχουσαν ἶριν, καὶ διὰ τῆς ἀνακλάσεως ἐκείνην ὁρώσης· διὸ καὶ ἀμαυρότερα τὰ χρώματα τὰ τῆς δευτέρας, ἅτε ἀπὸ ἀνακλάσεως γινόμενα δευτέρας.

space between the rainbows was not red, though it was nearer the larger band of the internal rainbow than the first band of the external rainbow.[15] Would not the reflection of the visual rays also show the same color in this part of the clouds? The little information given by Alexander may suggest that once more the difficulty should be dealt with on the basis of optical science, which taught that reflection should not occur from just any position: view, reflecting surface, and bright body ought to have specific positions and distances.[16]

Alexander of Aphrodisias's work was very influential in the Renaissance, as can be deduced from the frequent reprinting of Piccolomini's Latin translation. However, no solution could be found in it for the difficulties raised by Aristotle's text.

Olympiodorus's *Commentaries* on the *Meteorologica* are a different case, as they introduced a new element in the explanation of the formation of the rainbow's colors: they placed the clouds reached by the visual rays at different distances. The appearance of the three colors of the rainbow would depend both on the distance travelled by the visual rays and on the length of the distance covered inside the cloud. According to this point of view, when our vision meets the nearest clouds, it would have travelled a shorter distance and therefore would be stronger, whereas at the same time it would absorb a small quantity of the darkness of the water, thus causing the appearance of red.[17]

---

15 Ibid.: "Quaeret autem aliquis, nam si minor peripheria exterioris iridis phoeniceum colorem habet, quae prope ampliorem primae iridis peripheriam est, quae et ipsa per simile colorem retinet; propterea quia ab ambabus his peripheriis, fortior fit refractio visus ad Solem; quid nam, non et quod intermedium istarum peripheriarum est omne phoeniceum etiam habet colorem?" *CAG*, vol. 3,2, 160, 21–26: ἐπιζητήσαι τις ἄν, εἰ ἡ ἐλάττων περιφέρεια τῆς ἐξωτέρας ἴριδος φοινικοῦν ἔχει τὸ χρῶμα, οὖσα πλησίον τῆς μείζονος περιφερείας τῆς πρώτης ἴριδος, ἢ καὶ αὐτὴ τοιοῦτον ἔχει τὸ χρῶμα τῷ πλείους ὄψεις ἀπὸ τούτων ἀνακλᾶσθαι πρὸς τὸν ἥλιον, τί δήποτε οὐχὶ καὶ τὸ μεταξὺ τῶν περιφερειῶν τῶν εἰρημένων πᾶν φοινικοῦν ἔχει τὸ χρῶμα.

16 Ibid., 81: "An neque ex qualibet quidem parte refrangi habet visus, secundum opinionem eorum qui sic visionem fieri existimant; neque ab omni parte refrangi habet lumen ad visum, secundum aliorum opinionem, qui sic fieri visionem arbitrantur? Verum determinata ac definita sunt loca refractionum, ac praefinitam distantiam esse oportet luminosi ipsius corporis a speculis ipsis, quae suspicere eius lumen habeant. Quapropter ab his quidem refractio accidit fieri, quae huiusmodi determinatum habeant situm." *CAG*, vol. 3,2, 160, 28–33: ἢ οὔτε ἀπὸ παντὸς μορίου ἡ ὄψις ἀνακλᾶται καθ' οὓς οὕτως τὰ τοιαῦτα ὁρᾶται, οὔτε ἐπὶ τὴν ὄψιν ἀνακλᾶται τὸ φῶς ἀπὸ παντὸς μορίου καθ' οὓς οὕτω τὸ ὁρᾶν, ἀλλ' ὡρισμένοι οἱ τόποι τῶν ἀνακλάσεων καὶ δεδομένην χρὴ τὴν ἀπόστασιν εἶναι τοῦ τὸ φῶς ποιοῦντος σώματος ἀπὸ τῶν δεχομένων αὐτὸ κατόπτρων· ἀπὸ τούτων οὖν γίνεται, ἃ ταύτην τὴν θέσιν ἔχει.

17 Olympiodorus, *In meteora Aristotelis*, 65r: "Nubes enim, in quibus iris faciem ostendit suam, quia in exilia corpuscula minutaque stillicidia divisae discerptaeque sunt; et quaedam ipsarum longe ab oculis abductae iacent, quaedam vero propius consistunt; radii oculorum qui plurimi sunt, foras emissi ad regionem nubis quidam incursu suo

Though interesting, this new explanation would introduce an idea of *mixtio*, which in this case depends on the portion of the cloud traversed by the visual rays. Would it not be possible to solve the problem by explaining the phenomenon with a changing angle of incidence? According to Olympiodorus, vision became ever more weak the more the rays of the visual cone departed from its axis, i.e. from the ray that met the reflecting surface along a perpendicular line. Now in the case of the double rainbow the perpendicular rays and those nearest to it fell precisely between the two bands of red, that is in the space where no color was perceived. The strength of these rays could make it possible to perceive sunlight without any alteration. By moving away from this space, the visual rays were making the angle of incidence wider and wider, so that the perception of the different colors of the rainbow placed at the right, and the left of the space taken up by the rays near the perpendicular, became weaker. In this way, both the contrary order of the arrangement of the colors in the two rainbows and the space without color could be explained.[18]

---

propioribus nubibus obvii ad id quod apparet, reflectuntur hoc est ad Solem; quidam autem radii quibusdam a conspectu procul summotis nubibus incidunt; alii autem in alias ab oculis multo quoque adhuc remotiores nubes incurrunt. Sed radii quidem qui ad propiores nubes perveniunt, et quasi per exiguum nigrum, idest per nubilium aera ipsum aspiciunt Solem, non multum falluntur, quippe qui neque multum via defessi languent, et per brevia aeris nubilia ipsum cernunt. Ideoque phaeniceam Solis faciem in nubibus iis intuentur ...” CAG, vol. 12,2, 236, 24–33: ἐπειδὴ γὰρ τὰ νέφη, ἐν οἷς ἡ ἶρις ἐκφαίνεται, κατακεκερματισμένα ἐστὶν εἰς μικρὰς ῥανίδας καὶ τὰ μὲν αὐτῶν πόρρω κεῖνται τῆς ὄψεως, τὰ δ’ἐγγυτέρω, αἱ ἐκ τοῦ ὄμματος ἐπ’ αὐτὸ ἐκπεμπόμεναι ἀκτῖνες πλεῖσται οὖσαι αἱ μὲν εἰς τὰ πλησίον νέφη προσπίπτουσαι ἀνακλῶνται πρὸς τὸ ὁρατόν, τουτέστι τὸν ἥλιον, αἱ δὲ εἰς τὰ πορρωτέρω, αἱ δὲ ἔτι εἰς τὰ πορρωτέρω. ἀλλ’αἱ ἀλλ’ αἱ μὲν εἰς τὰ πλησίον νέφη προσπίπτουσαι ἀκτῖνες ὡς ἂν δι’ὀλίγου μέλανος ὁρῶσαι αὐτόν, τουτέστι ἀχλυώδους ἀέρος, οὐ πάσχουσι πολλὴν τὴν ἀπάτην ἅτε δὴ μὴ πάνυ ἀσθενήσασαι καὶ δι’ὀλίγης ἀχλύος αὐτὸν ὁρῶσαι. ὅθεν φοινικοῦν χρῶμα ὁρῶσιν αὐτὸ τοῦ ἡλίου ἐν τοῖς νέφεσιν ἐκείνοις.

18 Olympiodorus, *In meteora Aristotelis*, 65v: “Quum enim a visu nostro radii multi defluant, qui in rectam lineam ad rem spectabilem immittitur radius, qui itidem axis est cuiuslibet coni geniti, valentiorem et perspicaciorem videndi vim habet quam reliqui radii, qui non in rectam sed in obliquam partem perferantur. Et ex his rursum radiis qui recto et perpendiculari radio propiores sunt, videndo magis pollent quam qui a perpendiculari longius decidunt; ex quo fit, ut radius ad libramentum immissus quum validissimus omnium existat, nullum in videndo mendatium patiatur. [...] Caeteros vero radios perpendiculari confines mendatium et fraudem pati certe contigit, sed exiguam. Hos vero qui longius ab eo radio qui axis cuislibet coni est, absistunt, in magnum mendacium et errorem incurrere. Iis rebus ita constitutis in iride nubes multi ab oculis emissi radii circumquaque oberrare videntur; quorum quidem radiorum unus in rectum emissus perpendicularis existit, aliqui autem huic proximi adiacent, alii procul a recto decidunt. Sed radius ad libramentum iniectus medio inter utrunque arcum spatio incidit in eum scilicet locum, qui inter utrasque phaeniceas lineas media regione interiacet, quo quidem in loco nullius mendax omnino coloris similitudo apparet.” CAG, vol. 12,2, 238, 20–30: ἐπειδὴ γὰρ ἐκ

However, would it have been possible to combine the two theories on the formation of color, one that refers to the portion of the cloud crossed by the visual rays and the other that uses the variation of the angle of incidence of the same rays? It would seem a simple affair, but Olympiodorus did not explore the problem further.

3 Piccolomini's and Vimercati's Assessments of Aristotle's Rainbow Doctrine

The commentaries by Alexander of Aphrodisias and Olympiodorus became the main reference point during the Renaissance, though they did not completely replace the contributions by medieval commentators on the *Meteorology*. Alessandro Piccolomini adopted an extreme position: in his "Tractatus de Iride", he declared all studies made by the *Latini* to be utterly useless. Piccolomini's work, which was structured as a mathematical treatise, tackled the question of how colors changed towards black in relation to increasing distance and weakening vision, as well as other questions.

Piccolomini indicated that the changing distance, weakening visual power, and reflection were the main causes of the changing colors and of their formation through different reflections of the visual rays towards the sun. This was the conclusion Alexander of Aphrodisias had already reached, but it was possible to go further and relate the formation of the colors of the rainbow to the angle of incidence of the visual rays in the cloud, as Olympiodorus had pointed out. In the external band of the first rainbow the angle of incidence was greater, and therefore the penetration and the *mixtio* of the visual ray with the darkness of water was less. However, while reflection alone could be sufficient to cause the altered perception of the color of the sun, it seemed that this could not happen with much greater angles of incidence, where the *mixtio* could not take place. By reducing the angle of incidence the penetration of the visual rays increased, and as a consequence the *mixtio* of the visual rays with the color of the small drops of water also increased. Thus red, green, and blue

τοῦ ὄμματος πιπτουσῶν ἀκτίνων ἡ κατὰ κάθετον φερομένη πρὸς τὸ ὁρατόν, ἥτις καὶ ἄξων ἐστὶ τῶν γινομένων κώ-νων, ἰσχυροτέρα ἐστὶ τῶν μὴ κατὰ κάθετον, ἀλλὰ πλαγίων φερομένων (καὶ τούτων αἱ πρὸς τῇ καθέτῳ ἰσχυρότεραί εἰσι τῶν πόρρω τῆς καθέτου), συμβαίνει τὴν μὲν κάθετον ὡς ἰσχυροτάτην οὖσαν μὴ πάσχειν ἀπάτην ἢ σπανίως ἔχειν, τὰς δὲ πρὸς τῇ καθέτῳ πάσχειν μὲν ἀπάτην, ὀλίγην δέ, τὰς δὲ πόρρω πολλὴν πάσχειν ἀπάτην. τούτων οὕτως ἐχόντων ἐπὶ τῆς ἴριδος φαίνονται ὄψεις πολλαὶ παρὰ τὰ νέφη, ὧν ἡ μέν ἐστι κάθετος, αἱ δὲ παρὰ τὴν κάθετον πίπτουσιν, αἱ δὲ πόρρω τῆς καθέτου. ἀλλ' ἡ μὲν κάθετος ἐμπίπτει ἐπὶ τὸ μεταξὺ τῶν δύο ἰρίδων, ἐπὶ τὸ τῶν μεταξὺ δύο φοινικῶν περιφερειῶν, ἔνθα οὐδὲν ὅλως ἀπατηλὸν φαίνεται χρῶμα·

(violet) were formed. Blue (violet) was the last perceivable color because the visual rays nearest to the perpendicular, though they could most deeply penetrate and mix with the cloud, did not have a sufficient angle of incidence to cause the altered perception of the color of the sun. To cause the appearance of colors, the angles of incidence had to be between a maximum and a minimum inclination.[19]

It is obvious that if the change of the angle of incidence were the only cause of the appearance of colors, it would not be possible to explain the contrary order of their arrangement in the second rainbow. Another element ought to be considered, which through being changed would counteract the effect of the increasing angle of incidence: distance. In other words, red would continue to appear up to the maximum value of the angle of incidence, but the increased distance would weaken the strength of the vision of this red, making it appear first green and then blue (violet).[20]

But how to explain the colorless space between the two red bands? For Piccolomini this was due to the juxtaposition of two red colors with very different intensities: the red of the first rainbow would be much stronger then the red of the second, which would cause a change of the color towards white in the space between the two colors. To explain this phenomenon in the formation of the color red, Piccolomini referred to a presumed diverse structure or

---

19 Piccolomini, "Tractatus de iride," 124: "Radius enim visualis, si nimis forti extiterit, tunc aut nubem penitus pertransit, sicut radius perpendicularis, aut valde penetrans, quamvis maxima fiat dicta permixtio, debiliter tamen valde refrangetur, cum propinquior sit ipsi perpendiculari, ac naturam ipsius nimis sapiat, et ex hoc coloris phantasiam non causabit. Atqui e contra si radius magis quam necesse sit distabit a perpendiculari, tunc quamvis ad maximum angulum refrangantur, tamen modica fiet talis permixtio quam diximus luminis cum nigro nubis, et propter hoc etiam coloris emphasim non produxerit. Necesse est igitur quod radius ipse, nec nimis accedat ad perpendicularem, nec etiam nimis elongetur ab ea. Nam ad coloris productionem, non solum requiritur sufficiens ac debita permixtio luminis cum nigro nubis quae ex sufficienti penetratione causatur, quod non nimis longe a perpendiculari contingit fieri, sed etiam requiritur quod sufficiens refractio fiat, ad sufficientem, scilicet angulum; adeo quod non in tantum penetret, quod ad nimis parvum angulum reflectatur."

20 Ibid., 125: "Distantiam enim pro colore puniceo generando sufficiens est, quare inferior peripheria secundae iridis punicea est […]. Secunda vero peripheria, cum iam determinata ac proportionalis illa distantia defecerit, ex qua talis refractio fieri habet, ut color punicens generetur; tunc quidem cum refractio ex nimia distantia debilis iam fiat, (ex nimis longa enim et nimis brevi distantia, debilitatur refractio, ut diximus) fulgidum ipsum tendit magis ad nigrum, ac viridem colorem producet; et consequenter alurgum in extima peripheria secundum eadem rationem, extra quam peripheriam nullus amplius color apparet, propter elongationem partium nubis a debita distantia pro refractione sufficienti ad colorum generationem."

constitution of the external part of the cloud in which the rainbow is formed: that part would be less dense and its little drops would be 'badly' placed.[21]

Frequent references to ancient commentators were also made by Francesco Vimercati in his *Commentaries on the Meteorologica*, which was the most important edition with commentary on Aristotle's work published in the 16th century. Telesio certainly knew it, as he used Vimercati's translation, with few changes, in the first chapters of his *De iride*.

In his commentary on the Aristotelian passage concerning the double rainbow, Vimercati pointed out the difficulty of explaining the contrary order of the arrangement of the colors; it seemed evident to him that, if one strictly followed the laws of optical science, the arrangement of the colors in the internal rainbow should also be inverted since the visual rays nearer the perpendicular are always stronger than those departing from it.[22] Olympiodorus's solution should be rejected since he placed the perpendicular visual ray in the space between the two red bands of the rainbows, whereas according to the last part of Aristotle's treatment of the question that ray fell in the centre of the cloud.[23]

---

21 Ibid., 124: "Cum igitur e regione Solis rorida nubes constiterit, atque id iridis phantasiam secundum stillas disposita fuerit, tunc quaedam determinata distantia est inter nubes et Solem, ac inter nubem et visum, secundum quam non solum luminis cum nigro nubis permixtio ac penetratio, sed etiam refractio sufficiens est, ad hoc quod fulgidum ipsum non multum ab albedine deficere videatur, adeo ut puniceus, color producatur. Et haec determinata distantia incipit in exteriori iridis peripheria, ac perdurat extra ipsam, usque ad aliquam nubis partem, quod totum intervallum ex sui natura puniceum apparere debet. Sed quoniam, ut superius explanavimus, quilibet color iuxta nigrum positum, albior videtur, iccirco cum puniceus hic color, qui in dicto intervallo est, iuxta partem illam nubis valde remotam, situs est, a qua propter hoc quod nimium distantia superexcedit, refractio nobilissima est [...] propter hanc, inquam, iuxtappositionem albus apparet, et etiam in coloribus iridis quae ab ipsa Luna fit, est videre. Cum igitur nubes ipsa, in illa quidem parte ad quam intervallum dictum terminatur, non multum densa sit, et bene secundum stillas disposita, tunc quidem secunda fit iris."

22 Vimercati, *In quatuor libros*, 332: "Nunc ea dubitatio diluatur, qua obiici contra Aristotelem solitum est, si ex aspectu validiori color puniceus, minus valido viridis et purpureus appareant, rationi consonum, imo vero necessarium esse, ut intimus ambitus puniceus, extimus purpureus vedeatur. Aspectus enim radios ab intimo ad Solem, quam ab extimo validiores referri; quandoquidem perpendiculari radio, qui ad centrum arcus fertur, sunt propiores, monstratumque sit a perspectivis radium perpendicularem validissimum esse, nec unquam reflecti aut frangi; eos autem, qui ab illo recedunt, quo minus abducuntur, validiores esse, quo magis, imbecilliores."

23 Ibid., 333: "An huic dubitationi occorrendum est, illud tradendo, quod Olympiodorus, utriusque arcus colorum diversitatem assignans, ex Ammonio commemoravit, nempe radium perpendicularem ad illud spatium ferri, quod inter utrunque arcum positum est. Illud itaque spatium, quod radio valentiori conspicitur, absque errore ullo a nobis apprehendi, tum id, quod sequitur, puniceum, qui color a Solis colore minus quam caeteri

What then was the cause of the appearance of the color, which was the farthest away from the bright sun, in the place of the strongest reflection? According to Vimercati, some authors thought that this inversion was only accidental, and essentially due to two obstructing factors: the narrow internal space and the greater density of the cloud in its central part than in the external one.[24] Other authors denied that these factors could solve the difficulty, since the greater density of the central part of the cloud would have suggested placing the color red in the internal band.[25] The observed order of the arrangement of the colors could be explained by the fact that the visual rays near the perpendicular would penetrate more deeply into the cloud and for that reason would absorb more darkness of the water,[26] whereas the mixture would gradually lessen along with the increasing distance from the perpendicular. Reflection would thus take place at different levels of depth, and the greater strength from the optical point of view would become a greater weakness of the preservation of color.

How to solve the difficulties raised by the commentators, and especially how to explain the colorless space between the two rainbows? To answer these questions, Vimercati also turned to Alexander of Aphrodisias, but unlike Piccolomini, he did not consider the geometrical aspects of the problem. Rather, he thought that one should not understand Alexander's argument as based on the distances of points from a reflecting surface, so that the statement "a reflection does not take place from just any point in a mirror" became "a reflection does not take place from just any part of the cloud in which the

---

recedit, utpote minori errore conspectus. An prorsus falsum est, radium perpendicularem ad spatium id ferri, quandoquidem (ut post docebitur) ad nubis centrum fertur."

24 Ibid.: "Hanc igitur dibitationem aliqui aliter sustulerunt, concedentes, per se quidem colorem, qui ad candidum magis accedit, in intimo ambitu apparere debuisse, ob eamque causam puniceum, nigriorem autem veluti purpureum in extimo, ex accidenti tamen ob duo impedimenta, candidiorem, qui est puniceus in extimo, et purpureum in intimo apparuisse; ac impedimenta quidem esse ambitus illius interioris parvitatem, atque nubis, in qua apparet, crassitiem et densitatem, quae longe maior est, quam in exteriori. His ergo duabus de causis Solis colorem in interiori ambitu minus perfecte repraesentari."

25 Ibid.: "Sed si ex radiis validioribus, quales sunt, qui iuxta perpendicularem habentur, color Solis in nube perfectius apparere per se debeat, illis profecto impedimentis non tolletur, quo minus appareat; nam et a nube densiori magis reflectentur, utpote eam minus penetrantes ..."

26 Ibid.: "An vero potius dicendum est, radios perpendiculari proximos, quoniam caeteris validiores sunt, debiliter admodum, et ad angulos parvos reflecti, imo vero ipsam nubem magis penetrare, illique magis admisceri, ob eamque causam Solis colorem debiliter valde repraesentare, ac quo magis a perpendiculari recedunt, eo debiliores esse, validiusque et ad angulos maiores referri, ideo colores ad candidum propius accedentes, et a nubis nigredine remotiores ostendere."

rainbow appears". As a consequence, between the two bands of red color there would be a discontinuity only because that part of the cloud was too far away.[27] This solution supported the explanation that the second rainbow was nothing else than an image of the first. This explanation, however, raised the essential difficulty of the mirror image's turning over from concave to convex.

Vimercati's work offered an overview of past opinions but the challenge of finding the true cause of this complex natural phenomenon was still open. This challenge was taken up by Bernardino Telesio.

4 Telesio's *De iride*

If we now analyzeTelesio's *De iride* we must first point out that it removes an ambiguity which was always present in the Aristotelian tradition. In the *Meteorologica*, Aristotle had accepted the theory of the visual rays issuing from the eye, giving up his own theory of vision. Alexander of Aphrodisias had tried to justify this way of proceeding, pointing out that from the point of view of the geometrical explanation of optical phenomena it was a matter of indifference whether vision took place through a visual ray issuing from the eye travelling towards the object that was seen or whether the eye passively received it from outside.[28] Medieval optical science had rejected this ancient theory, and Telesio accepted the general opinion on this point.

27 Ibid., 333, 338: "An vero, inquit ille [Alexander], non ab omni nubis parte aspectus ad Solem, aut lumen Solis ad aspectum reflectitur, sed reflexionem loca definita sunt et certa, definitamque et certam splendidi lumen mittentis corporis a speculo distantiam esse oportet? Ab his igitur speculis ita distantibus, situmque certum habentibus, arcum et colores repraesentari. Quibus in verbis videtur Alexander docere, ideo colorem nullum inter utrunque arcum apparere, quia nulla ibi reflexio ad Solem seu ad aspectum efficiatur [...] Neque enim haec (ut mihi videtur) est Alexandri (quemadmodum nonnulli crediderunt) sententia, sed quod in spatio illo nubes nimis distet; siquidem ait, certam luminosi corporis et speculi distantiam esse oportere, et ab his speculis, quae ita distant, reflexionem fieri; quasi dicere vellet, ab hac nube media non fieri, quoniam longius distet, quam ut possit reflectere. Nec vero ait Alexander, ab omni puncto speculi cuiusvis reflexionem non fieri, ut quidam putarunt, sed ab omni nubis, in qua arcus apparet, parte."

28 Alexander of Aphrodisias, *In quatuor libros meteorologicorum*, 72: "Quoniam vero, quantum ad praesentem rationem attinet, nihil refert sive dicatur, quod visus ipse ad speculum ad aequales angulos refractus, in rem visibilem incidens, cum sub huiusmodi refractione res ipsa contigerit, illam videat; an dicatur potius quod res ipsa quae videri habet, propter aliqualem habitudinem, vel situm ad speculum per intermedium diaphanum patiens quidem atque affectum, emphasim faciat in speculo, quod quidem taliter diaphanum existat, ut non solum a colori pati possit, adeo ut alteri diaphano acceptam qualitatem elargiri

However, this seems to be the only time that Telesio followed the tradition of geometrical optics. In fact, in his critical discussion of the Aristotelian conception he rejects the fundamental assumption which explained the cause of the rainbow by means of optics, i.e. the assumption that the observer must be placed in the middle of the straight line joining the sun and the center of the mirror consisting of a great amount of small drops forming a cloud. To reject this assumption Telesio resorted to the same examples mentioned by Aristotle himself. In the case of the rainbow, which can be seen in the water drops raised by oars when rowing or in the drops splashed by hand, the former assumption is not verified. The same must be said for the rainbow which, in some particular conditions, is formed around the flame of an oil lamp. If we then add the experiences made with a transparent prism of glass to the examples mentioned by Aristotle a different explanation will obviously be needed.[29]

In Telesio's view, the rainbow should be explained on the basis of the assumption that light travels from the sun to the clouds and subsequently shines towards the eye. Light spreads from its source in all directions. In thin bodies such as air, it permeates them and can be perceived even when its source is not directly visible; in dense bodies, smooth and shining, light becomes more intense and while it doesn't penetrate them it is very bright and its color is

valeat, verumetiam et conservare, propter politiem ac splendorem, emphasim possit; ita quod ab ipso dehinc tanquam ab aliquo colorato patiatur rursus atque efficiatur diaphanum ipsum quod intermedium est. Quoniam, inquam, nihil refert in praesenti negotio, sive hoc dicatur sive illud, opinionem sequitur modo, quae emissionem radiorum ponit, quam quidem mathematici approbant." *CAG*, 3,2, 151, 20–30: ἐπεὶ δὲ οὐδὲν ὅσον ἐπὶ τῷ λόγῳ διαφέρει ἢ τὴν ὄψιν λέγειν ἀπὸ τοῦ κατόπτρου ἀνακλωμένην πρὸς ἴσας γωνίας, ὅταν ὑπὸ τὴν τοιαύτην ἀνάκλασιν τύχῃ τὸ ὁρατὸν ὄν, προσπίπτουσαν αὐτῷ ὁρᾶν αὐτό, ἢ αὐτὸ τὸ ὁρατὸν διὰ τὴν ποιὰν σχέσιν πρὸς τὸ κάτοπτρον διὰ τοῦ μεταξὺ διαφανοῦς πάσχοντος ἐμφαινόμενον ἐν ἐκείνῳ, ὄντι τοιούτῳ [διαφανεῖ], ὡς μὴ μόνον πάσχειν ὑπὸ τοῦ χρώματος δύνασθαι οὕτως, ὡς διαδιδόναι τὴν ἀπ' αὐτοῦ ποιότητα ἄλλῳ διαφανεῖ, ἀλλὰ καὶ φυλάσσειν δυναμένῳ τὴν ἔμφασιν διὰ λειότητά τε καὶ στιλπνότητα, ὡς ἀπ' αὐτοῦ πάλιν τὸ μεταξὺ αὐτοῦ τε καὶ τῆς ὄψεως διαφανὲς πάσχειν ὡς ἀπὸ κεχρωσμένου, τῇ δόξῃ τῇ τῶν ἀκτίνων καθωμιλημένῃ τε οὔσῃ καὶ τοῖς μαθηματικοῖς ἀρεσκούσῃ προσχρῆται.

29 Telesio, *De iride*, chap. 8, 5v–6r: "Quod igitur dictum est, vel eo una reflexione iridem fieri statuens Aristoteles, quod ibi modo fiat, ubi solum speculum sit nubes, et aspectum nostrum reflectere potest, probandus omnino videdur, minime vero e Solis illam regione tantum constitui decernes, oportere itaque aspectum nostrum medium inter Solem, nubemque fieri, et in eadem omnino linea Solem, aspectumque nostrum et iridis centrum polumque esse; passim enim irides intueri licet, quas inter, Solemque medii nos minime sumus […] et quae ex aqua Soli exposita, vel e guttis a remis sublatis, aut manu sparsis fiunt, nequaquam nobis inter eas, Solemque mediis fiunt, nec quae e serratili spectantur vitro, multoque etiam minus, quae circa lucernam fiunt."

not altered.[30] However, that does not happen when the light goes through something colored, or when, by illuminating a body with a certain density and depth, the light permeates it in a variable manner, making it shine with different colors. This was the case with water and with the clouds, which changed color from their natural whiteness to an increasingly greater darkness according to their greater depth or density. The cause of this alteration was the black color of matter, which became more notable when depth and density were more considerable.[31]

One also had to take into account the direction of the light, which could be either perpendicular or inclined. In the former case, light, reflecting on itself and acquiring strength, would have managed to overcome the black structure of matter, whereas in the latter case it would have mixed with matter more and more deeply, becoming altered into different colors.[32]

Hence, the question of the rainbow shifted from a discussion concerning the problem of the formation of images in small mirrors to an analysis of the variation of light in more or less dense bodies.

---

30 At the beginning of chap. 16 of *De iride*, Telesio refers to his *De rerum natura*. Here in book 4, chap. 10, 145 we read: "Itaque et ubi nullus conspicitur Sol, quo scilicet recta, qua sola progredi lux videtur, deferri non potest, a crasso quopiam retardata, reiectaque, et Sole non dum exorto, et penitus iam abdito, aliquantis per tamen universo in aere, et imis etiam in terris si non fulgida, at bene certe visilis, beneque spectatur clara. Non quidem id accidat nisi ab aere etiam, a se ipsa nimirum, vel summe exili in eo facta, reluceat. Nam quae a densis, aequabilibusque, et nitidis refulget rebus, a quibus, quod nihil eas subeat, integra relucet, et continua amplius, unitaque, nihil ab earum tumoribus, nec a maculis etiam [...] intercepta, intercisaque ullis, nihilo, quam a Sole ipso minus fulgida, minusque relucet ingens. [...] Nihil imminuitur ab ullo, quin in singulis bene in se ipsam colligitur, proindeque veluti alter Sol facta, a singulis, veluti a Sole ipso effulget, seseque effundit." For a *general* account of Telesio's light theory, see Mulsow, *Frühneuzeitliche Selbsterhaltung*, 104–139.

31 Telesio, *De iride*, chap. 16, 14v–15r: "Et aquas, nubesque permeans, et relucens etiam ab iis, si paulo profundiores, densioresve sint, non albo amplius, qualis, et lucis, et illarum utriusque est color, sed longe pluribus, et omnibus propemodum, qui album, nigrumque intermedii sunt, et ipso etiam nigro colorata relucet, quod nimirum penitus eas subiens, earum materiae nigredinem attingit, et prout maiori, minorive eius portioni immiscetur, eo magis, minusve ab ea exuperatur, ad nigrumque agitur. Itaque ubi humile est mare, album, ubi paulo viride, et ceraleum ubi amplius, et nigrum ubi profundissimum, eo scilicet magis obscurata, ad nigumque acta, quae ab eo relucet lux, quo ampliori ipsius materiae nigredine immista est."

32 Ibid., chap. 16, 15r: "Itaque aquam in vitro contentam matutina, vespertinaque lux, quae scilicet, quod bene obliqua advenit, nihil reflexa inseipsam colligitur. Itaque ab inexistente aquae nigredine esuperata irinis coloribus intingitur omnibus, minime vero et meridiana, quae nimirum bene directa inseipsam reflectitur, proinde copiosa, robustaque facta, materiae nigredinem penitus esuperat."

When the sunlight reached the cloud suitable for showing the rainbow by the shortest line, its strength would have prevented it from undergoing any alteration; thereby it was seen without any particular color. With increased inclination, the light would have been increasingly affected by the darkness of the cloud and would subsequently have formed the red, green, and blue (violet) bands. After the formation of this last color the inclination of the light would have increased too much and its variation would have been too faint to be perceived.[33]

Once more the inclination would reach a maximum value and a minimum value, but this time it did not depend on the theory of reflection but rather on a somewhat original idea of the emanation of light from the body of the sun.

Although Telesio acknowledged it was a fact that each part of the things that were lit up received light from every point of the surface of the sun, he thought it possible to establish a special relationship between some parts of the cloud and some parts of its surface. The single parts of the cloud would have only shown that alteration of light which was predominant over the other, and that predominance would have depended on the way in which illumination was taking place according to the greater or lesser inclination. Thus Telesio could spot those parts on the surface of the sun which were, in his opinion, mostly responsible for such variation. These parts show two extensive bands symmetrically placed in the two hemispheres. According to him, it is not the whole surface of the sun that spreads the light that causes the rainbow. The outermost parts do not, since the inclination of the light's rays coming from them is too great, and even the central part does not since the light from it reaches the cloud by the shortest line.[34] Through this division of the surface of the sun, Telesio could treat the problem of the double rainbow with great surety.[35]

33 Ibid., chap. 17, 15*v*: "In nube omnino bene in se ipsam conspissatam, et a luce iridem fieri existimare licet, nec maxime directa, maximeque robusta, nec maxime obliqua, languidaque, sed ab ea quae harum quasi media sit [...]."

34 Ibid., chap. 17, 15*v*: "et ab ea forte, quae nequaquam a Solis parte emanet, quae nubi proximior, earumque, quae nubi expositae sunt, media est omnium; eam enim nubi directam imminere existimare licet; neque ab iis, quae maxime ab illa absunt, maximeque obliquam ad nubem emittunt lucem, sed ab iis, quae utrarumque veluti mediae sunt. Sphericus enim cum sit Sol, assidueque eius superficies immutatur, singulae eius partes proprium ad nubium partes quasvis situm obtineant oportet, eoque singulas a reliquis magis diversarum, quo magis ab iis absunt, obtineant oportet."

35 Ibid., chap. 9, 8v–9v, Telesio had exactly noticed the contradiction in the passage where Aristotle had tied to explain the inversion of the arrangement of colors in the double rainbow. He had also carefully considered the solution of the problem offered in Olympiodorus' passage and had shown that it was untenable by briefly referring to the theories *perspectivorum*. Those explanations were similarly to be reject which had been

The illumination coming from the central part explained the missing color between the red bands of the two rainbows, whereas these last two were the result of illumination by those parts of the extensive symmetrical bands nearer the central zone. This was in fact the light which was striking the cloud in a less inclined way. The more it travelled towards the outermost part of the extensive bands, the inclination of the light increased, and thus in the corresponding part of the cloud the green color appeared first and then the blue (violet) color.[36]

To summarize, the lower hemisphere—the one turned towards the surface of the earth—was responsible for the appearance of the internal rainbow, whereas the other hemisphere was responsible for the external rainbow.

The problem which had so strained the minds of the Aristotelian commentators seemed finally to have been resolved, although the premise on which the solution was based was far from sound. Telesio's attempt, however, remained outside the scientific tradition, since this tradition continued to refer mainly to the laws of geometrical optics until it finally found the essential precondition for any further research into the law of refraction.

5 Concluding Remarks

In this essay I have read Telesio's *De iride* in connection with a scholarly tradition that can be traced back to Aristotle's *Meteorologica* (Book 3) and other sources of the Aristotelian corpus dealing with the formation of colors

---

offered by more recent commentators of the *Meteorologica*, who "ab antiquioribus acceperant acquiescere impotentes, dictarumque diversitatum rationem reddere disperantes, si quomodo Aristoteli placet, iris utraque aspectus ad Solem reflexione fiat exteriorem minime eo pacto exoriri contendunt, sed interioris iridis imaginem esse". This last criticism seems to be directed to Vimercati.

36 Telesio, *De iride*, chap. 18, 16r–16v: "A luce porro, quam diximus iridem fieri, non ratio tantum, sed eius colorum ordo aperte quidem in simplici, at multo etiam in duplici amplius manifestat. Propterea enim ubi duplex fit iris, non altera alteri contigua fit, proximaque, sed spatium inter utramque album spectatur. [...]. Utraque nimirum iris, prout ab albo spatio magis recedit, magis ad nigrum, obscurumque, et aeque utraque, eodem que tendit modo, in eo tantum ab altera differens altera, quod superioris colores paulo, languidiores apparent, quod iris utraque, et quod utriusque medium est spatium ab universo quidem Sole, at non ab universo simul singuli iridis utriusque ambitus, intermediumque spatium, sed et hoc, et singuli illi a certa illustrantur Solis parte, et inter medium quidem spatium a media iridum ambitus, pro ut ab albo spatio magis recedunt, ita a Soli partibus, quae a media magis absunt; et interior quidem iris a Solis parte, quae infra eam, exterior vero ab ea, quae supra mediam est, illustrari videtur."

(*De anima* and *De sensu et sensibilibus*). As I have endeavored to show, Telesio's attempt to provide an adequate explanation of the phenomenon of the rainbow still operates within an essentially Aristotelian framework. Aristotle's treatment of the rainbow is particularly interesting as a case in which mathematical disciplines such as optics can help to comprehend the phenomenon. However, at the same time, mathematics is incapable of adequately accounting for one of the essential features of the phenomenon, namely its color. In spite of Telesio's disavowal of Aristotle, he cannot be placed outside the Aristotelian tradition of scholars and their explanation of the rainbow and its colors. Within this tradition I paid particular attention to Alexander of Aphrodisias and Olympiodorus. Although Telesio breaks with the extromissive theory of visual rays put forward by ancient writers on optics, this does not radically alter the framework of his explanation: in fact, this break would seem to bring him even closer to a genuinely Aristotelian theory of vision.

The instance of the double rainbow is a crucial example. The Aristotelian solution de facto entails taking two separate mathematical components into consideration: on the one hand, in relation to the internal rainbow, the magnitude of the arches of the various colors; on the other, in relation to the external rainbow, the observer's distance from the reflecting surface formed by suspended droplets. Ancient and Renaissance commentators, including Telesio, tried to reunite these two mathematical components by employing geometrical analysis and a theory of perception. Further, they addressed the related question of why the space between the two rainbows is colorless.

In my view, it is only through such contextualization that we can understand Telesio's *De iride*. Telesio's theory of the formation of colors, which is so closely linked to the idea of matter, does not greatly differ from some of the solutions proposed in previous centuries, for instance by Olympiodorus. The alteration of the color of natural light depends on the density and depth of the illumined body. The cause of this alteration is the blackness of matter, which only becomes perceivable when this depth and density is substantial. It can hardly be denied that, given these assumptions, the appearance of a rainbow no longer has to do with the problem of the formation of images in tiny mirrors, but rather becomes a problem related to the variation of light in bodies of varying density. Although these ideas point beyond the Aristotelian conception, Telesio does not take the actual step taken by later writers on optics. He does not argue for the refraction of light rays as one of the causes of the formation of colors. While Telesio thought that the variation of the obliqueness of these rays plays a crucial role in the appearance of the colors of the rainbow, this element is never further explored through an in-depth study of optics.

This observation provides us with an improved appreciation of the importance of the work of other Renaissance scholars for Telesio, among whom are Alessandro Piccolomini and Francesco Vimercati. Piccolomini translated Alexander of Aphrodisias' commentary on the *Meteorologica* and was the author of a *Tractatus de iride*, which are fundamental sources for Telesio's treatment of the rainbow. Far more relevant for the study of Telesio's views is Francesco Vimercati's *Commentaries on the Meteorologica*, which was the most important commented edition of Aristotle's work to have been published in the second half of the 16th century. In this text he reopened the discussion on the phenomenon of the double rainbow.

Telesio derived his discussion of the double rainbow from these sources, and not from any mathematical enquiry—less still from any experimental study. Once he was aware of the fact that it is impossible to come up with a convincing interpretation of the phenomenon within a strictly Aristotelian framework, Telesio departs from it and develops a new explanation. Telesio proceeded using a process of 'elimination' of all those elements which inevitably led to unsolvable contradictions. Although this resulted in a very different interpretation of the phenomenon than Aristotle's, it nevertheless directly derived from it. Telesio's *De Iride* is representative of the wider context of the Renaissance debates about the reception, transformation or refutation of Aristotelian themes.

CHAPTER 8

A Conversation by Telesio: Sensualism, Criticism of Aristotle, and the Theory of Light in the Late Renaissance

Martin Mulsow

How often do we have the chance to listen in on a disputation between a Renaissance philosopher of nature and his opponent? There has to be a third party to document it, someone capable of capturing the conversation with clarity. This article is about the discovery of one such document, and the disputation it records will afford us deeper insight into the intellectual culture of the new philosophy of nature in the Late Renaissance—in Neapolitan culture in 1570, to be precise.[1] Biographical material available about Telesio (1509–1588), who led the conversation in question and who was described by Francis Bacon as *hominum novorum primus*, is not exactly abundant.[2] All too little is known about the life and research practice of this philosopher who belonged to the generation of Italian scholars—which also included Cardano

1 On the reconstruction of conversation groups and their significance for the history of ideas and philosophy cf. Mulsow, *Die unanständige Gelehrtenrepublik*, esp. Chap. 5, and Mulsow and Stamm, *Konstellationsforschung*. This article appeared originally in German as: "Ein unbekanntes Gespräch Telesios: Aristoteleskritik, Sensualismus und Theorie des Lichts in der Spätrenaissance". My gratitude to Eckhard Keßler who introduced me to the appreciation of Telesio.

2 Fundamental sources are the speech made by Giovanni Paolo d'Aquino upon the death of Telesio: *Oratione ... in morte di Bernardino Telesio filosofo eccelentissimo agli Academici Cosentini* (Cosenza 1596), as well as Bartelli, *Note biografiche*, 7–73; furthermore De Chiara, *Bricciche telesiane*, Scipioni, "Lettere inedite di Bernardino Telesio," Mercati, "Appunti telesiani," Egidi and Boretti, *Telesio: Regesto dei documenti*. Concerning Telesio cf. Lotter, *De vita et philosophia Bernardini Telesii commentarius*, Brucker, *Historia critica philosophiae*, vol. IV, 449–460; Bartholmes, *De Bernardino Telesio*, Fiorentino, *B. Telesio, ossia studi storici su l'idea della natura*, Abbagnano, *Telesio*, Soleri, *Telesio*, Saitta, *Il pensiero italiano*, 1–77; Kristeller, *Eight Philosophers*, 91–109; De Franco, *Bernardino Telesio*, Galasso et al. *Bernardino Telesio e la cultura napoletana*, Keßler, "Selbstorganisation in der Naturphilosophie der Renaissance," De Franco, *Introduzione a Bernardino Telesio*, Bondì, *Introduzione a Telesio*, Mulsow, *Frühneuzeitliche Selbsterhaltung*, Pupo, *L'anima immortale in Telesio*, Leijenhorst, "Bernardino Telesio," Boenke, *Körper, Geist, Spiritus, Psychologie*, 120–170. The following books appeared after the publication of the original German version of this article: Mocchi, Plastina and Sergio, *Bernardino Telesio tra filosofia naturale e scienza moderna*, Bondì et al., *Bernardino Telesio y la nueva imagen de la naturaleza.*

 | DOI:10.1163/9789004352643_010

or Fracastoro—who from their breadth of textual knowledge and new empirical research were the first to forge ahead with comprehensive non-Aristotelian concepts. Telesio studied in Padua, most probably from the end of the 1520s until 1535, although there is no documentary evidence of his matriculation or graduation. As he was born into the aristocracy, he had no need to strive for all the official documentation normally required for a university career. His uncle, the humanist Antonio Telesio, was of independent means and so took his education in hand. After this Padua period, the trail goes almost completely cold. What we do know is that he had long periods in a monastic environment where he taught himself, that later in life he started a family in his native Calabria, and that he held representational roles in his home city of Cosenza. He had good links to the Pope and a number of cardinals at the papal court. It was the death of his wife in 1560 that seems to have given Telesio the drive to devote his energies to research for the rest of his life. However, there is little detailed information about that later period.

It is this paucity of information which makes the brief report of a conversation discussed and transcribed in this article useful. This report belongs more than anywhere in the context of the break with Aristotelian science and the problems associated with it.[3] Given this context, the report can lay claim to an interest which goes beyond the purely biographical. The major point is that Telesio has been repeatedly represented as a critic of Aristotle but this has resulted in a failure to take into consideration the actual experience of the contention with the contemporary forms and representatives of Aristotelian academic knowledge, i.e. with the very milieu which was the direct opponent of the new philosophers of nature. Clearly, Telesio did not carry on a long distance conversation over about two thousand years with Aristotle, as he appears to, but with contemporary Aristotelians, and yet he is not entirely blameless for the problems surrounding his reception. His idiosyncratic style and paucity of references, which solely demonstrates criticism of the classical authors Aristotle, Hippocrates, and Galen, pretends a new beginning that is free of context, is not concerned with contemporaries and sometimes rather imitates the style of pre-Socratics like Parmenides or of the Stoics. But we should not fail to recognize that this text is constructed with extreme artificiality and was deliberately expurgated of references to contemporary authors. Telesio's silence

3 On the diversity and vividness of Renaissance Aristotelianism see Schmitt, *Aristotle and the Renaissance*. See, as well, Des Chene, *Physiologia*, Leijenhorst et al., *The Dynamics of Aristotelian Natural Philosophy*, Kusukawa, *Transformation of Natural Philosophy*, Gaukroger, *Francis Bacon and the Transformation of Early-Modern Philosophy*, id., *Emergence of a Scientific Culture*.

about his contemporaries is part of his breakaway from established science and, when viewed sociologically, establishes him as *distinct* from the academic conventions of his time.

Even if he was in isolation—*magnis solitidinibus*[4]—of academic debate, Telesio is characterized through and through as looking at problems which dominated Aristotelian science at the time of his education in Padua. Former Padua students who were his contemporaries, for example Alessandro Piccolomini or Lucilio Maggi,[5] later gave attention—within the Aristotelian framework—to optics, light reflection, and the generation of heat, which were exactly the questions that lead Telesio to his own ideas. The philosopher's scruples and caution are known from his visit in 1563 to Vincenzo Maggi, a fellow student from the Padua days, during which time he wanted to test the validity of his critique of Aristotelianism; it was actually Maggi's positive reaction that gave him the courage to go public two years later with the draft of *De natura*. Publications came late in the day: in 1565, just as Telesio turned fifty, he presented a first, slim version of his newly developed philosophy of nature. Next he built on this version and in 1570 published a second, slightly modified edition. Then, after a long period of research during which new theorems were constantly being accumulated, the complete *De rerum natura* was published in nine volumes in 1586.[6] Two years later, he died.

The report published here is the transcript from a manuscript left by Antonio Persio. Persio is considered to be the most important of Telesio's immediate pupils but, according to more recent findings, should rather be viewed as a collaborator of his tutor, and after his death even as an original mind of Late Renaissance philosophy.[7] The printing of the manuscript *De natura ignis et caloris*[8] was blocked after 1612 by church circles[9] and awaits its edition until today. Persio's work probably came into being between 1587 und 1590,[10] immediately before his publication of short writings left by Telesio, *Varii de*

4 Telesio, *De natura iuxta propria principia* (1565), Prooemium.

5 On this cf. Lohr, *Renaissance Latin Aristotle Commentaries II*.

6 Telesio, *De rerum natura iuxta propria principia* [1586] (1965, 1974, and 1976).

7 In essence Garin, *Umanisti, artisti, scienziati*. On Persio cf. Gabrieli, "Notizia della vita," Garin, "Nota telesiana," Artese, "Antonio Persio," id., "Filosofia telesiana e ramismo," id., "Il rapporto Parmenide-Telesio, dal Persio al Maranta," Mulsow, "Reaktionärer Hermetismus vor 1600?," id., "Philosophia italica als reduzierte Prisca-Sapientia-Ideologie, 253–280.

8 Persio, *De natura ignis et caloris*.

9 On the mood of this period cf. Redondi, *Galileo eretico*. About the context of the Accademia dei Lincei, *Federico Cesi e la fondazione dell'accademia dei Lincei*, Freedberg, *The Eye of the Lynx*.

10 On dating cf. Mulsow, "Philosophia italica als reduzierte Prisca-Sapientia-Ideologie."

naturalibus rebus libelli.[11] At that time he was in Venice and Padua and tried to facilitate the introduction of Telesian philosophy into academic teaching. The report given by Persio, however, However, dates back to an earlier time. He remembers back then and tells:

1 Persio's Report

[Antonio Persio: *De natura ignis et caloris* XI, 29; MS. Linceo VII, Biblioteca Corsiniana, Rome, 351*r* f. For greater readability the spoken parts of the report have been italicised, unlike in the manuscript.]

Memini cum ad Philosophum Bernardinum Telesium Neapolitanum dum in cuius degeret immortali gloria vir dignissimus quotidie visendi eum et cum eo commendandi de Philosophia ventilabam. Quadamdie aestatis tempore cum magnus regnaret calor solque esset flagrantissimus, cum nobis solis Telesius sermoenaretur. Ecce Quintius Bonianus Telesii quidem conterraneus Philosophus et ipse ac Medicus sed Aristotelicae sectae nimis addictus et Eristicis ad Telesium visendi ac percontandi seu potius consentendi gratia venit, assurrexit placide Telesius, et post mutuam salutationem senex resedit atque allata mox sella est in qua etiam sedit Quintius ad nos statim reversus qua de re loquebamur interrogavit ut caeptus sermo ac disputatio si quae esset non relinqueretur.

I remember when I was discussing philosophy at the home of the Neapolitan philosopher, Bernardino Telesio, at a time when I lived in the shadow of his glorious reputation. Here was a man so noble that I would seek him out daily and put my whole trust in him. One summer day it was extremely hot and the sun was blazing down, and Telesio himself fell into conversation with us. Along came Quintius Bonianus, a fellow countryman of Telesio, and himself a philosopher and physician but who had committed himself very much to the Aristotelian sect and was much of a squabbler to see Telesio and ask for information, let alone to mix with us. Telesio stood up calmly and, after they had greeted one another, the old man sat down again and as soon as a chair had been brought over for Quintius, the new arrival turned to us and asked what we were talking about so that the conversation already underway and the discussion, if that was what it was, was not interrupted.

11 Telesio, *Varii de naturalibus rebus libelli* [1590] (1981).

De solis respondi *calore et colore nempe albedine disserebat* [*, non*][12] *Telesius qui soli utrumque inesse reapte et non virtute et potestate quadam ut nos Peripatetici* (nam et ego Peripateticos partes saepissime suscipiebam non tam altercandi quam addiscendi ac veritatis ad ipsis cendae gratia id faciebam) *tenemus Quinte.*

I replied: *Telesio has been discoursing on the warmth and color of the sun, namely its whiteness, as real properties inherent to the sun and not simply a capability or a specific capacity, as we Peripatetics believe, Quintius.* (Because I, too, often took the part of the Peripatetics and liked this not so much for the sake of the argument but to learn and to attain truth).

Tum Quintius ad Telesium *utrumque quod docis falsum Telesi, cum nec calor, nec color ullus nedum albedo soli inquam inesse poterit.*

Then Quintius said to Telesio: *as far as both of these are concerned, you are teaching the wrong thing, Telesio my friend, because neither warmth nor any color—and certainly not white—could be inherent to the sun.*

Cur ita inquit Telesius

Why so? said Telesio.

quia replicavit Quintius *corpora coelestia nullas cum his inferioribus qualitate similes vel pares habent, cum coelum sit natura quinta seu quintum corpus, ideo fit ut non communicet cum his nostratibus corporibus ortui et interitui subjectis.* et alia his similia congerebat.

Because, answered Quintius, *the heavenly bodies have nothing to do with these lower things in terms of sameness or similarity, and because heaven is a fifth nature or a fifth body; from this comes the fact that it does not communicate with these common bodies which are subject to coming into being and passing away.* And he brought together other arguments of a similar nature.

At Telesius ut erat amoeno ingenio et alterandi inimicus, *Vir tibi dicam* inquit *Quinte Tu ut Aristoteles operationem et sensum tuearis / sensum negas et sensum ipsum amittis quia vera non sentis nec vides quae alii et vident et sentiunt /*

But Telesio, although friendly, was not inclined to change his view, and said: *My good man, I want to say to you: Quintius, like Aristotle you should pay attention to the effectiveness and your perception and instead of doing that you repudiate the perception of the*

12 Possible slip of the pen for 'nos.'

senses and go so far as to lose this because you neither feel nor see what others see and feel.

fac igitur repetit Quintius *ut sensum hunc agnoscam quem me negare ais.*

Then your attention should be given, retorted Quintius, *to my recognizing this perception of the senses, the one you say I repudiate.*

Quod suscipiens Telesius *non sentis tu e solis calorem, non vides in ipso alberem, etsi unquam tempore hoc in quo inquam sol est ornatissimus cum sol calorem emittat ad nos ardentissimum ac candidissimus appareat vel caecis.*

Taking up this point, Telesio said: *My friend, you feel neither the warmth which comes from the sun nor do you see in it the white, and of all times now, when—and I repeat—the sun is at its most brilliant because the sun sends out its greatest warmth to us and appears at its most brilliant white, even to those who cannot see!*

Certe inquit Quinius *id nunquam animadverti*

Of course, said Quintius, *I've never noticed that before.*

probi indignum facimus inquit Telesius surgens et apprehensa Quintii manu e cubiculo exiens uterque in silatio quod [351recto / 351verso] soli erat obiectum. *Videas Quinte* inquit Telesius *quam sit candidissimus sol iste ut calorem ipsum taceram, quem si negares te non amplius vivum sed mortuum esse pronunciarem, at si candorem negabis saltem caesum*[13] *te esse sentiam.*

That is a state of affairs so poor as to be unworthy of support, said Telesio, who stood up, took Quintius by the hand and went with him out of the living room to the outside so that the sun was facing him. *Look, Quintius,* said Telesio, *at how this sun is a very brilliant white, not to mention how very warm, and if you were to dispute this, I shall declare you no longer alive but dead; but if you dispute the brilliant white, I shall think you are, at the very least, blind.*

13 Slip of the pen for 'caecum.'

Quintius adhuc clamitans negabat se videre solem album sed illum candorem non esse candorem sed fulgurem *itaque sol non est albus sed fulgidus* ex illo principio quod iam dixerat, *quin nec coelum* aiebat *viderent aut solem sed nubes pro coelo* et alia id genus.	Quintius, his voice still raised, denied that he saw the sun as being white and said that this brilliant white was not whiteness but a sparkle; *this is why the sun is, not white, but gleaming,* according to that principle that he had already referred to, *because people,* he said, *would neither see the heaven nor the sun but clouds instead of sky* and other such things.
male igitur Poeta ille? ego in excandescentem Quintium protuli ridens *Pronaque cum spectent animalia caetera Terram* *Os homini sublime dedit coelumque tueri* *Iussit, et erecto et ad sydera tollere vultus*	*So does this mean the poet had described the situation poorly?,* I smilingly proposed to the agitated Quintius: *While the remaining animals bowed down and looked at the Earth /* *He gave man an upturned countenance* *And the sky to look at / He ordered him* *To stand tall and raise his eyes to the stars*
Itaque risus dissolvit pervicaces quique in sua opinione discessit Quintius neque persuasus sed infaestus Telesio semper &c.	And this is what made those stubborn mules break into laughter, each one being left with his own opinion. Quintius left but Telesio was not convinced, still rather angry, and so on.

2 The Context in Which the Conversation Occurred

This conversation probably took place in the period pre-1570 during which Telesio was preparing the second edition of his book *De rerum natura* and staying in Naples, where Persio had virtually daily contact with him.[14] In his *Liber*

14 Between 1563 and 1570, Persio was in Naples and there, as tutor to the Orsini family, got to know Telesio. The period in which he had contact with him seems to be after the first publication of *De natura*, which came out in 1565. For dates cf. Gabrieli, "Notizia della vita," 482 and 489.

novarum positionum Persio also described this time as an intensive exchange between the older man, Telesio, and his young pupil.[15] In that text he praised his tutor's *benignitas*, which is expressed as *aemoenitas* in the present text. Both of these were virtues greatly valued at that time. In the culture of *civil conversazione* in the Late Renaissance, the characteristics of good temper, kindness, helpfulness, and friendliness of behavior held a special, central position in dealings with other people. That these virtues were not merely a traditional theme in literature but reflected in the art of conversation is conveyed in the report that has been left to us. Thanks to its detail and clarity, it allows us to draw a number of conclusions as to Telesio's character.

It actually paints a picture of a gentle man but one who was emphatic when it came to presenting his opinions. The 'Socratic nature', which Tasso claimed was seen in Telesio,[16] is evidenced in the conversation recorded by Persio and, above all, in the dialogue-like development of his philosophy. Telesio is in conversation with Persio and the latter takes on the role of the traditional Aristotelian philosopher. In these Socratic role-plays, Persio clearly assumes more than the role of pupil. He formulates variations and objections quite independently. But then a true Aristotelian by the name of Quintius comes along and the dispute takes on a sharper edge. This brings about a situation in which three Calabrians are having a discussion in Naples, three fellow countrymen displaying a shared solidarity, so as to remain friendly in their conversation despite different opinions, and in spite of an opponent spoiling for a disagreement. For Quintius is introduced as someone who *had committed himself very much to the Aristotelian sect* and was eristic.

15 Persio, *Liber novarum positionum*, 12v: "Sed iam tempus postulat, ut ad id, quod de Telesio nostro narrare occoeperam, revertamur. Is igitur cum secundam editionem adornaret, egoque interea temporis essem Neapolis, mirum quam illum erga omnes humanitate insigni praeditum, et munitum expertus sim; neque enim erubuit, quae viri comitas, adque veritatem potius, quam ad laudem inveniendam, propensio est, quotidiano sermone meum de suis scriptis iudicium postulare; imo de cuiusvis, etiam [...] sententiam sibi reliquam facere. Quid! etiam nunc [...] etiam nunc equidem, cum tertia, eamque postremam cogitat editionem, novis de animalium generatione libris acutam, nullum neque impensae locum dat, quo doctorum de suis scriptis calculi compos fiat. Singularem vero eius animi benignitatem quo scelere praetermittam qui tam largus mihi fuit suorum dogmatum, consiliorumque, ut aliquid, quodipse sciret, quod me fugeret, latere me velle, nefas putaret; corrigique a me, senex a iuvene, si quid erraret, discereque in illa aetate preclarum existimaret." The fact that the third edition of *De rerum natura* had nonetheless appeared sixteen years after the second had, according to research carried out by Bartelli, probably something to do with Telesio's financial difficulties.

16 Tasso, *Opere ... colle controversie sulla Gerusalemme*, 264: "Fu Bernardino Telesio uomo di acuto ingegno, di profonda dottrina e di socratici costumi [...]."

Persio begins his narrative with the information—not unimportant for the latter part of the conversation—that the incident occurred on a hot summer's day when the sun was blazing down from the sky. He paints a lively picture of the setting in which the conversation took place: the way Telesio calmly gets to his feet, the way the two men greet one another, the way a chair is pulled up for the guest and people sit down again. Then comes the conversation in which Quintius follows his combative nature: Persio had barely explained Telesio's argument about the *candiditas* of the sun when the opposing argument is set out. Quintius says that the sun has no inherent whiteness because heavenly bodies have qualities over and above the four sublunar elements and no direct connection to them. He cites the Aristotelian theory that the sun is neither hot nor white nor warming by means of its light but simply creates warmth in the sublunar world through friction generated by its movement.[17] Behind this is, for the Aristotelian world view, a decisive separation of the transient world of the elements and the unvarying world of the heavenly bodies, a separation of symbolic dimensions still to be discussed. Telesio responds to this expected objection in a gentle but unyielding manner (*amoeno ingenio et alterandi inimicus*); he advises his Aristotelian opponent, like Aristotle himself, to pay attention to the effect (*operatio*) and above all to the perception of the senses (*sensus*) instead of denying these theories at the back of his mind. Telesio reinforces his argument, just as Quintius persists with his own viewpoint, going so far as the provocative statement that his opponent must be blind, even dead, if he persists in this manner with his sensuous impressions. He goes on to say that people must feel (*sentire, sense*) the warmth and recognize the brilliant white of the sun when it is high in the sky. Quintius, equally provocatively, remains calm, while flatly denying having this sensuous impression of whiteness. It is at this point in the conversation that something very interesting happens. Telesio stands up, takes Quintius by the hand and leads him outside. In doing this, he is leaving the theoretical discussion behind and literally reaching for real evidence of the immediate impression, in the way people have been wont to do since long time in the face of skepticism of direct recognition by the senses. Yet the way the conversation continues shows very well that consensus is still a long way off. Quintius now abandons his position of the simple naysayer and offers his interpretation of the perception of the senses:

17 Cf. Aristotle, *De coelo* 289 a 20ff and Aristotle, *Meteorologica*, 340 to 10. Aristotle develops his theory of the sky in *De coelo* in constant disagreement with the pre-Socratics, especially the Pythagoreans. For a harsh critique of the distortions of this latter tradition cf. Kingsley, *Ancient Philosophy, Mystery, and Magic*. On Aristotle's *De coelo* cf. in general Leggatt, *Aristotle: On the Heavens I–II* (1995).

that the sun is not white but gleaming. This theorem from the Aristotelian tradition[18] actually makes it possible to explain away the appearance to the naked eye while choosing a different word. According to the Aristotelian theory color is created by the darkening of light on a body which is not transparent. If this is not the case, then one cannot refer to it as color but only as something shiny, brilliant, or gleaming.[19] This is why the sun cannot have its own actual color, and if we nonetheless have an impression of color then it is, or must be, due to this darkening of the sun because of something not completely transparent, or it can only be called a gleam—or both. This made Quintius' claim that they were only seeing cloud instead of the actual sky comprehensible.

Once again the conversation takes a surprising turn. Persio intervenes in the dispute in order to come to his tutor's aid, perhaps to relax the tension which has arisen between the two colleagues and their undecided, apparently undecidable, exchanges of claim and counterclaim. Before Telesio can respond to Quintius' attacking argument about the 'gleaming' of the sun, Persio recites a passage from Ovid's *Metamorphoses*: "While the remaining animals bowed down and looked at the earth / He gave man an upturned countenance and the sky to look at / He ordered him to stand tall and raise his eyes to the stars."[20] This humorous quotation, which provides a temporary solution to the controversy, points not only to the Socratic-friendly context and to a possible quandary for the Telesians but also gives an indication of the constant use, in this non-academic setting, of quotations and aphorisms from the classical poets.

Of course this aphorism is not viewed as a formal argument—otherwise people would hardly have laughed—but it would surely also be wrong to reduce it to mere ornamentation. This is because the centrality of the symbolic dimension, which is always present when the accessibility or the inaccessibility

---

18 For terminology cf. for example Simplicius, *Commentarius in libro de caelo Aristotelis*, 15r: "lumen coeli & fulgidum". It is not entirely clear to me where Quintius got his argument about the 'gleaming' of the sun. 'Gleam' is mostly used to describe the stars, or clouds when the sun shines through so that its rays are visible (cf. Plutarch, *Moralia*, III), or views of nature which so delight philosophers. It is possible that Quintius or one of his predecessors developed an argument as to why sunlight cannot be directly seen. Then, exactly as Quintius indicates, even on a clear day there is always a thin veil of cloud between the sun and the earth and this darkens the sun's light and makes it 'gleam' but does not allow for pure whiteness.

19 Aristotle, *De coloribus* 793 to 14ff.

20 Ovid, *Metamorphoses* I, 84–86. The context is that of teachings on how the world came into being and starts with the creation of man; cf. 76–83: "Sanctius his animal mentisque capacius altae / Deerat adhuc, et quod dominari in cetera posset. / Natus homo est, sive hunc divino semine fecit / Ille opifex rerum, mundi melioris origo, / Sive recens tellus seductaque nuper ab alto / Aethere cognati retinebat semina coeli. / Quam satus Iapeto, mixtam fluvialibus undis, / Finxit in effigiem moderantum cuncta deorum."

of the sky or heaven is discussed, is clearly visible in the quotation from Ovid. Man's upright posture, his dignity—*dignitas hominis*—is expressed by the fact that man *looks up to the sky* and is able to do so. First-hand knowledge of the sky is not, as the Aristotelians believe, obscured by the theoretically required 'clouds', but rather, as the wisdom of ancient poets has far more genuinely acknowledged, is entirely possible to obtain and not unreasonable.[21]

Thus the presence of poetic texts, above all Latin verse—Persio often refers to Lucretius and Virgil in *De natura ignis*—was certainly not always simply for decorative purposes. The texts, because of their age and the eminence of the writers, had the authority of classical poetic wisdom. For example, Persio mentions Ennius in *De natura ignis* as the *poeta antiquissimus* in order to shore up the argument that light shines white, using his words *Prodibant Famul, cum candida lumina lucent*.[22]

We know the role played by Ovid's description of the beginning of the world in Agostino Steuco's *Cosmopoeia* in 1535.[23] Even if Telesio's sensualistic natural science and the learned speculations of the Roman librarian seem to be worlds apart—he would hardly have laughed about the Ovid quotation—one can recall that Telesio, who was in contact with Giovanni della Casa,[24] was familiar with the intellectual world of the courts and their preference for poetic and allegoric treatments of scientific questions. Persio's referencing practices are reminiscent of Steuco, who considered the oldest sources of wisdom as an expression of the *perennis philosophia*, the philosophy which was and is valid for all time. In fact, he views an author like Ennius as original and authoritative because his language was not distorted.

The fact that the conversation between Telesio and his Aristotelian counterpart is about the sun is by no means coincidental. During the Renaissance the sun was bestowed with symbolism in many different ways, so much so that it took on semi-religious identification with the principle of unity.[25] In 1550, Innocenzio Ringhieri composed, in the humanist, courtly spirit, a

21 This also had religious implications, namely the accessibility of the heavenly for the naturally religious side of all human beings. In this sense, Vincenzo Cartari writes in *Immagini delli dei de gl'antichi* [1556] (1647), 1: "L'homo alza gli occhi al Cielo, e spesso anco le mani insieme giunte, quasi che naturalmente senta, che di là su viene ogni bene." Cf. Mulsow, "The *Historia* of Religions in the Seventeenth Century."

22 Persio, *De natura ignis*, II, f. 350*v*. Cf. Quintus Ennius, *Poesis reliquiae*.

23 Steuco, *Cosmopoeia*; cf. Williams, *The Common Expositor*.

24 Cf. Imperiale, *Musaeum historicum et physicum*, 78; also see, Lupi: "Il sublime secondo Telesio," especially 53f. Cf. Márquez Villanueva, "Bernardino Telesio y el 'antiguo sacerdote' de *La Galatea*."

25 For the significance of the sun in the Renaissance cf. Fédération Internationale des Sociétés et Instituts pour l'Étude de la Renaissance, *Le Soleil à la Renaissance*, Pissavino, "L'altro sole," especially 53f.

poetic dialogue between *Otio* and *Diligenza* about the sun.[26] Here, and subsequently in Francesco de' Vieri, the eulogy to the sun is shaped mostly by neo-Platonism.[27] For a natural scientist like Telesio, who made efforts to break away from the Aristotelian 'normal science' and to strive for new concepts more fitting for nature, alternative traditions like the neo-Platonic must have been welcome sources of inspiration.

And yet there are clear differences from courtly culture when it comes to dealing with the sun. In Ringhieri there is no real confrontation between differing scientific positions. By contrast, the conversation referred to by Persio very clearly documents the *direct* confrontation between the new philosophy of nature and Aristotelianism. That Persio later regretted his suspension of the controversy using the Ovid quotation and the ensuing merriment is indicated in the continuation of the chapter in *De natura ignis* where he had recorded his recollection. "In fact," he said twenty years after the conversation, "Now I shall [...] show evidence taken from Aristotle and other Peripatetics of high repute that the sun and the fire are white, and that light is whiteness itself."[28] It is understandable that Persio still considers the requirement for evidence

26 Ringhieri, *Il Sole*, no pagination, G1v: "[Otio:] [...] per tanto se non ti fosse moesto di sapere come siano i corpi celesti di foco disiarei, & come colà sù questo corruttibile, & materiale elemento possa trovar si, che cosi poco distanzi essere il Sole, della sua sustanza favellandomi dicevi: & se da lui questo soavissimo caldo, che io nel suo lume sento, o pur altronde, com'egli è fecondissimo nasce. [Diligenza:] [...] Poi quando il foco delle sphere celesti considerasi, che egli sia quasi un quinto elemento piu nobile de gli altri quattro, per natura purissimo, considerar si deve, il quale come l'infernale non affligga, o tormenti: come questo nostro materiale non arda: come l'elementale scaldando non corrompa, o distrugga: come l'intelligibile de i Seraphim, & l'individuo, & trino d'iddio non ami: ma ben che lucido, liquido, labile, lubrico, agile, polito, piacevole, uguale, quieto, ogni cosa illumini, & con la virtu vivifica dolcemente scaldi: & per questo dal nostro, & da gli altri sia diversissimo assai. La onde il Cielo è caldo, ma non abruscia, et il lume del Sole con la sola repercussione intensissimamente abbrusciare si vede, et se i raggi della Luna ripercossi nulla scaldano, è perche l'humida Natura signoreggia in lei: scaldano adunque per la Natura ignea, et vivifica loro, i rai salutiferamente del Sole. Ben che quei duo alteri mostri di Natura, Aristotele, & Alberto dicano: che la calidita, e il lume, convenienti cagione non le appareno, ond'alcuno alle stelle, & alle Cielo, il foco sia: anzi per certissimo tengono, che non per loro naturale caldezza, ma per virtu del moto, che il cielo per quelle coseistesse, a quelle istesse, circa quelle istesse, in quelle istesse, da principio sanza principio, a fine sanza fine, ordinatamente avvolgendo, il tutto a scaldare, & conservare sia nato, il caldo nel mondo celeste, & in questo elementare si generi, & si conservi."

27 On the role of the sun in neo-Platonism cf. above all Macrobius' praise of the sun; cf. Liebeschütz, "The Significance of the Speech of Praetextatus," Fauth, *Helios Megistos*.

28 Persio, *De natura ignis*, vol. II, 351v: "Verum ego nunc ut gratiam aliqua in causa a manibus Telesii nolente aut seclamante Quintio demonstrabo ex Aristoteles et aliis clarissimis Peripateticis solem ac ignem esse albos et lucem esse ipsam albedinem."

with regard to Quintius unfulfilled. It was only with the edition of *De coloribus*, Telesio's short text which dealt specifically with color and which Persio published in 1590, that a follower of Telesio could find an answer to the problem of colorfulness which was actually an alternative to Aristotelianism. But before I go into the factual content of this answer there is still one point to clear up, the point which really makes it possible to restore a historical profile to the confrontation.

It is essential to identify Telesio's partner in the dialogue. Who is this Quintius, this fellow from the 'eristic' Aristotelians? He is Quinzio Buongiovanni (Bongiovanni, Bonjohannes, Bonioannes, Bonianus), professor at the University of Naples, born in Tropea in Calabria and, as such, a fellow countryman of Telesio. Buongiovanni was first a teacher of philosophy and later, during the course of his university career, of medical theory and practice.[29] In 1571, probably only a short time after the documented conversation, his *Peripateticarum disputationum de principiis naturae sectiones tres*[30] appeared. It is a book which clearly represents the standpoint of traditional Aristotelianism, obviously in acquaintance with Telesio's *De rerum natura* of 1565 and 1570. This text is worth closer scrutiny because it could bear some interesting traces of the disputation with Telesio.

Yet the expectation is not fulfilled; at least, not at first sight. Buongiovanni disagrees with Simone Porzio instead of with Telesio. In Naples in 1571, Porzio was clearly seen in academic circles as the most dangerous of the new thinkers. Porzio had a significant influence on the intellectual climate in Naples, and not only during his time there teaching which came to an end as early as 1547 with his departure for Pisa, but also in the years which followed. Just as Averroistic Aristotelianism had been firmly established in Pisa as a result of Agostino Nifo's professorship, it was through Porzio that a rival view grew in strength, a view which went back to the original Greek texts and in this way taught a very naturalistic—and undogmatic—Aristotle.[31] Aristotelian

29 Cf. Minieri Riccio, *Notizie biografiche e bibliografiche*, vol. II, 188; after Kristeller, *Iter Italicum* II, a number of handwritten medical descriptions by Buongiovanni are in MS 3 Oq E82, 201–206 of the Biblioteca communale in Palermo.

30 Buongiovanni, *Peripateticarum disputationum*; the exemplar which I myself used: Biblioteca Corsiniana, Roma, Misc. in folio 205. As early as 1567 a piece of his writing appeared in Naples: *Quaestio de Divina Providentia iuxta Aristotelis mentem examinata publice in gymnasio Neapolitano* (Naples: Apud Matthiam Cancrum, 1567). Buongiovanni was elevated in 1587 to 'sostituto alla catedra della teorica della medicina und zum protomedico del Regno'; he died in 1612 in Naples. On philosophy in general see Cortese, "Età spagnola."

31 Cf. Cortese, "Età spagnola", 419f. On Porzio's considerable impact see for example Tasso's dialogue, *Il Porzio ovvero della virtù'* or Tansillo, *Capitoli giocosi e satirici*, 111: "il miglior di

basic concepts were interpreted anew and at the same time made flexible and accommodating again. The 1578 edition of Porzio's *Opuscula*, published by Giacomo Antonio Marta, shows to what extent the questions raised by Porzio were still present in the Naples of the 1570s. This edition was certainly made in order to bring underlying discussions to light and to decisively counter Porzio's attack on the immortality of the soul in his *De humana mente* with Marta's own *Apologia pro animae immortalitate*.[32] It is the same Marta who, in his campaign against Neapolitan naturalism, wrote in opposition to Della Porta and Telesio.[33] In any case, Buongiovanni had composed a treatment of *de immortalitate animae* well before Marta; it was also a disputation with Porzio and with Pomponazzi.[34] The long delayed publication of Pomponazzi's *De Fato* gave rise to further intensification of the dispute in 1567 even though the work had actually come into being around 1520. Buongiovanni reacted in the same year with a *De divina providentia* in which he defended the Christian idea of providence against Pomponazzi's stoic determinism.[35] In the second part of his book *De rerum naturalium principiis* (1553),[36] Simone Porzio had already emphasized the Stoic view of fate, and in

questa etate" and 156: "il maggior uomo che oggi si vegga". Also see Facca, "Humana mens corruptibilis," Keßler, "Von der Psychologie zur Methodenlehre," Lohr, *Renaissance Latin Aristotle Commentators*; Vasoli, "Tra Aristotele, Alessandro di Afrodisia e Juan de Valdés," Castelli, "Tra ricerca empirica e osservazione scientifica," id. "Il *De' sensi* e il *Del sentire* di Simone Porzio," id., "Tra aristotelismo, naturalismo e critica," id. "Il *De conflagratione* di Simone Porzio."

32 Marta, *Opuscula eccelentiss. Simonis Portii Neapol.*; the copy in the Bibl. Corsiniana, Rome, is tied up with Buongiovannis *Peripateticarum disputationum*; one can also look at the document, contemporary in time and location to that of Buongiovanni, which appeared in 1571 in Venice, *Peripateticarum quaestionum libri quinque* (Venice: Apud Iuntas, 1571) by Porzio's pupil Andrea Cesalpino, in order to see how open to a new empirical Naturalism the Aristotelianism had become in Pomponazzi's succession.

33 Cf. Buongiovanni, *De divina providentia*, 15v: "in nostra quaestione de Anima Immortalitate"; the text, which was probably never printed, seems to be lost; it is not referred to in Di Napoli, *L'immortalità dell'anima nel Rinascimento*.

34 Marta, *Pugnaculum Aristotelis adversus principia Bernardini Telesii*.

35 Cf. Pomponazzi, *Opera*. Cf. the edition of *De fato* by Richard Lemay (1957), as well as the Italian translation by Compagni (2004). Buongiovanni starts the disagreement as follows on 10r: "Petrus Pomponatius huius opinionis de Fato accerrimus defensor, contendit si fatum ponatur nec humanam vitam auferri, ut putabat Alex. nec eius rationis cogunt tum etiam audet dicere quod philosophi rationes in hoc sunt frivolae, patet ex earum clarissima solutione, cum enim Arist. ait quod si Fatum daretur, tunc propositio de futuro contingenti esset terminatae veritatis, inquit Petrus (…)." Cf. Scribano, "Il problema de libero arbitrio nel *De fato*."

36 Porzio, *De rerum naturalium principiis*; cf. also his MS. Liber De Fato (Milan: Biblioteca Ambrosiana), 197. sup. (XVI), IV, ref. 75r–94r, as well as *An homo bonus vel malus volens fiat* [1551] (2005).

Naples this text was now read in the light of his tutor's *De Fato*, published posthumously in Protestant Basel. Although Bernardino Longo—Buongiovanni's teacher and an Averroist—succeeded Porzio,[37] salt was rubbed into the old wound during 1567 and the threat posed by the undermining of religious providence was more acute than ever.

With knowledge of this pre-history the reader of Buongiovanni's *De principiis naturae* has no further difficulty recognising the constant—but never explicit—relationship to Porzio's *De rerum naturalium principiis*. The questions raised by Porzio's idiosyncratic interpretation of Aristotle with reference to the concepts of matter and privation are all dealt with. From the perspective of this opposition, Averroist-oriented philosophers such as Buongiovanni or Longo should be seen more as the protectors of orthodoxy, while the thinkers schooled by the Stoics and Alexander of Aphrodisias were potential heretics.

With his seemingly eccentric arguments, Telesio was less suited to the established mold of university disputation. It was extremely difficult for anyone to pigeonhole him in any known direction of thinking. As the conversation shows, Telesio's deviations from customary thinking were so radical that any commensurability in the disputation was barely possible (a problem which I shall examine further). In any case the questions concerning Telesio in Buongiovanni's treatise are at most discussed in a special part of the text: the problem of the matter of the sky.[38] This is precisely the point in the horizon that Buongiovanni had in mind when he joined the conversation between Telesio and Persio and he sensed the issues under discussion. His attempt to categorize Telesio's views revolves around the problem that the sky, if accepted as fiery, must therefore be transient.[39] "It's clear to me," he says in his book, certainly against Telesio as well as Porzio, "that the path which the grammarian [i.e. Johannes Philoponus] tries to suggest with reference to the first book of the *Meteorology* and the first book of *De coelo*

37 Cf. the letter of dedication in Longo in *De divina providentia*, 2r: "Praeceptor suavissime;" on Longo cf. Lohr, *Renaissance Latin Aristotle Commentaries II*: "Longus, Johannes Bernardinus;" Cortese, "Età spagnola," 327 and passim. For his summary of Aristotelian Physics cf. *Lectiones in VIII Physicorum*, Ms. Vaticana, Rome, Reg. lat. 1968 (XVI), item 1.

38 Cf. the MS of the reading by Porzio *Quaestio de materia caeli*, Bibl. Ambrosiana, Milan, MS. p. 197. sup. IV, 33r–40r. In this question, too, Buongiovanni seems primarily to disagree with Porzio.

39 Buongiovanni's tutor Giovanni Bernardino Longo had already looked at questions to do with the sky. Whether, for example, Longo's *De cometis disputatio* of 1578 is related to a latent discussion connecting to the coming into being of earlier versions of Telesio's *De cometis et lacteo circulo* (1590) is not to be decided here. The only thing to be determined is that such relationships are now no longer to be excluded as improbable.

(according to Simplicius's interpretation), is no real possibility. According to this path, the matter in the heavens is of the same sort as ours [on earth] [...] and it is fire because it heats."[40] Philoponus, we learn, has been identified as the one who originally represented an argument of this type.

A flammable and thus transient sky was perceived as undermining the very foundations of the Christian faith. It is, in fact, known that early critics of Telesio, including those in church circles, saw above all an assault on the accepted picture of the sky, or heaven, alongside the supposed materialism concerning the soul. Telesio, they lamented, does not recognize any *intelligentiae* in the sky, spiritual unities that direct the celestial spheres.[41]

Buongiovanni postponed his own contention with these problems until another publication. Several times in the text he indicates a commentary on Aristotle's *De coelo* where he would go into the questions further.[42] This document has, however, vanished, if indeed it ever existed. The text could far better have explained the ideas behind Buongiovanni's heated exchanges with Telesio than the few passages in *De principiis naturae* do.

40 Buongiovanni: *Peripateticarum discussionum*, 41c/d: "nec tamen me latet aliam esse viam, qua Grammaticus tum 1. Meteor. tum primo coeli referente Simplicio, conatur suadere, in coelo, esse materiam eiusdem rationis, cum hac nostra, & sua natura, defecturum corpus coeleste, s. ipsum esse ignem, quoniam calefacit." Here is the entire related text: "Ad quartum, distinguit Simplicius, propositionem omne id quod aliquo indiget, ut sit, corruptibile est, posse bifariam intellegi. illud enim quo res aliqua indiget, vel non est de essentia illius rei, & tunc res illa sua natura, defectura est, quo pacto non sic coelum materia indiget, vel illud est de essentia rei, quo pacto res sua natura, non est defectura, quemadmodum sunt corpora coelestia, quibus indigent necessaria, & de essentia sunt illorum. postremum diluit Simplicius, atque totum corpus coeleste, esse virtutis finitae in vigore. infinitae autem in duratione. similiter & partes eius infinitas in duratione atque finitas in vigore asserit, nec tamen me latet aliam esse viam, qua Grammaticus tum 1. Meteor. tum primo coeli referente Simplicio, conatur suadere, in coelo, esse materiam eiusdem rationis, cum hac nostra, & sua natura, defecturum corpus coeleste, s. ipsum esse ignem, quoniam calefacit, non contentus Arist. sententia, quintam naturam ab his, quae sunt hic, longe remotam asserentis, quam postea sententiam ex Neotericis multi, mordicus defendunt. quorum quidem rationes, & sententias num naturae rei consentiant simul Arist. ad illud astruentium rationum solutiones examinare, ac perpendere diligentissime, in comm. de coelo huic rei commodissimo loco, differe operae pretium duxi. Nunc vero satis sit vidisse corporum in gyrum latorum materiam, eandem cum hac nostra caduca non esse, nisi identitate ab uno, & ad unum, quam analogicam nuncupamus, in qua plus aequo forsitan prolixiores fuimus, Grammatico, Egidio, & Avicenna, qui aperta mendacia in Arist. doctrina afferebant, vitium dabunt. Interea autem nostrum institutum prosequamur."

41 Cf. Marta, *Pugnaculum*.

42 Buongivanni, *Peripateticarum disputationum*, 35D: "quae omnia ego alias in Comm. de Coelo, simul expandam"; 40G: "demonstrabimus siquidem in Comm. de Coelo".

3 Problems with Light and Heat

At first sight the course of the conversation between Telesio and the Aristotelian Buongiovanni seems to fully confirm the familiar picture of the replacement of sterile Aristotelianism with the new, empirical science of nature. Of course, Telesio's act of escorting the Aristotelian outdoors to experience sensuous perception firsthand could immediately be seen as a sign of this new scientific spirit, just as the arch-Aristotelian Cremonini's refusal only a few decades later to look through Galileo's telescope became the symbol and icon of the intractable backwardness of the Aristotelian university world.[43] However, just as the Cremonini episode has been put in perspective in the meantime, in Telesio's case there are a few elements which do not fit the overall picture. For one thing, he quotes Aristotle when he demands for the methodology of empiricism. For another, Telesio's pupil, Persio, irritatingly describes himself as Aristotelian: *nos Peripatetici*. Thirdly, in the recorded conversation, Telesio does not manage to persuade the Aristotelian. What can be concluded from this?

The conclusion should be, more than anything, that the problem of 'anti-Aristotelianism' in the Late Renaissance is far more complex than people have often believed it to be. The allegation of empirical hostility towards Aristotelians is now seen within a greater perspective. We know, at least since Charles Schmitt, how vibrant the Aristotelian School was until around 1700, how it repeatedly yielded new resources of a 'modern' Aristotelianism in politics, economics, poetics, rhetoric, as well as natural science, and how close to empiricism such a new Aristotelianism could be. And today, after the merits of the 'first empiricists' have long been emphasized, one can be better persuaded of the *longue durée* of Aristotelianism. This also concernes the way the new philosophers of nature broke away from the 'old': the continuity and similarities are too great to be denied. In this way it can be seen and appreciated that Telesio—with his formula of 'Aristoteles oblitus sui', of an Aristotelianism which has forgotten its own methodological bases—could have appealed to Aristotle's original experience-based orientation.[44] Moreover it has been underlined that Telesio, in spite of his self-proclaimed openness to matters of the senses, has also been caught up in theoretical speculation more often than he admitted. On closer observation he feeds on speculative ideas from the

43 On Cremonini cf. Kuhn, *Venetischer Aristotelismus*, Riondato and Poppi, *Cesare Cremonini*. For the telescope incident cf. Drake, *Galileo at Work*, 162f.

44 Cf. Lerner, "Aristote 'oblieux de lui même' selon B. Telesio."

Averroists, the Avicennists, as well as from the aristotelian *Calculatores* of the fourteenth and fifteenth centuries.[45]

The situation in which Telesio and Quinzio could not mutually persuade one another also provides food for thought. This actually indicates that a pure appeal to perception alone is still not a sufficient criterion for deciding, since pure perception does not exist. That Quinzio, according to his own evidence, *does not see* the whiteness of the sun is a good example of the "theory-ladenness of observation";[46] in fact it precisely illustrates the situation in which people perceive only what fits the context of their assumptions at the time. This insight drawn from the philosophy of science exhorts us to maintain a distance from the familiar post-Bacon celebration of the new empirical spirit.[47]

Finally, there remains the irritating fact that Telesio's own pupil had categorized himself as an Aristotelian. Indeed, Buongiovanni is not the only Aristotelian in the conversation. In Persio's recording Persio is firmly included in the *nos Peripatetici* register. Persio explains this by saying that, in the discussions, he had mostly taken the traditional Aristotelian part. He was, as it were, the Aristotelian sparring partner in Telesio's Socratic role-plays. Later, in 1593, he said in a basic position statement that he would not want to be described as either a Telesian or an Aristotelian because his understanding of *libertas philosophandi* was counter to committing himself to an orientation of this kind.[48] Persio was never far away from Aristotelian debates and probably had more detailed knowledge of them than Telesio himself. In Persio's *Liber novarum positionum* the section entitled *Positiones pro Aristotele*[49] shows how nuanced Persio's view of the topic of "Anti-Aristotelianism" was, and in his great work *De natura ignis* he made up for what Telesio's 'mulishness' had not achieved, namely a cautious disputation with contemporary Aristotelianism in order to move it from detailed reinterpretation to Telesian persuasions, and thus to a work of transformation smarter than the insistence on both sides of apparently obvious positions by Quinzio and Telesio.

45 Cf. Mulsow, *Frühneuzeitliche Selbsterhaltung*.

46 For the concept cf. Russell Hanson, *Patterns of Discovery*.

47 Cf. for example Cleveland Van Deusen, "Telesio: The First of the Moderns" (1932).

48 Persio, *Draft letter to Andrea Chiocco*, Biblioteca Corsiniana, Rome, MS. Linceo I, 145r: "[…] exploris impietatibus multo in eo [Aristotele] probavi, ita in Platonem ita in Galeno […]. […]disputationes pro Telesio proposui et etiam pro Aristotele, sed sum non Aristotelicus, quia contra ipsum multa, ita nec Telesianus quia pro ipso pauca defendnda proposui, […]."

49 Persio, *Liber novarum positionum*, 176ff.

Incidentally, there was a reason that their conversation was about the sun, beyond the merely specific Renaissance interest in the sun's symbolism. The relationship between light and heat was one of the key problems of science at that time. It is certainly not as gripping and spectacular as the discussion about the Copernican Turn but it was hardly less decisive in its consequences for Aristotelian science. We have learnt from Buongiovanni's *De principiis naturae* that the problem of the flammable, and thus transient, nature of the sky was identified with the position of Philoponus. Perhaps including Telesio and Porzio among the followers of Johannes Philoponus, as Buongiovanni did, is not so wrong. Karl Schuhmann has indicated that Telesio's theory of space was influenced by Philoponus' commentary on physics.[50] Something similar can be proposed for the theory of light.

Now the case of Philoponus is certainly difficult as far as the profile of his philosophy is concerned.[51] The polemic raised against him, which Simplicius expresses in his commentary on *De coelo*, something which was available from 1535 and had considerable influence on Telesio, attacked Philoponus' later writings which came into being, or were redacted, after 529.[52] In these writings, Philoponus took up a position opposing neo-Platonism which, until that point, he had advocated himself; he developed a 'Christian' philosophy of nature which requires a divine creator who imbues their powers into the natural elements as well as a world which is not eternal but which did come into existence.[53] The thesis put forward by Philoponus, namely that the sky is of a fiery nature and that sunlight is warm, is obviously put forward as an interpretation of Plato's *Timaeus*, even if the fiery nature is thought of as a mixture of all elements.[54] In his commentary on *Meteorology*, which appeared before

50 Schuhmann, "Le concept de l'espace." There is also an essay on Telesio's theory of light written by De Franco, "La teoria della luce," but it does not really lead any further in my view.

51 Cf. Verrycken, "The Development of Philoponus' Thought." On Philoponus' theory of light cf. Sambursky, "Philoponus' Interpretation." On Philoponus' impact during the Renaissance cf. Schmitt, "Philoponus' Commentary on Aristotle's Physics."

52 Cf. the reconstruction of Philoponus' lost writings against Aristotle about the eternity of the world from the Simplicius passages in Wildberg, *John Philoponus' Criticism*.

53 It was already the case that Philoponus has God creating natural things, which then act and form the world, and this allows the start of a tradition which leads to Fracastoro and Telesio by way of Maimonides and Thomas Aquinas. Cf. in general Funkenstein, *Theology and the Scientific Imagination*.

54 Cf. Simplicius, *Commentarius in libro de caelo Aristotelis*, 11v ff. [on *De coelo* 270 to 3] the disputation with Themistius and then with Philoponus, esp. 14 recto ff. on the thesis of the flammability of the sky. See 14r: "Quoniam autem nescio qualiter dicta platonis placere huic videntur cum non habuerit doctores in ipsis, ut dicit [Grammaticus], neque ipse studiose inquisierit intellectum Platonis, et propter hoc putat quandoque suis phantasiis

the volte-face but was later revised, a warming quality is also attributed to the sun.[55] Philoponus develops a theory of light energies, *energeiai*, translated by Camutius as *operationes*. Referring to the passage of *Meteor.* 341 a 35 ('The way that the sun, considered above all other heavenly bodies for being warm, is in reality white and not fiery') he argues against the color criterion. Telesio, on the other hand, as we have seen, speaks in favor of whiteness and develops in preference a completely different theory of color from that of Aristotle.

Throughout there are great differences in detail, which still have to be further researched. This cannot be done here. However, the years after the conversation between Telesio and Buongiovanni provide two indications that our assumption that Philoponus and other late antique commentators played a role in the formation of Telesio's position cannot be entirely wrong. The first indication is the reaction in Venice to Telesio's philosophy after Persio had made it known in an open discussion there in 1575.[56] At that time, Nicolò Contarini and Alessandro Maranta saw in Telesio's theory, namely that the white of sunlight goes with extremely fine, hot matter, a new version of late antique

concordare dicta Platonis, quandoque autem dictis Aristotelis contradicere, videamus qualia etiam nunc de dictis Platonis proponit. Plato, ait, non ex igne solo coelestia corpora supposuit, sed plurimo maxime participare tali igne, qui et reliquorum elementorum mixturam melioris commixtionis facit. Omnis enim, ait, ab omnibus elementis subtilissima et purissima substantia, et obtinens rationem speciei ad reliquia, inconcretionem coelestium corporum segregata, a materialiori ipsorum et ut ita dicam rudiori parte substituta hic est astra quoque et solem ex tali igne Plato vult esse."

55 Philoponus, *In I. meteorum Aristotelis expositionum in tres libros* (1551) [printed together with Olympiodorus' commentary on *Meteorology*, 93v–139r]. Cf. 107r ff. [on Meteor. 341 to 35]; 112r f.: "Ex hoc igitur quoque loco satis planum fieri arbitror non esse Solis motum, qui aerem ipsum calefaciat, nempe qui ne ipsum quidem attingit aerem; sed huius esse qualitatem, quae huiusque omnis caloris effectrix existat, sicut ex igni manifesto apparet. Atque haec causa nempe est, quamobrem obductae obtectaeque partes aeris quod scilicet solis qualitas non transmittitur neque penetrat in ipsas ab obstruente corpore reiecta, non pari modo sicut aliae liberae partes aeris non incalescunt sed modicum tantum a propinquis partibus transmissum calorem recipiunt. Atque ita iis rebus tantum calor attribuitur, quae absque medi ullo libero campo solis radiorum splendorem ad se recipiunt. Ex quibus rebus intelligi potest cum lumine calore natura copulaverit atque coniunxerit. Qualitate ergo sol coelum totum calore accendit et non proprio suo motu, si quem peculiariter suum habet, neque motu, quo cum solari orbe circumfertur." The theory of a force of incalescence was later adopted—referring to Averroes' saying "Coelum est calefaciens per se motu et lumine" – mainly by those Averroists who adopted the Arabic optical theories as well. In this manner one could concede that at least in this way sunlight is warming. I have followed this path in my Telesio book and have seen in it a point on which Telesio did further work and whose inconsequence he tried to overcome by means of a more radical solution. Cf. Mulsow, *Frühneuzeitliche Selbsterhaltung*, 124–128.

56 Cf. for detail Mulsow, *Reaktionärer Hermetismus vor 1600?* esp. 167f.

neo-Platonic-Stoic theorems in Iamblichus and Proclus. Even if Philoponus had polemicized against Proclus' version of the eternal nature of the world, he is, as we have seen, not so far removed from the late neo-Platonists on the theory of light.[57] In Proclus, light is the purest form of fire and thus something physical which spreads and yet can be penetrated by something else.[58] With the infiltration of light into the earth it is weakened and its progress forward is slowed down. This thesis, too, could be taken from Simplicius' commentary on *De coelo*. It seems to have influenced Telesio's concept of a "species" of heat that is called light (*lux*), if it is visible and strong and can infiltrate matter, or is called whiteness (*albedo*), if it is invisible, weak, and is repelled by matter.[59] We shall see how this dual concept is related to Telesio's conversation with Buongiovanni.

The second indication is the reaction of Giacomo Zabarella to the theory of the hot or flammable nature of the sun and the sky.[60] In 1590, Zabarella, in his treatise *De calore coelesti*, polemicized against those philosophers who held the sky to be flammable and who "ascribed too much to the senses" (*qui sensum nimius tribuant*).[61] This characterization suited Telesio well—we have seen in the conversation with Buongiovanni how robustly Telesio persisted with the perception of the senses. Zabarella does not name Telesio anywhere. However, the fact that Persio dedicated a whole chapter in *De natura ignis* to the repudiation of Zabarella's argument allows us to conclude that he knew that Zabarella actually wrote in opposition to Telesio. Persio was at all times well informed by his brother Ascanio and his friend Federico Pendasio about the situation surrounding Zabarella.[62] In *De calore coelesti*, Zabarella

57 Cf. Philoponus, *Against Proclus*.

58 Cf. Sorabji, "Neoplatonists and Christians."

59 Cf. Telesio, *De coloribus*, in Telesio, *Varii de naturalibus rebus libelli*, 329; Telesio, *De rerum natura*, I, 44; cf. also Persio, *De natura ignis*, f. 306v–307v.

60 On Zabarella's Philosophy of Nature cf. Schmitt, "Experience and Experiment," 80–138. Cf. also Antonino Poppi, *La dottrina della scienza*.

61 Zabarella, *De calore coelesti*, in *De rebus naturalibus libri XXX*, 556 f. Cf. Chiocco, "Quomodo a solis corpore in his inferibus calorem gigni censuerit Aristoteles," in *Quaestionum philosophicarum et medicarum*, Quaestio XIIII; Piccolomini, *In libros Aristotelis de Coelo lucidissima expositio*; Cesalpino, *Peripateticarum quaestionum*, lib. III, quaest. 5–8.

62 Cf. also statements on the presence of Telesio's followers in Padua by Savile, cited in De Franco, *Bernardino Telesio*, 135f. A passage from *De Natura ignis*, II, 23*v* shows that Persio himself knew Zabarella from his time in Padua and Venice during the 1570s: "Inter Peripateticos vero quos viderim duos potissimum diligentiores fuisse in indignanda putredinis essentia; praesertim in Patavina Academia mihi visi sunt Archangelus Mercenarius et clarissimus Jacobo Zabarella." Persio's *Disputationes libri novarum positionum* (1576), 16 and passim show that Persio was carrying on debates with Mercenario at least and had done so since 1575. From 1568, Mercenario was professor philosophiae extraordinariae

officially addresses only the *prisci*, i.e. pre-Socratics such as Heraclitus and also, to a certain extent, Parmenides. But that does not conflict with a reference to the southern Italian innovator because Telesianism was soon recognized as the restoration of the early southern Italian tradition and of Parmenides in particular—first by the critics[63] but then increasingly by the propagators of Telesio himself.[64]

In any case, Persio reacted immediately to Zabarella's arguments and wrote a refutation, probably straight after their publication.[65] Zabarella included Johannes Philoponus as well as the *prisci* in the position under attack.[66] In this way we see the relationship between Telesio and Philoponus anew, however complex it may be to appreciate in its detail. Persio replied, without actually referring to Zabarella by name, in a special chapter of *De natura ignis*.[67]

Whether Telesio himself reacted to Buongiovanni's objections more elaborately than he did in the conversation is questionable. If he had, Persio would hardly have felt the necessity to respond in such detail in *De natura ignis*. After all it is noticeable that the later Telesio drew a distinction in the theory of the whiteness of light between *De coloribus* and the final edition of *De rerum natura*. He speaks now, as we have already seen, of *lux* on the one hand and

---

in primo loco; Zabarella was his rival in secundo loco; in 1577 he replaced Mercenario as associate Professor primo loco, while Mercenario rose to full professor in secundo loco; in 1585, after Mercenario's death, Zabarella succeeded him.

63 Cf. Artese: "Il rapporto Parmenide-Telesio."

64 In 1581 there was a turning point here in the form of Patrizi's rehabilitation of the pre-Socratics against Aristotle's criticism in his *Discussiones Peripateticae*. Cf. Mulsow, *Frühneuzeitliche Selbsterhaltung*, 319 ff; Muccillo, "La storia della filosofia presocratica."

65 This is spoken for by the fact that the chapter is added to the end of the manuscript (see footnote 67), as if it was added as an updated source after the other chapters had been completed. In general, I date *De natura ignis* as 1587–1590; cf. Mulsow, "Philosophia italica," 254f.

66 Zabarella, *De calore coelesti*, 557 B. Zabarella continues: "quare sive sublimior incedat, ut in aestate, sive humilior, ut in hyeme, eundem facere calorem debet: quoniam, ut in hoc igni nostro experimur, eadem eiusdem ignis distantia eundem calorem facit, resque ita est manifesta, ut omni dubio careat: quod si in radiorum repercussionem productionem caloris referamus, clara est huius temporum discriminis ratio, ut mox ostendemus."

67 Persio, *De natura ignis*, end. Because Persio's details are in the 11th book *de natura ignis et caloris*, it is possible to conclude that this chapter was envisaged for a 12th book. Zabarella deals with the question *Astra non ideo calefacere, quod ignea sint* as the first chapter but only to make clear that his own intention is the discussion of other problems. This debate is still to be reconstructed in detail and that is a matter for an essay of its own. Zabarella's first objection gives a rough validation to the idea that when something is warmed by sunlight it must be as warm on the earth in winter as in summer because in view of the curvature of the sky the sun must always be equally distant regardless of where it is. Persio had to deflate these and other objections.

albedo on the other.[68] He had understood that he could not speak, as was the case against Buongiovanni, of only *candor*, i.e. brilliant white. These semantic differences were always important. The conversation has given us insight into what a struggle these precise descriptions of sunlight induced. 'Whiteness', 'gleaming', 'brilliance', 'fieryness'—these are not all the same: rather each expression implies a different presumption in the philosophy of nature. The phrasing of a 'Science of Describing' used by Brian Ogilvie in connection with the Botany and Zoology of the Renaissance is, to a great extent, also valid for Telesio.[69] His works often go round and round in long, wearisome passages about the precise name to be used for a process: *ardentissimus, ornatissimus, candidissimus*—this is not merely enrichment through synonyms but an attempt at accurate description.

We can summarise our reconstruction of Telesio's conversation and its background by acknowledging two components. On the one hand Telesio had a genuine sensualism, with his steady practice and subtle attempts to find an accurate language of description, and on the other there was the demanding work on theory: the study of Simplicius, Philoponus, and many other theorists whose transformations of Aristotelianism Telesio followed and radicalized. The Christian theoreticians in late antiquity, such as Philoponus or Hiob of Edessa, were above all faced with the task of going beyond the Aristotelian and Galenist worldview without really being able to leave this conceptual framework.[70] That situation was similar to the one in which Telesio found himself in the late sixteenth century.[71] In more than one respect the movement

68 Telesio, *De coloribus*, p. 329, line 12 ff: "Itaque in longum sese effundit obviasque tenebras et species exuperat et inexistente oculis ipsum propria afficit specie, et per se omnino visilis est et lux dicitur. Haec vero quia summe exilis proptereaque et summe languida est multo in brevius sese effundit statimque vel ab obviis tenebris obscuratur vel in robustiore Solis albedine latet." Cf. Telesio, *De rerum natura*, vol. 1, 44, line 2 ff: "Patet itidem albedinem; nec eam modo, quae, quod sese amplificandi et quaque versus effundiendi potens est, et quae, quod sese assidue amplificat et quaque versus effundit, itaque animalium oculos subit iisque inexistenti spiritui se ipsam suasque affectiones manifestat omnes, per se visilis est et lux dicitur; sed quae veluti torpet et, siquidem sese et ipsa effundit, quoniam statim, prius omnino quam ad ocuos perveniat, ab obviis speciebus reicitur obscuraturque, invisilis per se est, et non lux sed albedo dicitur."

69 Ogilvie, *Science of Describing*.

70 Cf. Jacob of Edessa, *Book of Treasures*.

71 An indication of these connections is probably also Telesio's pointed—and in the 16th century less common—expression of warmth and cold as "naturae agentes". This talk of the primary qualities or even the elements as 'natures' is indeed also found in Cicero (*Tusculanae Disputationes* I, 22), but then above all in the Arabic tradition where it is very commonly used. From there, it may have been assigned to the Renaissance and to Telesio. For a position in the early Islamic world, which held both natures of warmth and cold

for liberation from the Aristotelian shackles repeated that of the sixth—ninth centuries which had, in part, been forgotten; but now, in the Renaissance, through the editions of Aristotle's commentators and also through the mediated reception of the Syrian and Arabic tradition, it regained its explosiveness.

Once the Aristotelian shackles had been broken, the way ahead was open for innovation. A little later, when in the context of the Genesis interpretation the problem of 'waters above the heavens' was seized upon by Patrizi and others, and then when it came to be seen under Telesio's new unified physics and his theory of 'liquid' heat, it suddenly became a burning issue once again. Now new ideas about celestial mechanics could emerge,[72] as could ideas like those of Galileo which transferred Archimedes' dynamic of the movement in liquids to the general dynamic of bodies. The theory of light and heat played a pivotal role in this transformation.[73]

However, at the same time these years were marked by rigid suppression of intellectual life by the Counter-Reformation. Had Persio still been composing *De natura ignis* at a time when many of the new philosophers of nature—Patrizi and Bruno come to mind—were hoping to be able to influence the views of the papal court, the climate would have deteriorated rapidly in the 1590s. Patrizi's *Nova de universis philosophia* was subjected to restrictions, Telesio's *De rerum natura* was placed on the Index in 1596, and in 1600 at the University of Padua there was an internal condemnation of Telesio's arguments. The prospects of a swift publication of *De natura ignis* became slimmer and after Persio's death in 1612 even his friends at the *Academia dei Lincei*, such as Federico Cesi and Galileo Galilei, could not manage to publish the huge manuscript.[74] This is why Persio's reply to Quinzio Buongiovanni did not go into print, and the transformation of Aristotelianism, which happened only implicitly with Telesio but was explicitly carried through by Persio, remained hidden from public view—until today.

Translated from the German by Deborah Langton

as measures for all the dynamic in the universe, cf. the opponent against whom Abu Ali Miskawayh writes: Arkoun, "Deux épîtres de Miskawayh."

72 Cf. Lerner, "Le problème de la matière céleste."

73 For the significance of the theory of heat for Galileo cf. Redondi, *Galileo eretico*; cf. further Mulsow, "Philosophia italica," 274ff.

74 See Gabrieli, "Notizia della vita" and Redondi, *Galileo eretico*.

CHAPTER 9

'*Haereticorum more leges refellendi suas proponit*'. At the Beginning of Telesian Censorship: an Annotated Copy of the 1565 Roman Edition

Alessandro Ottaviani

The library of the Accademia dei Lincei e Corsiniana in Rome has preserved an important copy of the 1565 Roman edition of Telesio's *De natura iuxta propria principia* at catalogue coordinates 31 A 9. Its importance lies in the critical annotations found throughout the entire volume which bear witness to the early reception of Telesio and the background of the later censure of his work by the Index. As far as can be ascertained from the stamp present on the frontispiece, this copy was part of the so-called *Bibliotheca Corsiniana vetus*, which is why, in all probability, it came to be a part of Cardinal Neri Corsini's collection between 1730 and 1750.[1] On the final page one finds an entry appended to the *verso*, which records the fact that Friar Angelo Baronio gave this copy to Medoro Patriarcha with permission to keep and read it. Baronio and Patriarcha were both part of the ecclesiastical and intellectual elite in the late sixteenth century. While our knowledge of their respective biographies is varied, it is at least enough to clarify the period in which this permit was granted. This allows us to complete the information contained in the entry, which is limited to the day and month (October 20), and also to put forward a hypothesis regarding the historical details of this particular copy, as well as the climate in which these annotations were made.[2]

1 I would like to thank Dr. Ebe Antetomaso for her help and expertise, and for providing data used to construct a possible scenario of what happened to this copy within the history of the Biblioteca Corsiniana. It also emerged from her analysis that the actual binding is definitely of a later date, and that the volume was trimmed. Traces of this trimming are found in the annotations themselves, where the writing appears to be slightly damaged and sometimes needs the addition of an initial or final character; an attempt was made to gather further data by consulting the documentation conserved in mss. *Cors*. 2628 and 2629, but the lists therein did not provide any information relating to whether it entered the library as a purchase or a donation. For the sake of comfort, the copy will, from here on, be referred to as *DNCors*, followed by an indication with reference to the numbering of the files being examined.

2 Here I transcribe in full the note on *DNCors*, Y4*v*: "Die 20 octobris/ conceditur facultas tenendi ac legendi hunc librum Ecc.mo D.no Medoro Patriarchae per triennium [per triennium *canc.*] conditionibus ut in indice Fr. Angelus Baronius Venetus Magister et Socius R.P.

 | DOI:10.1163/9789004352643_011

1 Angelo Baronio and Medoro Patriarcha

Angelo Baronio, born Francesco, the son of Tommaso Baronio and Faustina Molena, entered the Dominican Order, in the monastery of St. Dominic in the Castello quarter of Venice, on November 4, 1566; on November 5 of the following year, after changing his name to Angelo, he was performing the profession, overseen by Pietro Passamonti, the priest who had admitted him.[3] Then he went to the convent of St. Dominic in Bologna, where he rose to the position of professor of theology. We know for a fact that Baronio, under the office of Fra Pietro Martire Festa of Orzi, was nominated bachelor on May 5, 1600, an office which he soon left on September 8, 1600,[4] because he was called to Rome as *socius* to the Master of the Sacred Apostolic Palace, Giovanni Maria Guanzelli (in office from April 15, 1598, to June 25, 1607), performing well-documented activities as a revisor.[5]

However, his career as a revisor was cut short when Clement VIII made him a bishop in 1604 and entrusted him with the diocese of the city of Cattaro. In 1611, he moved on to the diocese of Chioggia, where he served until his death on September 12 of the same year.

Scarce information is available about Medoro Patriarcha, who applied for and received the loan. It is known that he originated from Grottammare, the town where Felice Peretti (later, Pope Sixtus V) was also born. Gaetano Marini credits him with being pontifical *archiatra* under Clement VIII and Paul V, as well as first physician in the years 1604 and 1616. He died on November 17, 1623, and is buried in Santa Maria in Aquiro.[6]

Many of his initiatives animated Roman medical society; the most well-known is his collaboration that appeared in Castore Durante's *Herbario Novo* (New Herbarium) in 1585 when Medoro Patriarcha must have still been relatively young.[7] His name also emerges during a discussion on the medicinal use

---

<...>"; special thanks go to Patrizia Landucci Ruffo, who made a valuable contribution to the deciphering of this section.

3 For biographical notes on Baronio, see Cornelio, *Ecclesiae Venetae antiquis monumentis*, 21 and 25; id., *Catharus Dalmatiae civitas*, 42; De Rubeis, *De rebus congregationis*, 232; Vianelli, *Nuova serie de' vescovi di Malamocco*, 239–242; Cicogna, *Delle iscrizioni veneziane*, 300.

4 Cf. D'Amato, *I Domenicani a Bologna*, vol. I, 527.

5 For Giovanni Maria Guanzelli (1556–1619) and his *Indice* published in 1607, see, in particular, Rebellato, "Il miraggio dell'espurgazione."

6 For Medoro Patriarcha see, initially, Marini, *Degli archiatri pontifici*, vol. I, XXXVIII, XL–XLI, 464; Mascaretti, *Memorie istoriche di Grottammare* [1841] (1994), 73.

7 Patriarcha's name is included in the list *Aliquot insignes Medici qui hac aetate in alma Urbe florent, et quorum consilium Auctor in Simplicium medicamentorum usu adhiberi admonet*, in Durante, *Herbario nuovo*, †4*r*.

of hot and cold drinks, which, between 1602 and 1607, involved Pietro Cassiani from Bologna, on the one side, and Giacomo and Giuseppe Castiglione, father and son, on the other.[8] This dispute takes on a specific importance for our purposes because one of the motives behind the entire affair derived from the wish to subject Antonio Persio's theses of 1593 to severe criticism. These theses called for the consumption of hot drinks as opposed to the conventionally accepted opinion, which also existed within the Peripatetic school, that giving cold or iced drinks to patients was preferable.[9]

Patriarcha was explicitly called into the dispute by Giuseppe Castiglioni, who, upon seeing himself in turn drawn into the argument in another work by Cassiani in 1606, responded through an *Apologeticus*, published in 1607, in which he presented a list of excellent physicians whose opinion he had requested. They declared themselves in favor of iced drinks.[10] However, Teodoro Ameyden, who returned to the dispute with a treatise on the argument in 1608, expressed the suspicion that Castiglioni had not been honest when dealing with Patriarcha's position. Ameyden indicated that he had had a lot of time to reason with Patriarcha on the subject and that he had never heard him express himself in favor of the use of very cold drinks, but rather, just like Bernardino Catellano who was also on Castiglioni's list, that he had completely banned the administering of drinks with added snow to patients.[11]

---

8 Here I limit myself to providing the biographical details of the works directly involved in the dispute: Castiglione, *Discorso sopra il bever fresco*; Cassiani, *Risposta ... al discorso sopra il bever*; Castiglione, *Observationum in criticos decas prima*, 26–32 (Chap. IX: *Frigida potione et ad aquam refrigerandam nive, et glacie veteres Romanos usos, Senecae et Tranquilli a Lipsio perperam accepta*); Cassiani, *De calidi potus apud Veteres usu*; Castiglione, *De frigido et calido potu*; once the question was opened, it raised a series of collateral interventions spread over time, amongst which, while the polemic was still underway, included one by Fuscone, *Trattato del bere caldo e freddo*, then by Teodoro Ameyden, see, below, Scacchi, *De salubri potu dissertatio*, and finally by Peccana, *Del bever freddo libro uno*; Artese dealt with the dispute in Antonio Persio, *Trattato dell'ingegno dell'huomo*, 99–103.

9 Cf. Persio, *Del bever caldo costumato da gli antichi Romani*; looking beyond Persio, Cassiani (cf. *Risposta*, 8–9) writes that Giacomo Castiglioni's *Discorso* was also aimed at Nicolò Masini's contemporaneous treatise, *De gelidi potus abusu libri tres*.

10 Castiglione, *Apologeticus*, B1v "Excellentes item medici Bernardinus Castellanus, Camillus Gorus, Demetrius Canavarius, Medorus Patriarcha, omnes denique, quibus cum hac de re locutus sum, potionem quatidiani victus frigidam probant, calidam damnant".

11 Ameyden, *Trattato della natura del vino, e del ber caldo e freddo*, 105–106: "L'haver noi visto quello che sentirono gli antichi circa il ber caldo o freddo farà più facile le risposte a gli argomenti de' quali il primo era che a tutti li più dotti medici d'Italia lodano l'uso di ber freddo, quali sono Marisilio Cagnati, Bernardino Castellano, Ascanio Mandosio, Angelo Vittorio, Iacomo Bonaventura, Cinthio Clementi, Antonio Porto, Camillo Goro, Medoro Patriarcha, Demetrio Canavario. Risponde a quest'argomento Piero Cassiano, rivocando in dubbio se li sopradetti siano i più dotti medici o no; ma per esser questo dubbio

We are not in a position to know what Patriarcha's conviction was, but it is of little importance compared with the fact that the context of the dispute also provided him with a reason to request the Telesian text; it must have happened between 1600 and 1604 because Baronio signed as *socius* to the Master of the Sacred Apostolic Palace.[12] However, if this can be indicated as a specific reason, which can be linked to the more general ones that caused physicians to deal with the *De natura* and the *Libelli* after they had been placed on the *Index*,[13] then we should consider it unlikely that Patriarcha wrote the annotations, because no great interest in medical-biological matters is found in them. Moreover, something that provides greater motivation for his exclusion as the author of the annotations is the fact that the manner of intervention seems to hold no trace of any reference whatsoever to the difficult points which had been placed in the *Index*;[14] instead, they reflect problems and instances that are only comprehensible if placed within the context of the years immediately following the publication of the Roman edition, or more probably immediately before the Neapolitan one of 1570, which ignited the Telesian *affaire*.[15]

---

per molti capi difficile a discutere, lasciatolo da parte, verrò ad esaminare la verità di quella propositione, cioè ch'essi lodano il ber freddo. [...] Havendone io ragionato seco sopra quella materia, mai l'intesi lodare quest'uso di ber freddissimo, cioè con la neve. Bernardino Castellano e Medoro Patriarche, non solo non lodano la neve, ma l'hanno bandita di casa loro, onde non poco mi maraviglio dell'opinio del Castiglione, ch'egli ponga costoro in questo numero, perché non credo ch'eglino fossero mai si empii, che volessero lodar ad altrui per buono quel che interiormemte tengono per malo. Giacomo Buonaventuri per il ber freddo credo che crepasse qualch'anno prima, che fatto non harebbe, s'havesse bevuto temprato. De gli altri medici non havendo io cognitione, né per scritti, né per parole, non ne posso dar giudicio, ma se d'una bugia si può inferire l'altra et havendo il Castiglione citato per se li sopranominati doi, Bernardino Castellano e Medoro Patriarcha, i quali veramente sono contra esso, verrò anche dubitando di tutti gli altri, se sono tali quali egli li fa".

12 What is more, this testimonial is not without a certain interest if referred to the complex question of the issuing of the reading permit, which occurred at a time of particular tension between the Master of the Sacred Apostolic and the Congregation for the List of Prohibited Books (Indice); see Fragnito, "La censura libraria," Baldini, "Il pubblico della scienza," Fragnito, "Un archivio conteso," Savelli, "Allo scrittoio del censore," id., "La biblioteca disciplinata."

13 Cf. Baldini, "Il pubblico", 195, n. 79, from which it emerges that, after being placed on the *Indice*, one of the two reading requests presented over the time period that the scholar considers was forwarded to Giovanni Talentoni, the physician and natural philosopher.

14 See Martínez de Bujanda et al., *Index de Rome*, 477–78: "Traités absents de Rome 1590, à expurger pour la première fois dans Rome 1593," Firpo, "Filosofia italiana e Controriforma. IV," Bondì, "Expurgatio impossibilis."

15 In correspondence to the passage in which Telesio comments on the Aristotelian theory of seawater salinity and refers to *proprius commentarius* (Book 2 Chapter 12, see *DNCors*, pp. 94–95), the *scholium* notes: "Telesius composuit librum de salsedine maris"; certainly,

2 A Preliminary Analysis of the *scholia*

In order to accurately understand the *scholia* I will now present an annotation which, in terms of its breadth, can be considered a synthesis (in many ways effective) of the main themes of *De natura*. An entire transcription of this interesting annotation can be found in the appendix to this paper, where the reader can note the list of sources which will be used to philosophically consider the Telesian notion of the total pervasiveness of the *anima vegetativa*.[16]

First it is necessary to mention the reference to the Egyptian *prisca sapientia*, which is summarized here in its Hermetic *facies*. The reference is interesting as it reveals a concordistic vision of philosophy. Although Aristotle's primacy is repeatedly stated and defended in the annotations, the emergence of his philosophy is regarded as the synthesis of a tradition inaugurated by Hipparchus the Pythagorean. The development of this tradition constantly strived to rationally express, however imperfectly, the knowledge that was hidden behind the *symbola* and and *occultissimae notae* of the Egyptian, Chaldean, and Indian wisdom (*sapientia*).[17]

---

the vagueness with which the title of the *libellus* was restored does not permit us to infer anything about the need to use 1570 as *terminus post quem*. However, it should at least be noted that this vagueness may appear even less comprehensible given the hypothesis that the annotations date from after the inclusion in the *Indice*, which, as is known, also included the edition of the *Libelli* edited by Antonio Persio.

16 Cf. *DNCors*, c. *1*v*: "II lib. cap. XLIII, pag. 148, cap. LVI pag. 169 ita ut omnia quae illustrentur a sole non solum agant anima vegetativa, sed etiam sensitiva, ut habes apud Mercurium, II, IX et XI, *Pimand.* Aristotelem VIII de divina sapientia secundum Aegyptios, cap. II et III ex aliis I *de anima* tex. 86 M. T. C. II *de natura deorum* in persona Lucilii Balbi philosophi stoici pag. 66, V *Tuscul.* Pag. pag 517 quam in contrarium dixerit 1 *de nat. deor.* in persona C. Cottae adversus C. Velleium, pag. 46. Laërtium in *Thalete*; Senecam philosophum VI *natural. queastiones*, cap, XIV, pag. 12 <.>, cap. XVI, pag. 140"; for the entire transcription of this long, interesting annotation, see the Appendix.

17 Cf. *DNCors*, c. *2*v*: "Scientias primi sapientes apud Aegyptios Chaldeos et <I>ndos semper occultarunt <e>as vel symbolis vel occultissimis notis posteris declarantes. Quas tenebras Hiparchus discipulus Pythagoras, primus de medio sustulit, deinde Plato et clarius Aristoteles et philosophi instarum familiarum, ut graves auctores observaverunt; relicta tamen semper rerum ipsarum difficultas, quam nemo potuit de medio tollere. Quid igitur garris Telesi? Quod priscis symbola, tibi rerum ipsarum difficultas, quam, etsi methodice ab Hiparcho ad nostra usque tempora quasi per manus explicata, non percipis, sed in meridie caligas, et dicacitate tua graves auctores offendis"; the *scholium* is written in the left-hand margin and referred to by the letter "h" in connection with the following part of the text which is found between cc. *2*v*–3*r*: "Et neque propterea tot sua tenebris occultantem illum existimare potentes, quo, ut suis placet, ignavos deterreret, quibus nimirum pulcherrimarum rerum invideret cognitionem, non eadem omnia obvolventem caligine videntes, at obscuriora abstrusioraque quae sunt, quae igitur illustranda aperiendaque

By contrasting Telesio with the tradition that links the *primi sapientes* to Aristotle, Galen and the most direct followers of Aristotle, it is possible to understand the most salient character of his philosophy. The annotations explicitly attack him as typically heretical. Telesio's shares the *hybris* of heretics which, on a dogmatic level, made them totally reject laws that were established and generally accepted at that time. His thought is their *nova fides* transposed onto the level of *philosophia.*[18]

Once the *outlaw* nature of Telesio's philosophy is taken as established, the annotation still has to clarify the terms of dissent established by *De natura* vis-à-vis Aristotelian philosophy and the Peripatetic tradition. The annotator is absolutely convinced that this comparison will shed light on the falsity of Telesian philosophy and operates in such a way as to convince the eventual readers of this fact. They are supposed to recognize what is declared in the frontispiece of the *scholium*, namely that the expression *iuxta propria principia* of the title refers to principles that are "false, as the following scholia will reveal" (*eaque falsa, ut progressus scholiarum docebit*).

---

essent maxime, adeo profunda ut linceus nullus superare et pervadere illam queat, aperta magis et quae penitus innotuisse visa sunt, nulla plerunque, valde exili interdum, ut suspicari liceat, propterea id esse ab eo factum, ut ne sui penitus dissimilis in dissimilium traditione visus, non aeque omnium sciens videri queat, est et quae in nimia ponere velit luce, ut suis etiam ambitiosius circa quaedam revolvi videatur, sua ostentans et pluribus quam opus firmans illa rationibus. [...] Id volentem omnino, non ignota nimirum sibi illa fuisse homines suspicari, sed abstrusiora quam quae omnibus innotescere et manifestari queant omnibus, se ipsos igitur damnare omnes, illum admirari semper".

18 Cfr. *DNCors*, c. **4r*: "Haereticorum more leges refellendi sua<s> proponit; ut enim illi, negatis divinis scripturis, patrum dogmatis et vetustissimis traditionibus iam inde ab initio quasi per manus ad haec usque tempora in ecclesia Dei acceptis, ita etiam et Telesius Calaber, negata universa philosophia [*add. in marg. sin.* philosophia] Mercurii Trismegisti, Phytagoreae Samii, Platonis Atheniensis, Aristotelis Stagiritae et philosophorum istarum familiarum ad nostra usque tempora quasi per manus accepta, nobis leges, tamquam alter Lycurgus, cum eo de naeniis suis disputantibus proponit. Negatis principiis, bone Calaber, non est disputandum, ut docet philosophus I *phy*. a tex. 8 usque 12, sed respondendum": the *scholium* is referred to by the letter "o" in reference to this part of the text: "Tum, ne ut nobis notas illius afferant distinctiones terminosque quas ingenue fateor percipere me nunquam satis posuisse, propterea, reor, quod non sensui expositas, nec huiusmodi similes continent res, se summe a sensu remotas, et ab his etiam, quae percaepit sensus, quales tardiore qui sunt crassioreque ingenio, cuiusmodi, mihi ipsi et nulla animi molestia esse videor, percipere haud queant; quae igitur contra nos afferent, exponant oportet et veluti in lucem ponant, tarditatis meae, si libet, commiserti et rebus agant, non ignotis vocibus quae, nisi res contineant, vanae sint inanesque. Illud pro certo habere omnes volumus nequaquam pervicaci nos esse ingenio, aut non unius amatores veritatis, et libenter itaque errores nostros animadversuros et summas illi gratias habituros qui quam solam quaerimus colimusque patefecerit veritatem".

The analysis begins with a criticism of the first *sententiae*, which, according to Telesio's intention, have the function of providing a synthesis of his thought.[19] At the point in which Telesio characterizes the nature of beings by means of a fundamental opposition (*contraria itaque inter se apparent entia*), the annotation (indicated by the letter 'p') stresses the totally inadequate way in which Telesio introduces the cornerstones of his philosophy, because he neglected to define the universal nature (*natura universa*) under which *contraria* would fall.[20] The criticism goes on to exhort Telesio to remedy the ambiguity of his formulation and indicates the two models he should conform to: Aristotle, here symbolized by *Categoriae* and the first book of *Topica*; and Cicero, introduced through *De officiis* and *De oratore*. The combination of Aristotle and Cicero should, though, be seen here in an almost paradoxical sense. Elsewhere, the annotator is as mocking as he is precise when discussing Cicero and his humanist and Renaissance followers from Lorenzo Valla to Mario Nizolio.[21] The list of the 'know-alls' displayed in this way deserves an accurate name-by-name examination. However, I consider it useful to first concentrate on the presence

19 Cf. *DNCors*, p. 1*r*: "Entia longe inter se diversissima longeque apparent dissimillima. At quae in sese mutuo agentia et mututo a se ipsis patientia, mutuo in se ipsa invertantur omnia, talia autem quae sunt, contraria dicuntur, contraria itaque inter se apparent entia [...]."

20 Ibid.: "Ut duorum istorum libellorum, ingeniose Calaber, inscriptio, et disputatio melius intelligerentur, tibi prius de natura universa, quam de contrariis quae eam supponunt disputandum esset. Ambiguitas enim a verbo, de quo est disputatio, in primis tollenda, ut sciat quisque quo conatus suos debeat dirigere, et ne andabatarum more in tenebris pugnet. Aristoteles cum Archyta Pythagorico initio *Praedicamentorum*, et I *top*. cap. XIII; M. T. C. [*scil.* Marcus Tullius Cicero] in *Officior*. et in lib. *De orat. ad Q. fratrem*. Quam multa enim natura significet, nosti. Tolle igitur in primis homonymiam istam, ut sciant lectores de qua illarum praesens disputatio sit"; the reference to the Pythagorean Archytas of Taranto is due to the fact that a work entitled *Dieci Categorie* was ascribed to him. Although this is in reality a Pythagorean work, it probably dates from the first century BC. The accusation of ambiguity is repeated in correspondence with *DNCors*, 2*r*, in the annotation referred to as "QQ.": "Hinc percipies satius fuisse ut a naturae significationibus disputationem istam incoharet, ut supra notavi cap. 1 fol. 1 pag. 1 Vide sequentes locos, ubi de natura loquitur et semper eadem ambiguitas relinquitur. Nota deinceps quam confuse in Aristotelem agat, ut vix intelligatur quomodo pro ipso contra ipsum agat" with reference to the final clause in the second chapter and the initial considerations in the following chapter.

21 Cf. for example the *scholium* in *DNCors*, c. *2*r*: "Scioli similes tibi refutarunt ut Graeci quidam Cornutu<s> et alii, quorum meminit Simplicii prooemium in *praedicamenta*, Laurentius Valla, Nizolius, [et *canc.*] alii huius farinae, et Iustinus philosophus stoicus et Martyr in lib. *De falsis dogmatis Aristotelis*," indicated with the letter *alfa* in reference to the part of the text: "[...] numinis instar hominis genus universum veneretur et veluti a Deo ipso edoctum et Dei ipsius interpretem, summa audit cum admiratione et cum religione etiam summa, novam ipse invehere tentem."

of Mario Nizolio, in whom the Ciceronian lesson, in an anti-Peripatetic and anti-Scholastic sense, finally emerged in his writing of *De veris principiis et vera ratione philosophandi contra pseudophilosophos libri IV* (*Four books about the true principles and the true way of philosophising against the pseudophilosophers*) in 1553.[22] What is striking is the intelligently ironic way in which the author of the *scholia* makes use of the category of *pseudophilosophus* and turns it against Telesio himself by conferring it upon a third person, Vincenzo Maggi, to whom Telesio (as he said in the preface) had turned to receive support regarding the quality of his *philosophia*.[23] The annotator then goes a little further by reiterating—still referring to Maggi—his strictly Ciceronian inclination.

22 The literature on Nizolio is vast; regardless—excluding the joint works on the relationship between rhetoric and dialectics in the Renaissance—it is worth consulting Rossi, "Il *De principiis* di Mario Nizolo," Del Vivo, "Valla, Vives e Nizolio," Vasoli, "Un episodio della disputa cinquecentesca su Cicerone e il ciceronianesimo," Nauta, "Anti-Essentialism and the Rhetoricization Knowledge," id., "De-essentializing the World."

23 Cf. the *scholium* indicated with the letter "l" in *DNCors*, c. *3*v*: "In lib. adversus pseudophilosophos, cum et is et sui similes pseudophilosophi sint; neque eius philosophia de rerum sed verborum cognitione"; it is worthwhile reporting the whole passage in which Telesio referred to the meeting with Maggi: "Facile igitur suspicari vererique potenti, et revera suspicanti interdum verentique deceptum me, neque enim fieri posse ut tot praestantissimi viri, tot nationes atque adeo humanum genus universum tot tam saecula Aristotelem coluerit, in tot errantem tantisque. Madium Brixianum adire et consulere visum est, quem et in philosophia excellere videbamus, et cuius mihi iamdiu animi ingenuitas innotuerat, ut si a praestantissimo viro cogitationes meae improbatae forent, nequaquam supprimerentur illae, sin minus, errores intuitus meos, quod reliquum vitae esset, et ipse Aristotelem suspicerem venerarerque. Brixiam itaque ad Madium profectus et itineris mei exposita ratione, nequaquam ille, quod multi fecerant et quod facturum et illum minitati fuerant, inauditum reiecit, at summa diligentia plures dies, quibus apud illum fui, et summa cum animi tranquillitate et audiit et perpendit omnia. Principia nihil improbavit et quod non e principiis flueret videre nihil potuit, Aristotelem in nullis certe satis defendere est visus. Damnavit etiam illum prima constituentem corpora, nequaquam res ipsas invitum tot illum taliaque posuisse affirmans, at proprias sequutum positiones. Neque igitur talia esse illa, qualia Aristoteli ponuntur, et ipsius positiones ab innumeris iisque inexplicabilibus excipi difficultatibus, quas a suis descriptas ostendi nobis curavit. Vir videlicet genere quidem nobilissimus, at multo animo magis et nihil, nisi ipsam colens suspiciensque veritatem, mihi, quem ipse interpretabatur, cui igitur veluti iuramento obstrictus videri poterat, veritus Aristotelem, quin, ubi parum placeret, oppugnaret illum et damnaret etiam defendi impotentem. Nihil itaque ab illo audiens, quod vel nostra labefactaret, vel quod Aristotelis positiones a nobis oppugnatas tueretur stabiliretque, et neque ab aliis ullis, quibus cum multis Romae et eximiis quidem viris communicare vel disserere illa licuit, et a multis, ut mea aederem impulsus, nihil id facere amplius veritus sum" (in all of the part, the *scholium* is, though, referred to the following passage: "Madium Brixianum adire et consulere visum est, quem et in philosophia excellere videbamus [...]").

I would say, however, that it is precisely this attack on Maggi and Nizolio, and their being painstakingly tied to this phase of the contrast between Ciceronian rhetoric and Aristotelian dialectic, which truly renders unconvincing the dating of the annotations around the turn of the sixteenth century into the seventeenth. In this time it would not have been very effective to refer to that climate, so much so that no reference emerges in the other annotations to people belonging to the generation after that of Maggi and Nizolio. Unless new research will discover new unpredictable scenarios, we can hypothesize that the compiler of the *scholia* was Baronio himself, who, having come into possession of this copy, added his comment during the years suggested above. In other words, this is when he might well have still been in the Dominican convent in Venice or have already been transferred to the one in Bologna and was thus at the beginning of his promising professorship in theology.

This hypothesis presents a reasonable level of probability, and opens up a scenario which is not without interest since, in the general discipline of issuing permits for a work which had finished in the *Index*, this is a truly unusual case because the person in the position to grant the license was the very owner of the copy in question. There is another detail that should be looked at and weighed in the light of this hypothesis: Baronio gave Patriarcha the *facultas tenendi et legendi* and set a temporal distinction, *per triennium*, which was, however, cancelled, clearly rendering the time limit ineffectual. It might be suggested that this cancellation was made *inter scribendum* or subsequently, at the hand of Baronio, following the very early return of the text, or for the opposite reason, in the sense that Baronio, once he had been elected bishop in 1604, may either not have been in a position to ask for the book to be returned or decided to leave it in the hands of Patriarcha.

Cf. the following *scholium*, referred to by the letter "n," in *DNCors*, c. *4*r*: "Immo veteribus symbolis sublatis, ut super littera h notavi, ea tibi inaccessibilis restat. Aristoteles II *metaphy.* tex. 1<...>, M. T. C. IV, *Acad. quaest.* 1 pag. 8 parvis voluminibus. Consulo igitur Brixianum tuum philosophum Ciceronianum", in reference to the following portion of the text: "[...] difficultates nullas, nec sibi dissentit unquam, sed penitus sibi ipsi cohaeret et una efficitur omnis; tum vel ignavissimis crassissimisque hominibus aperit manifestatque quaevis omnia, omnia sensui exponens apertissime, huiusmodi nulli, reor, Aristoteliana videri queant. Nostra ne sint, ii recte iudicabunt qui illa, quod Aristoteles faciendum praecepit, non ut adversariis, sed ut iudices arbitrique legerint consideraverintque religione, qua erga Aristotelem obstricti videntur, exoluti et tantisper illius positionum decretorumque obliti."

Appendix

The long significant *scholium* presented here extends over two non-adjoining pages: indeed it was begun on c. *4*v* and continued on c. *1*v*, as the writer had the presence of mind to write the word *retro* to indicate the jump backwards to be made in order to complete the reading.

I have followed a criterion of conservation in the transcription; I have distinguished "u" from "v"; I have re-extended the abbreviated forms; I have modernized the punctuation and reserved the use of capital letters for starting new sentences, for given names and for the titles of cited works; I have only used italics for the titles of sources; I have put the text into paragraphs to facilitate reading; I have indicated the change of page with the "//" sign; as is normal, I have adopted angle brackets for the additions introduced, except for where use was made of dots where the gap has not been corrected; finally I have set up an apparatus for the registering of corrections, cancellations, and interlineal additions.

De principiis rerum naturarum[24] adversus Aristotelem in hoc libro tractat, de quibus has conclusiones constituit:

I cap. II:[25] caelum summe calidum, terra summe frigida, quia maxime inter se distant loco; ergo maxima inter se distant essentia, ut de contrariis fatetur Aristoteles cap. de oppositis.[26]

II conclusio, cap. III: caelum et terra sunt duo principia, quae neque ex sese mutuo,[27] neque ex aliis et ex ipsis omnia fiunt in hoc mundo sensibili, ut definit Aristoteles I *phys.* tex. 42,[28] ex qua colligit cap. II: quia caelum et terra sunt summe contraria, mundus corporeus iis constans, sphaericus est et unus, caelum supremum, terra infimum locum tenet, illud mobile, haec immobilis, ut circa ipsam caelum volvatur, et aptius fiant in hoc mundo sensibili quae a sole fiunt.

III conclusio, cap. VI: caelum tenuissimum, remotum a terra, in varios orbes divisum, contrariis motibus distinctum, non iisdem polis innixum, ne terram frigidissima in perniciem traheret et mundum hunc inferiorem calore suo labefactaret; ex qua colligit cap. VII:[29] quia caelum summe calidum, omnibusque superstat, ut inde aequaliter vires suas iis inferioribus communicet, summe mobile, summe tenue, summe perspicuum, summe album, summe lucidum. Terra contra, quia summe

24 *Sequitur* in hoc libro *canc.*
25 cap. II *add. in interl.*
26 *Sequitur* cap. III (*fortasse*) libri 1 *canc.*
27 mutuo *add. in interl.*
28 ut definit I *phys.* tex. 42 *add. in interl.*
29 cap. VII *add. in interl.*

frigida, omnibus substat, summe immobilis, summe crassa, summe opaca, summe nigra et tenebricosa est.

Colligit secundo ibidem, caelo et terra hoc commune esse, ut se conservent fugiantque quae sibi advervantur, cognoscant quae ea, quae nocitura videantur, ut copiose declarat cap. VIII, docens omnia praedita esse cognitione sensitiva plus minus ve, ut plus minusve accedunt ad alterum istorum[30] extremorum, caelum, scilicet, et terram, ita ut caelum omnibus excellat iis quatuor sensibus, visu, tactu, gustatu et odoratu, alterum altero saltem constent omnia corpora necesse est; id tamen sine organo praestant, quemadmodum fuse explicavit per solis illustrationem II lib.[31] cap. XLIII, pag. 148, cap. LVI pag. 169 ita ut omnia quae illustrentur a sole non solum agant anima vegetativa, sed etiam sensitiva, ut habes apud Mercurium, II, IX et XI, *Pimand.*, Aristotelem VIII *De divina sapientia secundum Aegyptios*, cap. II et III ex aliis I *de anima* tex. 86 M. T. C. II *De natura deorum* in persona Lucilii Balbi philosophi stoici pag. 66, V *Tuscul.* pag. 517 quam in contrarium dixerit I *de nat. deor.* in persona C. Cottae adversus C. Velleium, pag. 46. Laërtium in *Thalete*; Senecam philosophum VI *natural. queastiones*, cap, XIV,[32] pag. 12 <.>, cap. XVI, pag. 140. Quibus etiam addit cap. VII. caelum et terram, tamquam partes universi, contra naturae suae vacuum densitare.

Deinde multa tumultuarie doce<t> ut probet a sole et terra tamquam ex principiis omnia // fieri, ut cap. XVI, ignis calore moderato fieri pullos ex ovis, Pyraustas in fornacibus Cipriis; cap. XXII: quomodo a sole et terra fiat aurum in visceribus terrae; quomodo aquae dulces et salsae a sole et terra, et non a vapore, ut putat Aristoteles, quia magis terra et sol distant quam vapor; cap. XXVII: // quomodo sol et terra faciat lapides, metalla, salsa, sulphurea, bitumina, plantas, et multa animantium genera; a vapore pluvias, nives, grandines et ventos; cap. XXVII: quomodo Plato ex iis quae per putrefactionem fiunt, idem colligerit, quo propterea merito reprehendit Aristotelem; cap. XXIX: quod motus sit per se caloris, quia a calore ens caelis est, et non per accidens, ut putat Aristoteles; cap. XXXVII: reprehendit Aristotelem, quod caelum ponat quintam essentiam et calorem et frigus nullius rei esse formas essentiales, ut pluribus ex Peripateticis expendit cap. LIV, LV usque ad finem libri; cap. XXXVII: colligit quomodo caeli sint eiusdem speciei sola raritate et densitate differentes. Ergo materia et forma et privatio non sunt principia rerum physicarum, sed caelum et terra; quia nullum illorum est activum, utrumque istorum activum; privatio non est ens, utrumque istorum ens.

---

30 istorum *add. in interl.*

31 II lib. *add. in interl.*

32 XIV *corr. ex* XLVI.

CHAPTER 10

Reformation, Naturalism, and Telesianism: the Case of Agostino Doni

Riccarda Suitner

1 A "heterodox" Telesian Philosopher

Only a single work by the philosopher and physician Agostino Doni survives: the *De natura hominis* (On Human Nature) published in Basle by Johann Froben in 1581. It is a testament to the author's closeness to the philosophy of Bernardino Telesio, whose relationship with Doni will be one of the main topics of this paper. Little is known about Doni's biography, especially his early years and the later part of his life. Doni is a very obscure author from both a theoretical and biographical point of view. He was virtually ignored by his contemporaries and is completely absent from modern research on sixteenth-century philosophy. Thus many details of his life, including the year of his birth and the place and year of his death, are surrounded with an aura of mystery. The few studies that do exist constantly reiterate two associations. First, with the publication in 1939 of Delio Cantimori's *Gli eretici italiani del Cinquecento* (Italian Heretics of the Sixteenth Century), which for the first time highlighted Doni's connection to the circles of Italian reformers in exile for religious reasons, the philosopher suddenly became associated with the events of the European Reformation. In a long footnote to his classic study Cantimori cited some letters sent by Doni in the 1580s depicting the Italian doctor's extreme state of poverty; he was living at the time in Basel and was even forced to beg the humanist and physician Theodor Zwinger to find him a second-hand silk beret, at the time the mark of his profession, since he could not afford a new one.[1]

In the second half of the twentieth century, Doni came to the attention of scholars interested in heretical movements such as Antonio Rotondò and,

1 Cantimori, *Gli eretici italiani del Cinquecento*, 341, n. 2. A previous version of this article was presented at the Max-Weber-Kolleg (University of Erfurt) at a seminar of the research group *Religious Individualization in Historical Perspective*. I thank all the participants for their valuable comments.

 | DOI:10.1163/9789004352643_012

more recently, Michaela Valente.[2] The 'entry' of the philosopher into the field of research on the European Reformation gave rise to the genesis of two different images of the philosopher. Some scholars, such as Rotondò, approached the figure of Doni from a merely biographical point of view, arguing that it is impossible to identify any affinity to reformed doctrines in his texts;[3] by contrast, other scholars have linked Doni to some of the main antagonists of Calvinist orthodoxy in Basel: Fausto Sozzini, Celio Secondo Curione, Giovanni Bernardino Bonifacio, and Francesco Pucci.[4]

Doni always seems to be at the fringes of the classic category of 'Italian heretics of the sixteenth century'. We do not know why he fled Italy, nor do we know the date and circumstances of his death. After a period in Basel, we find the philosopher in Leipzig, where he unsuccessfully attempted to meet the controversial physician Simone Simoni (1532–1602), a figure who came close to atheism and was 'on the fringes' of the Italian heretical movement. After passing through Breslau (Wrocław), Doni reached Krakow full of hope; here he attempted to enter the circles of the King of Poland Stefan Batory, the protector of many Italian exiles, to whom he had dedicated the *De natura hominis*. However he was met with hostility by the court physician, the Unitarian Niccolò Buccella (?–1599).[5]

Thus, from a purely biographical point of view, Doni always moves in the framework of reformed circles, in Basel and also in Poland. The king's entourage was in fact entirely composed of sympathizers to the so-called 'radical Reformation'.[6] One of them was Buccella, born into a family of religious dissenters, who had fled from Padua where he organized illegal dissections that were enormously popular, especially among the university's northern European students. Another figure was Fabio Nifo, who was also a professor of medicine at the Paduan university, who had come to Poland after a spectacular escape from the bishop's palace: he lowered himself down at nighttime with a rope and the help of his German students. He integrated so well into the Polish court that he joined the king in the Danzig campaign, as in 1575 the Baltic city refused to recognize his election. A third, Simone Simoni, had arrived

2 Rotondò, *Studi e ricerche di storia ereticale italiana*, Chap. 8, Valente, "*Libertas philosophandi*."
3 Rotondò, *Studi e ricerche*, 450.
4 See Firpo, *Juan de Valdés and the Italian Reformation*, 191. The monograph presents a detailed and fascinating picture of the dissemination and reception of the Reformation in southern Italy.
5 Rotondò, *Studi e ricerche*, 395–397 and 408–420.
6 On this phenomenon, fairly heterogeneous in terms of its theoretical and geographical boundaries, see Williams, *The Radical Reformation*, Biagioni and Felici, *La riforma radicale*, Wilbur, *A History of Unitarianism*.

in Krakow after stops in Geneva, Paris, Heidelberg (where he frequented the circle of Thomas Erastus), Leipzig, and Basel. In each of these places he managed to teach medicine or philosophy at the university and invariably aroused the suspicions of his colleagues. After repeated recantations, accusations of murder and Arianism (by some colleagues at Heidelberg, including the physician Antonio Francesco Pigafetta), and repeated scandals in which he became embroiled virtually everywhere, he, like Nifo, reconverted to Catholicism even though he continued to be suspected of Nicodemism, i.e. of merely formal adherence to orthodoxy.[7]

However, while Doni's work betrays a restless personality and an extremely radical naturalism it does not demonstrate explicit sympathies with Lutheran or Calvinist thought. More generally, the *De natura hominis* lacks any evangelical undertones, a feature which Luigi Firpo also emphasized in a figure who is in many ways very similar to Doni: Bernardino Telesio.[8] For biographical and theoretical reasons, it is impossible to deal quickly with the relationship between Doni and the so-called 'radical wing' of the Reformation, and I will return extensively to these issues in the course of this paper.

Doni's second recurring association—apart from religious reformers—is the one with Telesio. Doni was also from Cosenza; the *De natura hominis* appeared eleven years after the second edition of the *De rerum natura*, and four years before the third edition of Telesio's main work. Paradoxically, Doni never mentions either Telesio or any philosopher close to him, nor is he himself named in the writings of authors close to Telesio. Francis Bacon describes Doni in the *De dignitate et augmentis scientiarum* (The Dignity and Advancement of Learning) as a "disciple of Telesio".[9] Francesco Fiorentino, in his monograph on Telesio from the 1870s, accuses Doni of having "done nothing more than applying Telesian principles".[10] It is to Eugenio Garin that we owe some pages expressing a hope for a more in-depth study of Doni's philosophy, accompanied by an analysis of the events of late Telesianism and a comparison between the various notions of *spiritus* in the natural philosophy of this period.[11] Luigi

7 On these physicians, see Verdigi, *Simone Simoni*, Marchetti, "Figure di esuli italiani del Cinquecento." Stella, *Dall'anabattismo al socinianesimo*, Firpo, *Antitrinitari nell'Europa orientale*, Caccamo, *Eretici italiani in Moravia, Polonia, Transilvania*. On the specific relationship between medicine and the radical Reformation, see Suitner, "Radical Reformation."

8 See Firpo, "Filosofia italiana e Controriforma IV." On Telesio in general, see Abbagnano, *Telesio*; Soleri, *Telesio*; De Franco, *Bernardino Telesio*; id., *Introduzione a Bernardino Telesio*; Bondì, *Introduzione a Telesio*; Mulsow, *Frühneuzeitliche Selbsterhaltung*.

9 Bacon, *Opera*, IV, 3 [1623]. On the reception of Telesio in Bacon's writings, see the contribution of Roberto Bondì in this volume.

10 Fiorentino, *Bernardino Telesio*, vol. 1, 324.

11 Garin, "Il termine *spiritus*."

de Franco was the first scholar to devote a full-length study to Doni; the short final chapter of his edition of the *De natura hominis* (the first and, so far, only full-length study to be published) compares its doctrines with Telesio's *De rerum natura*, clearly showing that the philosopher cannot be dismissed as a mere imitator of Telesio.[12] Shortly afterwards, a chapter of Antonio Rotondò's *Studi di storia ereticale del Cinquecento* (Studies on the History of Heresies of the Sixteenth Century), focusing mainly on the reconstruction of the philosopher's 'heretical' network, also provided a crucial contribution to the 'discovery' of his relevance to the history of ideas.[13]

Doni's other works are lost today, probably because they were never published. From indirect sources we know that he completed a second edition of the *De natura hominis*, a treatise on colors—which would have been interesting to compare with Telesio's *De coloribus* and *De colorum generatione*—and a *Commento sul Petrarca*.[14] There is no proof that Telesio knew of Doni's work or vice versa. Yet (and this is also apparent from the few references to this matter by De Franco and Rotondò) comparing their doctrines is not only entirely legitimate; it is also necessary given the obvious similarities between the *De rerum natura* and the *De natura hominis* and, of course, their common origin from Cosenza, which is a further suggestion of possible connections between the two authors.

2 *De natura hominis* and *De rerum natura*: Some Remarks on the Relationship between the Two Texts

The issue of the relationship between Doni and the European Reformation and that of his links with the work of Telesio are both delicate problems that are difficult to resolve. Among other things, studies have appeared—I am thinking in particular of those by Roberto Bondì on the crucial issue of the two souls—that have shed light on the evolution of Telesio's ideas in the various editions of his masterpiece.[15] This was made possible by the studies from De Franco and Rotondò which responded to a need already stressed by Giovanni Gentile

---

12 De Franco, *L'eretico Agostino Doni*. De Franco's edition presents the Italian translation of Doni's text as well. On Doni's philosophical perspective, see also Plastina, "Un moderno eretico."

13 Rotondò, *Studi e ricerche*, Chap. 8.

14 See De Franco, *L'eretico Agostino Doni*, 34–38.

15 Gentile, *Bernardino Telesio*, 99–144; Garin, "Noterella telesiana;" Bondì, "*Spiritus* e *anima* in Bernardino Telesio." For a detailed analysis of Telesio's conception of the soul in the three editions of *De rerum natura*, see Miguel Á. Granada's paper in the present volume.

and Eugenio Garin. My first aim in this text is to note some divergences and similarities between the doctrines of Doni and those of Telesio that have hitherto not been remarked upon by interpreters, based in particular on the 1565 and 1570 editions of *De rerum natura* which predate the publication of Doni's work. I will then end with a reflection on the state of the sources. This may allow us to draw some conclusions on the relationship between the theories of the two philosophers and on Doni's place in late sixteenth-century natural philosophy. Furthermore, I will advance some considerations, based on the role of the concept of *spiritus* in the *De natura hominis*, on the relationship between Doni and the 'radical fringe' of the Reformation, in particular Michael Servetus.

Some of the basic premises of the *De natura hominis* are shared by Telesio's work. For both philosophers, nature is the only reality and has powers of its own. The vital principle is intrinsic to it, an active force upon which the movement of inert matter—representing the locus of the conflict between opposing forces—depends. The *De natura hominis* contains only three brief allusions to this subject, which nevertheless clearly refer to the Telesian philosophical milieu:

> The superior parts of earth, from which the matter which composes animated beings derives, consist of an even stronger dissimilarity. And this not because—as many have believed—the first four elements mix in it, but because in the middle of this space the acting natures fight with variable fortune; and while they are affected in various and diversified ways and are mutated, it also happens that matter, which occupies them [the superior parts of earth], contracts various and diversified forms, as will be explained in another and more appropriate section.[16]
>
> As soon as substances, which are occupied by the opposite nature—i.e. by cold, which otherwise loves immobility—are captured by heat, they move, which is not what they previously did; but in order to flee the motion and capture by heat they compress in the centre of the earth or in themselves; and this is the only way for them to aim at similar substances or to flee from the antithetical ones.[17]

16 Doni, *De natura hominis, libri duo*, I: "Suprema terrae, unde animantum materia desumitur, multo adhuc maiore constant dissimilitudine. Nec quia quatuor prima elementa, ut putarunt multi, misceantur, sed quia hic in medijs spatijs pugnant naturae agentes vario Marte varioque successu; dumque illas varie ac diverse affici mutarique contingit, fit ut materia quoque, quam illae occupant, varios diversosque contrahat habitus, ut alias, loco commodiore, exponetur."

17 Ibid., II, 8: "A contraria natura tenentur substantiae, a frigore nimirum, quod alioqui immobilitatem amat, a calore correptae, faciunt quod non faciebant moventurque, sed quo

> As substances, which are naturally different from one another, would always fight between themselves and agitate themselves mutually, in the human body it is necessary that there be one which is superior and leads the others. Chapter III.[18]

Furthermore, both in the *De rerum natura* and in the *De natura hominis* we see a universal power of perception[19] where there are no "abstract processes that come between things and spirit".[20] In Doni, just as in Telesio, sense is superior to intellect as far as the process of acquisition of knowledge is concerned; the imagination is conceived as a repetition of past movements of the spirit.[21] For Doni, and for Telesio from the 1570 edition onwards, heat precedes movement, with the only difference that movement is not described as an operation *operatio*) of heat in *De natura hominis*; rather, heat is defined simply as "motus localis auctor".[22]

However, the most interesting comparison concerns the nature of the spirit. Doni also describes a single, corporeal *spiritus*, which is neither a form nor an instrument of the body, in open disagreement (this long *pars destruens* does not appear in Telesio's text) with those philosophers who in varying ways had divided the soul into several parts with different functions (such as Plato, Aristotle, and Galen).[23] The proof of the corporeal nature of the soul is

---

fugiant motum raptumque caloris, ac cogunt sese intro ad centrum vel terrae vel etiam suum; qui modus unus est ipsarum vel petendi similia vel fugiendi contraria."

18 Ibid., II, 3: "Quia substantiae naturae inter se dissimiles semper inter se pugnarint seque invicem turbarint, oportere in corpore humano una aliqua superet, quae premat et corrigat alias. Caput III."

19 Ibid., II, 10: "omne quod natura constat vim sentiendi habet et sine ea esse non potest [...]. Omne simile petit simile et fugit dissimilia [...]. Nos enim fere nihil sentire putamus, nisi clamet ad tactum, nisi membra quatiat. Iniusti, qui prae nobis velimus de omni statuere natura et in omni exposcamus similes nostris motus, si eas facultates, quasi si habemus, illi velimus attribure, etiam si non habeat instrumenta, quae nos aliave animantia [...]. Nam aër, qui prope similis est substantia spiritibus animalium, (de eodem certe genere substantiae est), compressus vehementer collisusque, gemit fugiens compressionem et iam de ea conquerens; quae quidem querela est sonus; tamen vulgo minime omnium sic rem esse existimatur."

20 The quotation is from Rotondò, *Studi e ricerche*, 456.

21 Doni, *De natura hominis*, II, 18.

22 Ibid., II, 9.

23 Ibid., II, 4: "Considerantes conditiones huius crassi, quod est immotum, subinde transeuntes per collationem ad contraria, ex adverso ponemus substantiae prope perspectae conditiones. Hoc grave est, ergo illa levis erit; hoc non aeque quidem sed crassum omnino omne, igitur illa tenuis; hoc opacum, lumen neque reddens neque transmittens, illa igitur mobilis, per se semper mota, sicut natura per se a sua propria operatione cessare nequeunt substantiae; hoc natura frigidum per se ut apparet saltem sub mortem, illa igitur per

very similar in both philosophers. For Telesio, an incorporeal soul could not explain, for example, the fact that when an animal dies its parts are not all corrupted at the same time, nor could it explain temporary deaths; Doni views the crucial proof of the corporeal nature of the soul as lying in the weakening of the intellectual faculties of the sick. Just as one finds in *De rerum natura*, Doni's *spiritus* is spread throughout the body but present in larger quantities in the cavities of the brain.[24] However, this represents several clear-cut divergences from Telesio's ideas. Though stressing the incoherence of the Aristotelian definition of form, Doni, unlike Telesio, continues to use this term in several places in his work. Secondly, Doni approaches the issue of the spirit from a perspective that is at once more traditional and bolder than that of Telesio. Above all, he hypothesizes the existence of a single soul, identified with the spirit. Telesio did the same in the 1565 edition of the *De rerum natura*, before developing the distinction between the soul "a Deo immissa" and that "a semine educta".[25]

For Doni, then, there is just one soul.[26] However, it is this same soul/spirit that yearns to return to the heaven from which it comes:

> Our soul, excluded from heaven for some time and wrapped in this heavy, brutish, and lifeless body, is not able to rise and to direct its sight without difficulties.[27]

---

se natura calida; mobilem etiam eam a se esse, quando hoc est immotum, praedictum est [...]. Spiritum non incommode vocaris, usitato iam nomine. Nam spiritus mihi et corporea omnino substantia ea, quam quaerebamus, visa est et incorporea esse non posse." For the critique of Plato, Aristotle and Galen, see the whole Book I of the *De natura hominis*.

24 See ibid., II, 15.

25 The question of the genesis of the theory of the soul "a Deo immissa" is very controversial; diverse interpretations have arrived one after the other in the course of the decades. Roberto Bondì has convincingly hypothesized that the short reference to this second soul in the pamphlet *Quod animal universum ab unica anima substantia gubernatur* is a later addition, motivated merely by reasons of prudence. In the second edition of Telesio's text (1570) we find a brief reference to a soul infused by God: "Nec vero de humana etiam anima id dictum existimet quis, qua a Deo ipso in singula hominum corpora infusam esse non sacrae tantum divinae Literae; sed, ut suo amplius expositum est loco, humana etiam ratio nos dicet : sed de ea modo, quae e semine educitur, & quae agentis naturae opus videri potest, & de ea hoc in commentario perpetuo erit sermo" (Telesio, *De rerum natura iuxta propria principia* (1570), 22). The theory was included (very clumsily, as Giovanni Gentile pointed out) in the Telesian system only in the last edition of the text. See also Miguel Á. Granada's paper in the present volume.

26 See particularly Book II, § 7, *Demonstratio, qua ostenditur spiritum esse unum, sibi totum consimilem*.

27 Ibid., II, *Praefatio*: "Exclusus enim animus noster coelo per aliquod tempus obvolutusque corpore hoc gravi, bruto atque inerti, tollere sese in altum aciemque suam sine obstaculo intendere non potest."

> Certainly, from every side that force is exercised, and from every side its passage is precluded, therefore the spirit—which is a thin and light substance and would deserve a different kind of association—is forced to stay in its thick and heavy [body].[28]

In Telesio the discourse of the soul's aspiration to reconnect with the divine dimension only appears with the introduction of the concept of the incorporeal soul, beginning from the 1570 edition of Telesio's main work. The fifteenth-century conception of the *spiritus* as an intermediary between the soul and the body, still present in the *Trattato sull'ingegno dell'uomo* (Treatise on the Human Mind) by Antonio Persio (1576), was definitively overcome in both Doni's and Telesio's texts. The mystical tone that appears in the discussion of these issues in the *De natura hominis* makes Doni more similar to Persio than to Telesio.[29] Another very interesting feature of Doni's *spiritus* is its *brightness*:

> From the apparition of images, which takes place inside the spirit without penetration of any light from the outside, one can deduce that it [the spirit itself] is part of the light and that it would be able to produce light, if it would contract more. If the spirit would not be so, it [i.e. light] would not be produced; and it should be so, as nobody can see anything with natural means in darkness.[30]

Differences can also be found in the doctrine of generation. Both philosophers conceive the function of the body as being the prevention of the spirit's exit. However, Doni restricts himself to stating, fairly generically, that heat is found

---

28 Ibid., II, 5: "Iis vero modis undique vi facta et undique prohibita via, esse cogitur sub hoc grave et crassum spiritus, substantia tenuis et levis, digna alio consortio."

29 See Persio, *Trattato dell'ingegno dell'huomo*, 122: "Et quantunque piu e piu maniere d'Amore conoschino i dotti, chi cinque, chi tre, e chi piu, e chi meno; è non però cosa indubbitata che ciascuno di questi amori incominci dal mirare, ma sono questi mirari indirizzati a vari fini. Chi mira questa bellezza corporale per servirsene come a stormento alla contemplation della divina; chi mira la prefata bellezza per ivi fermar i vestigi, & far sosta di mirar piu avanti; chi la mira per isperanza di non voler toccandola goderla carnalmente [...]. L'amore aguzzante lo'ngegno per mezo della bellezza venghi da Dio nell'angiolo & nell'anima quasi trapassando per vitree finestre, da quella salti nel corpo apparecchiato a ricever quell'anima da questo corpo che frescamente ha alloggiato quello spirito che vi dimora."

30 Doni, *De natura hominis*, II, 18: "Ex apparitione autem imaginum, quae spiritui fit sub ipso, nihil luminis subeunte foris, colligi potest luminis participem esse illum et redditurum lumen, si densetur magis. Si enim non esset, non fieret illa; nihil enim omnino a cernente ullo secundum naturam cerni potest sub tenebras".

in the semen; it is the spirit, generated by heat itself, that destroys all the liquids present in the body by taking possession of it.[31] For Telesio, by contrast, heat is located in the uterus, which turns blood and semen into fluids that in turn give rise to all the parts of the body.[32]

Doni's treatment focuses exclusively on the human psyche and on the notion of *spiritus*, without touching on the cosmological and cosmogonical themes, the issues of pleasure and pain, and the ethics of conservation fundamental to Telesio's work.[33] A complete theory of the five senses is also lacking, and is found for the first time only in the definitive edition of Telesio's main work, in which taste, smell, and sight are essentially subordinate to touch. The only reference to the senses in the *De natura hominis* concerns taste, described as the main sense, the only one which allows the essence of things to be grasped fully: "flavor can, above all, reveal the nature of the thing to us; if it is possible to experience and taste it, it can be considered trustworthy, and we will be good judges of the process of knowing on the basis of flavor."[34]

Though an exhaustive comparison between the first two editions of the *De rerum natura* and the *De natura hominis* is needed to draw any truly concrete conclusions, it is certainly clear that the similarities between Telesio's and Doni's work are greater than their differences; to deny the virtual certainty that Doni had read the *De rerum natura* is a fairly bold statement. I fully concur with the theory proposed by Antonio Rotondò, according to which ignoring

31 "Ea autem tali massa comprehensa loco ubi procreamur, consequitur in tempore non valde longo ut de substantia, quae in ea massa proxima spiritui est, spiritus plane enascatur, similis, ut dicam breviter, spiritui parentum. Factus autem spiritus profecto statim elaberetur; nam nullum natura est libens cum dissimilibus, quae eadem cuncta sunt contraria; sed non unum obijcitur obstaculum. Per id enim temporis locus, in quo fit generatio, quasi studente natura parentis atque conante retinere fugitivum, ita clausus undique et ita constrictus est, ut merito res mira videri soleat [...]. Sed ipse etiam spiritus, ut factus est, ipse quoque sibi exitum usque arctiorem difficilioremque reddit. Nam, (quae est cupido et studium inexplebile omnium naturarum), dum sese multiplicat et sui copiam adauget, agens etiam vehementiore vi quam per ipsius exortum in semine agebatur, consumitque in eam rem non halitus modo consequentes, sed vapores atque liquida; crassitudines privatae liquentibus, quorum interventu mahis laxae magisque apertae tenebantur, cogunt constringuntunque sese, sicut videmus fieri in aperto, ubi liquida de medio crassiorum, quibus composita erant, exhauriuntur; inde vero postea difficilius adhuc spiritui fit exeundi negocium" (ibid., II, 5).

32 On the theory of generation in Telesio, I refer to Book VI of the 1586 edition of the *De rerum natura* (1586). The generative process is described there in a much more complex and extensive way than in Doni's brief treatment.

33 On this specific aspect of Telesian philosophy, see Mulsow, *Frühneuzeitliche Selbsterhaltung*.

34 Doni, *De natura hominis*, II, 2: "sapor super omnia potest prodere naturam rei, si liceat experiri, et gustatus syncerus haberi queat et simus boni cognoscendi ex sapore iudices".

the work of the other scholar was a safety measure on the part of both men.[35] This is especially true for the Telesian school: the works of Telesio, as is well known, were never condemned during the author's lifetime, although he did have some troubles with the authorities five years after his death; Doni, however, had already spent five years in prison "for a religious reason"[36] during his youth, perhaps in Cosenza, and was an exile in Protestant lands. Furthermore, if we restrict ourselves to the example of Germany, where Doni spent a long period before moving to Switzerland, we can note that there are still some copies of the first two editions of the *De rerum natura*, to which Doni may have had access when he was drafting the *De natura hominis*.[37]

Regarding copies of the *De natura hominis*, the few specialist studies on this text always include the adjective "rare". De Franco states that "the minimal or virtually non-existent dissemination of his thought was certainly conditioned by the rarity of copies of his work", hypothesizing that a wide European circulation "could occur, since Bacon read it in England, there was a copy held in Paris and a Dutchman, Van der Linden, mentioned it in his catalogue of medical works".[38] Rotondò defines the *De natura hominis* as a "rare" work, Enrico De Mas as "practically impossible to find", Francesco Fiorentino as "extremely rare".[39] However, these statements must be reconsidered in light, for example, of the number of copies of the *De natura hominis* held in German libraries. I know of three copies in Munich, and one each in Gotha, Berlin, Freiburg, Mainz, Wolfenbüttel, Dillingen, Dresden, Hannover, Göttingen, Rostock, Weimar, and Erlangen.[40] This is a larger number of copies than in Italy or any other European country. In some cases the owners were physicians, who were most probably interested in Doni's work for purely scientific reasons. This is true for example of the owner of one of the two copies in the Munich university library, Philipp Fischer, professor from 1789 at the University of Ingolstadt, or of Martin

35 Rotondò, *Studi e ricerche*, 397.

36 De Franco, *L'eretico Agostino Doni*, 18.

37 Staatsbibliothek zu Berlin; Universitäts- und Landesbibliothek Sachsen-Anhalt, Halle; Zentrale Hochschulbibliothek Lübeck; Niedersächsische Landesbibliothek Hannover.

38 De Franco, *L'eretico Agostino Doni*, 163.

39 Rotondò, *Studi e rirerche*, Chap. 8; De Mas, *Francesco Bacone*, 97; Fiorentino, *Bernardino Telesio*, 324.

40 Bayerische Staatsbibliothek München; Universitätsbibliothek der LMU München (2 copies); Forschungsbibliothek Gotha; Staatsbibliothek zu Berlin; Universitätsbibliothek Freiburg; Wissenschaftliche Stadtbibliothek Mainz; Herzog August Bibliothek Wolfenbüttel; Studienbibliothek Dillingen; Sächsische Landesbibliothek/Staats- und Universitätsbibliothek Dresden; Niedersächsische Landesbibliothek Hannover; Niedersächsische Staats- und Universitätsbibliothek Götingen; Universitätsbibliothek Rostock; Herzogin Anna Amalia Bibliothek Weimar; Universitätsbibliothek Erlangen-Nürnberg.

Fogel (1634–1675), who possessed the copy now held in the Niedersächsischen Landesbibliothek in Hannover. Other cases are far more interesting: one example is the copy in Rostock, whose owner (Jacobus Ebersperger, a physician who was Doni's contemporary) had it bound with a treatise by the Swiss physician and theologian Thomas Erastus (1524–1583), accused in the 1570s of being a Socinian and who formally introduced Doni to Stefan Batory's Polish court.[41] At this stage of my research on the German circulation of the *De natura hominis* it remains unclear if reconstructing the provenance of all these examples may help to shed some light on the true relations between Doni and Reformed circles and to clarify the aspect of his life that perhaps remains most obscure. In any case, in my opinion another line of enquiry may provide some evidence of a connection between Doni and the European Radical Reformation.

3 Some Hypotheses on the Relationship between Doni and Servetus and between Doni and Telesio

The few studies that exist on Agostino Doni, even though they are philologically accurate and tremendously relevant in historiographical terms for the discovery and understanding of this obscure philosopher, have given us an incomplete image of him. On the one hand, some researchers have 'included' him in a migration wave, mainly Italian, which scattered throughout northern Europe and into Poland and Transylvania after espousing the theories of the 'left wing' of the Reformation.[42] On the other hand, studies with a more philosophical orientation, such as De Franco's, focus on his naturalism *sui generis* which, to a much greater extent than Telesio's naturalism, turns on the doctrine of the *spiritus* and radically denies any sort of connection between Doni and the European Antitrinitarian movement of the sixteenth century.[43]

41 I thank Irene Friedl of the Universitätsbibliothek der LMU München, Sabine Brügmann of the Niedersächsische Landesbibliothek Hannover and Christiane Michaelis of the Universitätsbibliothek Rostock for this information. The Rostock copy contains numerous handwritten marginalia, which I plan to study in the near future. On Erastus as mediator between Doni and Batory, see Valente, "*Libertas philosophandi*," 126.

42 See notes. 2 and 4.

43 De Franco, *L'eretico Agostino Doni*, 46: "di certo dunque sappiamo che, quando veniva lottato dai 'papisti', non era né un calvinista né un antitrinitario; potremmo aggiungere anche che non fu sicuramente neanche un anabattista, dato che nessun accenno nelle sue lettere e nella sua opera potrebbe farlo sospettare."

From a biographical point of view, we can irrefutably demonstrate Doni's presence in numerous philosophical and medical milieus which at the time were under the influence of reformed ideas. When he was young he was imprisoned for religious reasons; unfortunately, the circumstances of this episode and of his later escape from Italy are unknown. In Germany, Doni tried to contact Simone Simoni, who at the time was a very controversial figure suspected of atheism; in Switzerland he sought the protection of the Paracelsist Theodor Zwinger; in Poland he tried, with the help of Thomas Erastus, to introduce himself to the circle of reformed physicians which inclued Niccolò Buccella, Fabio Nifo, and Simoni himself. In short, Doni's biography perfectly matches that of a sixteenth-century Italian 'heretic'. Unfortunately, no proof exists of his hypothetical heterodox religious convictions, neither in the surviving correspondence nor in the *De natura hominis*, his only extant work. Nonetheless, the philosopher has often been associated—due to a sort of historiographical misunderstanding—with opposition to Calvinist orthodoxy and generally with movements of open religious dissent (this is in relation to his Swiss period). However, all of this does not rule out the possibility of identifying other kinds of signs that might indicate a connection between Doni and European Antitrinitarianism.

As is well known, one of the founders of modern Antitrinitarianism was a physician, the Spaniard Michael Servetus.[44] His execution on the orders of Calvin and the publication of his great Antitrinitarian work, the *Christianismi restitutio* (The Restoration of Christianity), which both occurred in 1553, are two events that famously had an enormous impact on European heresy. Servetus' death led to the flight of many Antitrinitarians, especially Italians, from Switzerland to countries that were at the time more tolerant, such as Poland and Moravia.

In the *Christianismi restitutio*, Servetus posited a highly innovative theory on the passage of blood through the lungs. He argued that the soul consisted of bloody matter, which could pass throughout the body, and described the passage of blood from the right ventricle, through the kidneys, from the pulmonary artery to the pulmonary vein to the left ventricle, where its transformation into the *spiritus vitalis* was completed. In the lungs, the inhaled air was mixed with the refined blood, which then passed from the right to the left

44 There are a fair number of monographs on Servetus. See for example Zuber, *Michel Servet*; Bainton, *Michel Servet*; Birnstein, *Toleranz und Scheiterhaufen*; Domeyne, *Au risque de se perdre*; Manzoni, *Umanesimo ed eresia*.

ventricle.[45] Servetus' medical hypothesis had meaningful theological implications as well. His aim in this respect was to demonstrate how the soul could be concretely infused into man through the vital, divine breath, in accordance with the Biblical account: through the purification of blood by respiration described above. The involvement of the lungs was perfectly suited to justifying the contact between air and blood, which was refined and purified, thus allowing Servetus to explain the formation of the *spiritus vitalis* in the lungs and its passage into the left ventricle.[46]

In theoretical terms, the most interesting point is that, according to Servetus, the spirit manifests itself in the human body in three forms, which are three manifestations of the single Holy Spirit. There is thus an immediate connection with the three persons of the Trinity: in the same way, the three hypostases are but three different ways in which the one God manifests himself.[47]

The reception of Servetus' work in Italy is a highly controversial issue which cannot be treated in greater depth here.[48] As far as Switzerland is concerned, if the *Christianismi restitutio* was obliterated following Servetus' death there is no reason to doubt that less than three decades later—when Doni settled in Basel—the major theories of the *Christianismi restitutio* were still well-known in the city's heterodox milieus. For this reason, I think that we can connect the theory of the single, corporeal *spiritus* in Doni and in Servetus. This is surely the most innovative thesis of the *De natura hominis*, in comparison with Telesio's work; and, if we assume a connection between Servetus and Doni, it is a theory to which Telesio, living in southern Italy, could have had no access.

45 Servetus, *Christianismi restitutio*, 170: "Hinc dicitur anima esse in sanguine, et anima ipsa esse sanguis, sive sanguineus spiritus [...]. Ad quam rem est prius intelligenda substantialis generatio ipsius vitalis spiritus, qui ex aëre inspirato, et subtilissimo sanguine componitur, et nutritur. Vitalis spiritus in sinistro cordis ventriculo suam originem habet, iuvantibus maxime pulmonibus ad ipsius generationem [...]. Generatur ex facta in pulmonibus mixtione inspirati aëris cum elaborato subtili sanguine, quem dexter ventriculus cordis sinistro communicat. Fit autem communicatio haec non per parietem cordis medium, ut vulgo creditur, sed magno artificio a dextro cordis ventriculo, longo per pulmones ductu, agitatur sanguis subtilis: a pulmonibus praeparatur, flauus efficitur: et a vena arteriosa, in arteriam venosam transfunditur. Deinde in ipsa arteria venosa inspirato aëri miscetur, exspiratione a fuligine repurgatur. Atque ita tandem a sinistro cordis ventriculo totum mixtum per diastolem attrahitur, apta supellex, ut fiat spiritus vitalis."

46 Book 5, Chap. 5 of the *Christianismi restitutio* is devoted to the circulation of blood. These passages of Servetus' argument are described in detail in Ongaro, "La scoperta della circolazione polmonare". See also Simonutti, "Après Michel Servet."

47 Servetus, *Christianismi restitutio*, 169: "primus ergo est sanguis, cuius sedes est in hepate, et corporis venis. Secundus est spiritus vitalis, cuius sedes est in corde, et corporis arteriis. Tertius est spiritus animalis, quasi lucis radius, cuius sedes est in cerebro, et corporis nervis. In his omnibus est unius spiritus et lucis Dei energia."

48 See Ongaro, "La scoperta della circolazione polmonare."

I do not claim that we could define Doni as an Antitrinitarian on this basis; at least at the "official" level of his surviving printed work, Doni never made any Antitrinitarian statements. Nevertheless, he was in some ways very close to Antitrinitarian circles: this is proven by his journeys, his correspondence network, and the protectors who helped him during his European peregrinations. Furthermore, he was a religious dissenter, though we do not know exactly of what sort, but we do know it concerned an offence he was convicted of while still young. Finally, he was a physician and also probably studied in Padua.[49] In the sixteenth century, the Paduan university was one of the Italian environments most receptive to the ideas of the Reformation, due above all to its international nature, the relatively tolerant policies of the Venetian Republic and specific theoretical foundations: many intellectuals at the university had espoused radical Aristotelianism, skepticism, and materialism, in close connection with issues of medical theory.[50] All these conditions make it highly likely that Servetus' work, in particular the theory of the single *spiritus*, could have exercised considerable attraction for Doni, making him decide to include it in the Telesian framework of the *De natura hominis*. Suggesting that this connection is highly plausible does not imply a theorization of Doni's adhesion to Antitrinitarianism but rather that exposure to the same cultural context as other Italian exiles *religionis causa* may have influenced at least one of his philosophical theories.

More than one explanation is possible regarding his unsuccessful integration into the Polish court (apart from the usual explanations advanced in such cases, such as a very difficult personality). Even if we accept the hypothesis that Doni had a strong affection for certain Antitrinitarian ideas, some evidence of which could perhaps be provided by, say, lost writings of his, it was highly difficult to integrate his philosophical system into the milieu of the Italian exiles in Eastern Europe. It was a philosophy that was in no way compatible with the highly rationalistic approach of Fausto Sozzini or Ferenc Dávid. At the point when Doni went to Poland, Antitrinitarianism had already entered into a new phase which was completely different from the "Servetian" one. The "Paduan" form of Aristotelian criticism and ethical issues had won against the philological, highly complex neo-Platonism of the Spanish physician.[51] Even if the *De natura hominis* is the only one of Doni's writings to have survived, this text is fully sufficient to have us clearly understand the reasons for

49 See Rotondò, "Doni, Agostino."

50 On this, see Nardi, *Saggi sull'aristotelismo Padovano*; Randall, *The School of Padua*; Poppi, *Introduzione all'aristotelismo padovano*. For a general overview of the teaching of medicine at Padua, see Bylebyl, "The School of Padua." On the specific link between the radical Reformation and medicine in Padua see Suitner, "Radical Reformation and Medicine".

51 Balázs, *Early Transylvanian Antitrinitarianism*.

his intellectual marginalization by the Italian Reformers in Poland. Doni was neither a neo-Platonic nor an enthusiastic Aristotelian thinker. His world, like Telesio's, was pervaded by the struggle between hot and cold, by the absolute supremacy of nature as the only reality and by a material *spiritus*, which was identical to the corporeal soul. The fact that he is not mentioned in the early stories of Polish Antitrinitarianism is also proof *per se* of Doni's unfortunate fate in Poland. The spread of the legend that he was assassinated by Fausto Sozini, reported by the *nunzio* Alberto Bolognetti (and, in current research by Antonio Rotondò), is also a strong indication of the involvement of the philosopher in the circles of Venetian exiles in Poland and their mutual rivalries. Nevertheless, it is certain that more research still has to be done in order to shed light on the story of his presence in Eastern Europe, and hopefully this will lead to further understanding.[52]

As I have shown in the previous pages, this is not the only aspect of Doni's philosophy that is very difficult to clarify. The recurring association between Doni and Telesio is also not without contradictions. So far two opposite theses have prevailed. Some scholars have considered Doni a mere plagiarist of Telesian philosophy. By contrast, De Franco's study almost exclusively insists on the novelty of his work, omitting an incisive comparison between Doni's text and the first two editions of Telesio's main work. An accurate analysis of the concept of *spiritus* in Doni, in Persio and in the three editions of Telesio's *De rerum natura* would probably be the best starting point to find clarification on the relationship between the two philosophers from Cosenza.

The examples I have presented show that Doni's philosophy, despite the importance of De Franco and Rotondò's contributions, deserves far more attention than it has hitherto been given, in particular regarding these three aspects: a precise comparison between the *De natura hominis* and the first two editions of the *De rerum natura*; a rigorous contextualization of the philosopher's thought in the natural philosophy of the late sixteenth century, above all in relation to Antonio Persio; and an examination of the relationship between Doni and the European Antitrinitarian movement, in particular of his debt towards Michael Servetus' work. The integration of the Servetian theory of the single *spiritus* into Doni's *De natura hominis*—highly probable, as we have seen in the previous pages—is extraordinary proof of a connection between Telesio's natural philosophy and the major developments of sixteenth-century religious non-conformism.

52 There is no trace of Doni in Stanislas Lubieniecki, *Historia reformationis Polonicae*, English transl., *History of the Polish Reformation*; Lampe, *Historia Ecclesiae Reformatae*.

CHAPTER 11

Between Myth and Reality: the *Accademia Telesiana*

Giulia Giannini

Whether or not Telesio actually founded, or was affiliated to, a scientific academy is a question that has remained constantly ambiguous. Historiography seems to have expressed a tacit desire that has often led to discussion of an Accademia Telesiana or Accademia Cosentina, of which Telesio would have been the founder or re-founder. On the other hand, Telesio is a key figure of the Italian Renaissance, whose most important institutional characteristics include precisely the birth and proliferation of academies. Beginning above all with groups established towards 1440 around famous humanists such as Ottavio Rinuccini and Marsilio Ficino in Florence or Pomponio Leto and Cardinal Bessarion in Rome, hundreds of academies flourished throughout the Renaissance period, which were quite different from one another in terms of methods, structure and approaches. From Ficino to Galileo, the biographies of the greatest intellectuals of the time were marked by the foundation of or affiliation to an academy.

But there are more than just historical reasons connecting Telesio in historiography to what can be referred to as a true academic movement, which involved the founding of academies in nearly every medium or large Italian city. This is the fact that he belonged to that historical period deeply marked by the emergence of new associative forms.

For a long time, the genesis of academies was considered an organizational change driven exclusively by new intellectual needs and, in particular, by a reaction against the "stagnant" teaching system pursued by universities. Recent studies show that the relationship between traditional institutions (such as universities and colleges) and new forms of the organization of knowledge is much more complex. It is sufficient to consider the fact that many of the members of the new academies also held university positions, or the heterogeneity of "Aristotelianisms" taught within the university system and the occasional attempts to integrate non-Aristotelian ideas into the traditional system. Historiographically, however, the founding of academies in humanistic Italy has long been explained as the formation of intellectual groups outside the universities and as something fundamentally opposed to the Scholastic philosophy perpetuated by those institutions.

 | DOI:10.1163/9789004352643_013

In this sense, the open and continuous Telesian debate with Aristotle's philosophy and tradition, and his determined defense of the *libertas philosophandi*, based on which he rebelled against the dogmatically established *auctoritas*, would certainly make the scholar a worthy founder of an academy. No wonder, then, that a historian of philosophy such as Eugenio Garin tells how Telesio, upon finding the doors of the ancient universities closed, "had to be satisfied giving vent to his 'dreams' in an academy in his native Cosenza".[1]

But did an Accademia Telesiana actually exist? And if so, in what form?

1 Aulo Giano Parrasio and the first *Accademia Cosentina*

Much of the literature from the eighteenth and nineteenth centuries which concerns Telesio, academies, and the culture of southern Italy already makes reference to an academy in Cosenza during the first few years of Telesio's life. The academy, promoted and founded by Giovan Paolo Parisio (1470–1522), probably originated between 1511 and 1534 and would have represented not only an important reference point for the young Telesio but also the nucleus from which a second academy would later emerge: the Accademia Telesiana. Better known by his Arcadian name of Aulo Giano Parrasio, Parisio moved from Cosenza, first to Lecce and then to Corfu, where he perfected his study of the Greek language, then to Naples and Rome, where he assisted in Pomponio Leto's last lessons, and finally to Milan, Vicenza, Padua, and Venice. The Latin name Aulus Janus Parrhasius likely dates back to the scholar's stay in Naples, where he attended the Accademia Pontaniana. It was during one of his return trips to Cosenza that, according to the literature, Parisio founded the institution that would be known as the Accademia Cosentina or Accademia Parrasiana.

An initial problem concerns the actual dating. Parrasio in fact returned to Cosenza around 1511 and left again in 1514; he did this a second time between 1520 and 1521. He would then remain in Cosenza until his death, which was typically traced to 1534, but was more recently confirmed to have taken place in 1521 thanks to the rediscovery of Parrasio's will which indicated the date it was read.[2] But the dating—which is also discussed in more recent literature—is a collateral issue: conclusive documentary evidence of the existence of the academy during this so-called "first phase" is in fact lacking.

1 "The old universities closed their doors to Telesio. He had to be satisfied giving vent to his "dreams" in an academy in his native Cosenza". Garin, *Renaissance Characters*, 150.

2 See: Lepore, "Per la biografia di Aulo Giano Parrasio," Tristano, *La biblioteca di un umanista calabrese*, in particular 43 and onwards.

A classic source for the study of Italian academies is the monumental work by Michele Maylender.[3] During the 1920s, Maylender made the first real attempt to bring together all the evidence related to Italian academies in a single work, including those which were founded between the second half of the 15th century and the early 20th century. As reflected in his introduction to the volumes, Maylender views the academic movement as a fundamental prism through which one can understand every aspect of a period in Italy's cultural history, and not only from a social or literary standpoint.[4] Although sometimes incomplete in terms of the list of academies and sources, and often too hasty in its historical contextualization, Maylender's work remains a fundamental point of reference. The Accademia Parrasiana is classified as the "oldest" among the academies that emerged in Cosenza. Giano Parrasio would have founded it in imitation of the Accademia Pontaniana in Naples, to which he would have been an influential contributor. While not substantiating his claim with any relevant document and relying solely on 18th and 19th century literature—which, is must be said, is not entirely in agreement[5]—Maylender does not question the existence of the academy, stating only that "the Parrasiana carried the title of academy since its inception, while not having an academic structure".[6] Girolamo Tiraboschi, who was particularly attentive to the role of institutions in the production and transmission of knowledge, appears to share the same opinion, mentioning the academy in Cosenza as "established in some way by Giano Parrasio".[7] Francesco Fiorentino's thoughts

3 Maylender, *Storia delle accademie d'Italia*.

4 "Dall'illustrazione particolare delle vicende accademiche si ritraggono anche con somma facilità dei dati importantissimi intorno alla vita, abitudini, relazioni di famiglia, studi ed opere d'una gran parte degli scrittori d'Italia, e d'altra parte vi si legge quasi lo spirito dei tempi e l'influenza che esso esercitava sugli ingegni, sullo stile, sul modo d'avvisare le opere dell'ingegno, sui costumi e sul carattere. E soltanto le storie particolari delle Accademie possono farci vedere lo stato e la forma dell'istituzione nel corso de' tempi, cioè il periodo della sua nascita, sviluppo, fiore e decadenza con tutte le fasi intermedie". Michele Maylender, *Difesa* 4 (1900), published in Maylender, *Storia delle accademie d'Italia*, xv.

5 See for example Giacinto Gimma, Idea della storia dell'Italia letterata esposta coll'ordine cronologico dal suo principio fino all'ultimo secolo, vol. 2, 478; Francesco Saverio Quadrio, Della storia e della ragione d'ogni poesia, vol. 1, 65. Maylender himself is aware of the contrast, and in the conclusive part of his notes he writes: "[...] l'ignorarono, erroneamente asserendo il Gimma che il primo a formar in Cosenza una Accademia fu Bernardino Telesio, e dichiarando il Quadrio che non prima del 1580 ebbe Cosenza la sua virtuosa Adunanza", Maylender, Storia delle accademie d'Italia, vol. 4, 214.

6 "Portò la Parrasiana sin dal suo aprimento il titolo d'accademia, ma non ebbe assetto accademico". Maylender, *Storia delle accademie d'Italia*, vol. 4, 212.

7 "Dell'Accademia di Cosenza, cominciata in qualche modo da Giano Parrasio [...]".Tiraboschi, *Storia della letteratura italiana*, book 7.1, 124.

on the matter are not much different in his famous essay on the Accademia Cosentina and on Bernardino Telesio's philosophy. After addressing the dating problem, the lack of documents, and the absence of references to the academy in the literature of the time (for example in the works of Niccolò Salerni who wrote a long death ode dedicated to Parrasio), Fiorentino concludes: "I believe, rather, that those friendly get-togethers had not been given the name of *Academy*: the academic activity was there, but the name was missing".[8] More recently, Tobia Cornacchioli has sought evidence attesting to the existence of an academy in the works of authors from Parrasio's or Telesio's time;[9] he was perhaps the first to seriously address the problem of the lack of documented origins of a first Accademia Cosentina. Like Walter Lupi[10] after him, Cornacchioli notes how, in the account of his journey to Cosenza in 1526, Leandro Alberti does not mention an academy but only refers to "many noble minds", including Parrasio and Antonio Telesio.[11] Therefore, both come to the conclusion that, for a long period of time, it is not possible to speak of an actual academy which had a relative structure and rule system in Cosenza. Yet resolving the issue of the existence of the academy or its degree of informality solely on the basis of the absence of documents is slightly premature, and even Alberti's narrative proves to be in no way conclusive. Moreover, no academy corresponding to the particular period is mentioned in his *Descrittione di tutta Italia*, a work quite attentive to the historical and geographical study of the peninsula.

Conversely, Emilio Sergio's recent research takes a different direction, identifying the first "Accademia Cosentina" in a Greek and Latin school founded in Cosenza by Giano Parrasio.[12] Sergio's hypothesis is essentially based on the collection of letters edited by Parrasio, and in particular on a letter sent by the scholar to the Baron of Belmonte Vincenzo di Tarsia in 1512.[13] In his letter Parrasio implicitly refers to the existence of a school for the education of youth in Cosenza. Even assuming that Parrasio is referring to an actual Greek and

8 "Credo bensì che questo nome di Accademia non si fosse usato a qualificare quegli amichevoli ritrovi: ci era il fatto e mancava il nome", Fiorentino, *L'Accademia cosentina*, 34–35.

9 Tobia Cornacchioli, *Lineamenti di storia della cultura calabrese*. See also: Cornacchioli, *Nobili, borghesi e intellettuali*.

10 Lupi, *Alle origini della Accademia Telesiana*.

11 Alberti, *Descrittione di tutta Italia*, 187.

12 See in particular: Sergio, "Parrasio in Calabria," id., "L'Accademia Cosentina," 259–272, id., "Parrasio, Antonio Telesio e l'Accademia Cosentina."

13 "Remittiret aliquid de iudicio suo Lucilius, et, qui Lucilio subscripsit, Cicero, si viverent hac aetate: iuventutemque Cosentinam bonarum artium studiis cum quavis Italiae civitate certantem viderent." Parrasio, *De rebus per epistolam quaesitis*, 62; Mattei, *Quaesita per epistolam*, 110. See among others: Sergio, "Parrasio in Calabria," 428.

Latin school, nothing suggests that the school was founded by Parrasio nor—above all—that the school itself had at some point taken on the form of an academy or assumed the "Accademia Cosentina" name. More than the actual existence of the academy, Sergio seems interested in the problem of its dating which, on the basis of the letter in question, he establishes as 1511.

2 The Accademia Cosentina in Telesio's Time

There are two main reasons that prompted historiographers to insist on the idea of the Accademia Parrasiana. The first, attributable to reasons relating to local patriotism, is that the existence of a "first" academy with Parrasio in Cosenza would make the Accademia Cosentina one of the first Italian academies still in existence, founded even before the Accademia della Crusca, established in Florence in 1583.

The second and more interesting reason is the substantial lack of biographical information concerning Telesio during the period before the first publication of *De natura iuxta propria principia* (1565). This represents a long period in the life of the philosopher (more than half a century) in relation to which only scant and conflicting information is available, a period that includes his education, his studies, and the decisive gestation phase of his best known and most important work.

According to early biographers, Telesio had followed his uncle Antonio to Milan (1517) and to Rome (1523), and then made his way to Padua to study philosophy and mathematics under the guidance of Girolamo Amalteo and Federico Delfino. Yet claims of his studies at the University of Padua are rather inconsistent with the account provided by Telesio himself in the preface of the first edition of *De natura iuxta propria principia*, in which the scholar states that he was not permitted to attend any of the famous public Italian universities.[14] However, the scholarly literature ranging from Fiorentino to Bondì[15] agrees on one point: the role played in Telesio's education and studies by the philosopher's uncle Antonio, his first and perhaps only teacher.

14 "At neque adhuc mihi confisus, cui, ut dictum est, extremum modo vitae tempus philosophari licuit, et nequaquam in magno ocio magnaque animi tranquillitate, neque in publicis inclitisque Italiae Academiis a praestante aliquo viro edoceri, sed in magnis plerunque solitudinibus, molestissimis oppresso impedimentis, Graecorum monumenta evolvere, Latina non satis percipienti, ignotis referta vocibus." Telesio, "Proaemium," in *De natura iuxta propria principia* (1565).

15 Bondì, *Introduzione a Telesio*.

Although Parrasio and Bernardino Telesio probably had no significant direct contact,[16] Antonio was in Cosenza during Parrasio's first trip back to his hometown. It is therefore through his uncle that Bernardino would have benefited from Parrasio's teachings: it would have been nearly impossible for the founding of an academy during this period to have not engaged a scholar like Antonio Telesio in some way, as he is often mentioned by historians as one of Parrasio's students and sometimes even among the "members" of the alleged Accademia Parrasiana.

A line of continuity would thus connect Bernardino Telesio to Giano Parrasio through the mediation of his uncle Antonio, as well as through Telesio's assumption of the leadership of the Accademia Cosentina following Parrasio's death.

In the preface to the first edition of *De Natura*, Telesio provides the first direct reference concerning his life. In particular, he describes the journey he made in 1563 to Brescia to present his new doctrine to the Aristotelian Vincenzo Maggi, who was a Professor in Padua and Ferrara. Some scholars have detected the first trace of Telesio's succession of Parrasio in a letter he received in Bologna.[17] The letter was written by Sertorio Quattromani, who had transferred from Cosenza to Rome a few years earlier. Having received notice from Telesio regarding the outcome of meetings with Maggi, Quattromani does not show any surprise, stating that "that thing" which for others is new is "a thousand years old" for him.[18] Is this really an allusion to an Accademia Telesiana? It is true that in a letter dated August 5, 1571, and addressed to Giovan Battista Ardoino in Cosenza, Quattromani—a few kilometres away at the farmhouse in Cerisano (Cosenza)—inquires about local activities, asking, among other things, if anything was being done for an academy which he refers to as "ours" and defines as being "in ruin".[19] It is not clear if the reference is to an academy that was already operational or simply to a project; no further explanations are provided. The reply from Ardoino, who was a noble from Cosenza and Quattromani's brother-in-law, is missing. The testimonies are too insubstantial to establish the existence of an academy in Cosenza and, above all, to

16 Telesio was a child when Parrasio returned to Cosenza for the first time between 1511 and 1515; instead he was in Milan during Parrasio's second stay in Calabria between 1520 and 1521.

17 See in particular: Lupi, *Alle origini della Accademia Telesiana*, 23–24.

18 Quattromani a Telesio, 22 September 1563 in *Lettere di Sertorio Quattromani*, book 2, 67–69.

19 "[…]che si fa in cotesto mondo di Cosenza? Farsi nuovo preparare dai cieli in roina della nostra Accademia?" Sertorio Quattromani a Giovan Battista Ardoino. Cf. *Lettere di Sertorio Quattromani*, 128.

even vaguely define the circumstances. However, the knowledge of Telesian philosophy exhibited by Quattromani in his letter to Telesio provides a first important reference relating to the presence of a cultural and social environment in which Telesio's philosophy had already been long known.

The first appearance in a printed text of the expression *Academia Telesiana*, already noted by Fiorentino during the 1870s and more recently reiterated by Walter Lupi, appears to point in the same direction. The text *Disputationes libri novarum positionum* (1576)[20] made public the contents of the debate on Telesian philosophy organized by Antonio Persio in Venice. Persio, who was originally from Matera, had moved to Naples in 1560, which is where he met Telesio. Following a brief period in Rome and Perugia, his travels led him to Venice as a tutor for Andrea Corner, a son of the Venetian patrician Giorgio. In 1575, at the time of the debate, Persio had not only already written his *Apologia pro Bernardino Telesio adversus Franciscum Patritium*[21] but had also just published the *Liber novarum positionum*,[22] a doxographic collection of philosophical opinions on various branches of knowledge, including rhetoric, dialectics, ethics, law, and physics. Moreover, it is in the *Liber* that Persio clarified his relationship with Telesio, even describing how the philosopher from Cosenza had discussed the second edition of *De rerum natura* with him before its publication. The Venetian debate, based precisely on some of the arguments contained in the *Liber*, is the first public debate contrasting Telesio's philosophy to Paduan Aristotelianism. The content was published by one Andrea Aletino, who is also not identifiable. In his dedicatory letter, Aletino turns to Ugolino Gualteruzzi, Jerusalemite Knight and son of that Carlo to whom Monsignor della Casa had, in 1545, repeatedly declared his desire to want Telesio at his side as a secretary in Venice.[23] Little to nothing is known about the author. The name "Andreas Alethinus" is likely a pseudonym with possible Leccese origins.[24] In the letter the writer suggests he had known Telesio and defines Persio as the "*alterum columen*"[25] of that academy that Ugolino had known for some time and in the defense of which the contents of the debate were being published. Supporters of the two doctrines—Telesian and Aristotelian—had

20 Persio, *Disputationes libri novarum positionum*.

21 Id., *Apologia pro Bernardino Telesio*, in Telesio, *Varii de naturalibus rebus libelli*.

22 Persio, *Liber novarum positionum*.

23 The letters sent by Giovanni della Casa to Carlo Gualteruzzi on 26 March and 11 April 1545 can be seen in: *Lettere di Monsignor Giovanni Della Casa*, 13–14.

24 According to one of the accepted uses of Latin during the Renaissance, Alethinus refers to "a citizen of Lecce".

25 "[...] Antonius Persius, alterum Academiae nostrae columen cum a Telesio discessi [...]". Persio, *Disputationes*, 4.

confronted each other for three days in May of 1575 at the home of Giorgio Corner. Among Telesio's supporters, mention is made of the Venetian diplomat Giovanni Michiel, the young classicist Luigi Lollino with his two brothers Giovanni and Angelo—sons of the Venetian patriciate in Candia recently returned to Venice—and Ugolino Gualteruzzi. But the "witnesses" and "arbiters" of the dispute also included the Venetian Pietro Contarini, to whom Persio would dedicate his *Trattato sull'ingegno dell'uomo* (1576), Andrea Corner, Luigi (or Alvise) Gradenigo, librarian of the San Marco public library in Venice, the physician Daniele Ferulano, Arcangelo Mercenario, Professor of Philosophy at the University of Padua, and the Minorites Clemente Montefalco and Lelio Piacentini.

The discussion was anything but peaceful and on the last day a number of Paduan scholars even threatened to move "from words to action". The violent reactions had driven Persio to abandon the idea of repeating the debate in Padua[26] and Aletino to publish and publicise the contents, "*ne quid Academia Telesiana detrimenti patiatur*".[27]

Aletino's dedicatory letter was written in Rome, and the leaflet was printed in Florence. The participants in the dispute were philosophers, scholars, and aristocrats of the Venice area, professors and students from the University of Padua, and Venetian nobles or scholars in their service. If Fiorentino sees a clear reference to the Accademia Cosentina—not necessarily referred to as "Telesiana" by its own members, but known as such on account of Telesio's fame[28]—Lupi correctly emphasizes that nothing points to Calabria.[29]

Certainly one cannot but take into account the fact that the phenomenon of academies spread rapidly throughout Italy, changing form and adapting to different cultural needs and requirements, together with the ambiguity that had characterized the term throughout the Renaissance. It was used to identify a range of very different phenomena: formalized groups, public or private, with regulations and statutes, but above all—at least until the massive intervention of public power between the eighteenth and nineteenth centuries—simple informal gatherings among friends and scholars who would freely decide to discuss issues of common interest, without deadlines, fixed locations, rules, or regulations. Often, especially until the end of the sixteenth century, the term was also used to indicate casual encounters, lectures, and debates on a specific issue.

---

26 Persio, *Liber novarum positionum*, 8.

27 Id., *Disputationes*, 7–8.

28 Fiorentino, *L'Accademia cosentina*, 102–103.

29 Lupi, *Alle origini della Accademia Telesiana*, 56.

Nevertheless, some references still make it possible to link together that quantitatively formidable and qualitatively diverse series of events that deliberately took on the title 'academy': (1) An academy took a name, which was often strange, curious, excessive, and not infrequently the target of irony, especially from abroad, but which was, nevertheless, the first sign of group identity; (2) An academy had an emblem (*imprese*) or motto, which symbolically communicated intentions, objectives, and approaches; (3) Less frequently, during the Renaissance, there was the assumption of a stable and organized form with the adoption of a legislative system and written regulations; (4) Above all, an academy had its very own production. An academy would print its *acta*, or proceedings, whether from the contribution of individual members or in the form of collective texts, and select and disseminate academic discourses. These were works which primarily comprised literary texts, orations, prose and poems related to special occasions, acclamations meant to celebrate specific events but also "arguments" and "discourses" which were more philosophical in nature and dedicated to specific issues or problems. Often, it is only possible to trace the existence of a specific academy through produced materials or through membership. Association with an academy also soon became a question of social consideration and prestige, a mark of identity to affix on the cover page of one's own works.

Within this context, Andrea Aletino's and Antonio Persio's references to an "Academia Telesiana" become easier to interpret. The term may not only refer to the existence in Italy and abroad of a group of Telesio's followers but may even refer to the very same debate organized in Venice.

It is this aspect that is most surprising in relation to the Accademia Telesiana and the Accademia Parrasiana: not so much the absence of a founding action—common to most Renaissance and modern academies—but rather the lack of "academic" works, the lack of specific references to the two academies not only in correspondence but also on the cover pages of works authored by their alleged members. And this imbalance is even more striking when analyzing documents relating to the so-called "third phase" of the Accademia Cosentina.

3 "Third phase", Sertorio Quattromani and the Academy

In 1591, three years after Telesio's death, the *Ragionamento chiamato l'Academico, overo della Bellezza*,[30] a treatise on aesthetics written in vernacular by Jacopo di Gaeta, was published in Naples. Very little information

30 Iacopo di Gaeta, *Ragionamento chiamato l'Academico*.

is available on the author, who is repeatedly cited by Tommaso Campanella in his works. In the brief biographical index included at the end of *Rime in lode della Illustrissima et eccellentissima Donna Giovanna*, Giovan Giacomo de Rossi mentions that he was from Cosenza and a Doctor of Law. The name of a jurist Jacopo di Gaeta also appears in the list of consultants of the Diocesan Congregation in Cosenza, who in 1600 would have had to review a series of texts, including Telesio's *De rerum natura*, which were entered in the Index of Clement VIII as "*donec expurgentur*".[31]

The *Ragionamento*, a "brief treatise on beauty"[32] which has only recently been rediscovered, is the first publication that can be directly attributed to the activities of an Accademia Cosentina. In fact, not only is the text explicitly addressed "to the academics of Cosenza", but when introducing the theme the author turns to the "Prince" who—he writes—gave him the task to "reason [*ragionare*] on beauty [...] in front of such a noble and scholarly group".[33] Not only was the treatise most likely discussed within the academy, but it had been commissioned by the Prince. The first references to a system of norms or practices also emerges in the short preamble preceding the discourse. Rhetorically downplaying his abilities in relation to the superiority of the argument and the audience, Jacopo justifies its "undertaking [*impresa*]" by appealing to the authority of the Prince and by making reference to established practices:

> If it were up to me, I would gladly give up this task to people certainly more capable than myself. But as a universal law has been established among us over time and is accepted by all, without acknowledging any excuse in relation to that which from time to time is ordained by our Prince, [and] despite knowing that I shall not receive any praise, I bow to the weight of this task and dedicate myself to it with a good heart..[34]

Thus a picture emerges of an academy whose members are alternately called upon by the Prince to "reason [*ragionare*]" on particular themes, as confirmed by the last few lines of the text in which it is announced that Fabio Cicala is

31 Ms. Vaticano Latino 6529, ff. 28r–29v. See: Lupi, *Alle origini della Accademia Telesiana*, 82.

32 "Un picciolo trattatuzzo della bellezza", Gaeta, *Ragionamento*, n. 650, 48.

33 "ragionare della bellezza [...] appresso così nobile et intendente brigata". Gaeta, *Ragionamento*, 53.

34 "Io per me cederei volentieri a coloro che di me vagliono certamente assai più. Ma da che legge universale è tra noi ricevuta, prima in commune da tutti e poi ancora per costume lungo fermata, di non ammettersi scusa niuna a quello che pur di volta in volta ci venga ordinato dal nostro Prencipe, et io, tutto che poca o niuna loda me ne sia per seguire, m'inchino al peso e sotto entro di buon cuore alla impresa." Gaeta, *Ragionamento*, 53–54.

already preparing to "reason [*ragionare*] on love".[35] Someone would perhaps lecture on Dante[36] and discussions had already been held, likely by Sertorio Quattromani and Giulio Cavalcanti, from whom Di Gaeta cites a number of works in reference to arguments previously made by others. Whether the arguments had been discussed in front of the academy or not, the two authors were clearly distinguishable as members of a group, even if that group was probably unofficially established.[37]

Thus, when reading between the lines of the *Ragionamento*, a veritable Renaissance academy comes to light, together with a name, a Prince, specific practices, and members. There is also a first hint of what will later be recognized by Tommaso Campanella[38] as the academy's symbol: the painting of a "valiant and noble Woman" presented a few days earlier to the academics and depicting Danae impregnated by Jupiter in the form of golden rain.

Dante, Petrarch, but above all Telesio emerge as the examples and writers dear to the academy. And it is precisely "according to the art" of Telesio, lauded as "master [*maestro*]" and "father [*padre*] of the Accademia Cosentina", that Jacopo di Gaeta seeks to address the theme of beauty:

> I shall not be able to vividly portray and to represent for your eyes this noble image with the vagueness of various colors, but I shall nevertheless strive with a few lines, as best as I can, according to the art of our never sufficiently praised master, and the father of the Academia Cosentina.[39]

Although the academic interests of the Accademia Cosentina were mainly encyclopedic in nature, this was typical of most Renaissance academies and not only of Telesio's work. On the other hand, the first work signed as such by an academic is precisely *La philosophia di Bernardino Telesio ristretta in brevità et scritta in lingua volgare* (1589) by Sertorio Quattromani, who adds "Cosentian academic [*Accademico cosentino*]" on the cover page.

The first reference to an Accademia "Cosentina" had already appeared in the annotations of a work from 1585, when Telesio was still alive. The work in

35 "Veggio ben io troppo altamente apparecchiarsi il Cicala a ragionar dell'amore." Ibid., 84.

36 "Ma delle grandezze e maraviglie di questo veramente divino Poeta si tenerà, forse, altre volte particolare ragionamento." Ibid., 69.

37 Cf. ibid., 71–72: "[…] questa è parte già trattata da i nostri […]".

38 Campanella, *Poetica*, 426.

39 "Non potrò con la vaghezza dei varii colori ritrarre al vivo e rappresentare a gli occhi vostri questa nobilissima imagine, m'ingegnerò nondimeno con alcune picciole linee tirate, al più che potrò, secondo l'arte del nostro non mai a bastanza lodato maestro, e padre della Cosentina Academia". Gaeta, *Ragionamento*, 54.

question is the aforementioned *Rime et versi in lode della illustrissima et eccellentissima Signora Donna Giovanna Castriota Carrafa*,[40] a collection in celebration of the Duchess of Nocera. A number of celebrated southern lyric poets contributed, such as Angelo di Costanzo, Bernardino Rota, and Luigi Tansillo, as well as many Calabrian authors including Jacopo di Gaeta and Bernardino Telesio himself. Commissioned by Scipione de Monti upon his departure to war, the work was entrusted to Sertorio Quattromani and Giacomo de Rossi; at the conclusion of the text de Rossi drew up a short biographical index. It is in this "table of authors" that, referring to Fabrizio della Valle, the nephew of Quattromani, De Rossi added: "this man is from the good Accademia Cosentina".[41] It is difficult, however, to consider the collection the direct result of an academic activity. It is an incidental work in which the academy's name appears only in reference to Fabrizio della Valle (there is no mention in the short biographies of other Calabrian scholars such as Telesio or Jacopo di Gaeta). With regard to the name of the academy, a letter from Quattromani to Giovanni d'Aquino illustrates how four years later, in 1589, the matter was still being discussed.[42] It was perhaps d'Aquino who proposed it as the official name of the academy shortly after the death of Telesio. It was, on the other hand, precisely d'Aquino who edited the *Oratione in morte di Berardino Telesio*, written to be read publicly in Cosenza in front of the academics. Only published in 1596, the work closes with a dedicatory letter to the Cosentian academics, dated January 1 of that year. Two other works from the same period signed by "Cosentian academics" are the *Rime del sig. Giovanni Battista Ardoino Academico Cosentino* (1590) and *Le historie de monsig. Gio. Battista Cantalicio vescovo di Civita di Penna, et d'Atri, delle guerre fatte in Italia da Consalvo Ferrando di Aylar, di Cordoua, detto il Gran Capitano, tradotte in lingua toscana dall'Incognito Academico Cosentino* (1595, 1597).

These documents represent the only evidence of the academy during the 16th century. Although Telesio is an important point of reference, the most important documents date after his death. Moreover, the academy maintained a strictly Renaissance structure, with a major interest in literature, especially in

40 Castriota Carrafa, *Rime et versi*.

41 "[…] costui è della buona accademia cosentina." Ibid.

42 "Il nome, che Vostra Signoria ha posto a i nostri Academici è così proprio, et così honorato, che non potrebbe desiderarsi migliore. Pure quando piacesse a lei, et a gli altri, parrebbe a me, che i nostri Academici non havessero a chiamarsi altramente che 'Academici Cosentini'. Perché questo nome così semplice, et così puro ha del più grande, che tutti gli altri nomi ricercati, et investigati con ogni studio, et con ogni industria, et mostrerebbeci più modesti et meno ambitiosi". Quattromani to Giovanni d'Aquino, 13 March 1589 in Quattromani, *Scritti*, 66.

the vernacular, and in humanistic culture in general. Therefore, even if Telesio's role in the foundation of the Academy cannot be dismissed, it is also not particularly likely.

4 Conclusions

If one can speak of the Accademia Telesiana during Telesio's lifetime, it is only in reference to the diffusion and reception of his philosophy in Italy and abroad. There is no evidence of the actual foundation or creation of an academy with a more or less regular location and activity. The situation changes following Telesio's death. Despite its informality, the Accademia Cosentina presents the typical traits of the Renaissance academy; however, with this in mind it is appropriate to scale back Telesio's role. Defined in *Ragionamento* as "the father of the Accademia Cosentina", Telesio was undoubtedly an important source of inspiration; however, keeping in mind chronological considerations, the expression should be interpreted broadly, i.e. in terms of an ideal rather than an actual institutional founding. And, as much as Telesian philosophy served as an important point of reference for its members, the academy did not have Telesio as its only model. Other important points of reference included Dante, Boccaccio, and Petrarch, and the use of vernacular language was considerable.

The Accademia Cosentina, which was the first academy in Cosenza based on existing evidence, takes the form of an academy of a primarily encyclopedic nature: arguments related to the philosophy of nature are present but always as part of an organic relationship with the classic and literary forms of knowledge. In this sense, even if we want to attribute a somewhat "Telesian" character to the Accademia Cosentina, it is certainly impossible to speak of a scientific academy in the strict sense of the words.

The testimonials and documents in our possession exclude the hypothesis that Telesio may have founded an academy or that he even took part in one. However, one should not overlook Telesio's symbolic value in the academic (and academic-scientific) culture of the late Renaissance, from the presence of "Telesians" among the Lynceans[43] up to the references to and adoption of Telesian ideas by Francis Bacon, the mediator and innovator of scientific and

43 In a recent publication, Paolo Galluzzi, expanding on a number of considerations already put forward by Eugenio Garin, identifies a Telesian style in the Cesian project by the Lyncean Academy. See: Garin, "Fra '500 e '600: scienze nuove, metodi nuovi, nuove accademie," Galluzzi, *Libertà di filosofare in naturalibus*, Chapter 2.

philosophical culture in the transition from the philosophical era of the Italian Renaissance to the birth of modern scientific society in Britain.

Therefore, if Telesio was not directly the founder or a member of a scientific academy, it is not because his philosophy was not compatible with the academic form. The rather complex and varied phenomenon relating to the foundation of academies was above all a social phenomenon and it was only from the early years of the seventeenth century, long after Telesio's death, that the first academies that could be defined as "scientific" began to make actual progress.

CHAPTER 12

The Transformation of Final Causation: Telesio's Theories of Self-Preservation and Motion

Rodolfo Garau

In an attempt to overcome the widespread narratives concerning Telesio as a precursor of Newton (Cassirer) or as a prisoner of an unscientific form of hylozoism (Gentile),[1] current scholarship has increasingly emphasized the anti-Aristotelian traits of Telesio's philosophy as his most historically relevant contribution to the development of early modern natural philosophy. In a paper tellingly entitled *The First of the Moderns or the Last of the Ancients?*, Guido Giglioni argued that the notion of sentience, far from representing an outmoded vestige of a naive animism, provided the basis for a radical shift from the Aristotelian notion of movement, and thus must be seen as the most original trait of Telesio's natural philosophy. Giglioni writes,

> [...] the Aristotelian notion of nature suffered from an ineliminable residue of Platonism, in that all natural beings, in their irresistible tendency to be actualized by the form, appeared to be drawn to a principle that in the end transcended nature itself, regardless of whether that principle was the unmoveable mover, the pure actuality of the ultimate form or the active intellect. Put in a nutshell, the actions of nature were supposed to terminate in something other than nature itself.[2]

Detaching himself from the fundamental kernel of Aristotelian natural philosophy, Telesio saw the Aristotelian notion of nature as an insufferable deviation from his project of understanding nature *iuxta propria principia*. While he did not reject the project of understanding nature within a teleological framework or declare teleology as present but unintelligible (as many seventeenth-century inquirers would do), Telesio rejected the idea that the *telos* was somehow situated outside of nature. As Giglioni observes,

1 See Cassirer, *Das Erkenntnisproblem*, Gentile, *Bernardino Telesio*, in *Opere complete di Giovanni Gentile*, 133–206.

2 Giglioni, "The First of the Moderns or the Last of the Ancients?," 85.

 | DOI:10.1163/9789004352643_014

> [Telesio first] dismissed the notion of nature as an unconscious source of activity by emphasizing instead its fully sentient character; second, he demonstrated that the view of unintentional finalism was a pure abstraction of the intellect, for nature could not be regarded as occupying an autonomous sphere of activity without adding the decisive qualification that it has to be aware of its ends and purposes.[3]

The reaction of Telesio's contemporaries to the publication of *De rerum natura* seems to confirm that what struck its readers was the proposal of a system of natural philosophy that represented a coherent and all-encompassing alternative to Aristotelianism.[4] Its novelty was not necessarily seen in the alternative physical theses propounded by Telesio but rather in its ambitious attempt to replace the very fundaments of Aristotelianism. In 1572, the Platonic philosopher Francesco Patrizi was asked to draft some of his major objections to Telesio's theories. On that occasion he was certainly not very lenient with Telesio's work. Patrizi accused Telesio of reviving the vision of nature of the pre-Socratic philosopher Parmenides. Second, Patrizi observed that Telesio largely based his theories on abstract reasoning (excluding his well-known claim that all true knowledge either derives from sensation or is developed by means of an analogy with sensory experience). According to Patrizi, Telesio's scarce reference to direct observation is also exemplified in glaringly counterfactual claims, such as the claim that heat cannot derive from motion but, vice versa, that heat always precedes motion as a substance precedes its operation. On the contrary, the most coherent aspect of *De rerum natura* seemed to Patrizi to be the second book, devoted to the confutation of concurrent Aristotelian theories—therefore the *pars destruens* more than the *pars costruens* of Telesio's system. Patrizi described this part of *De rerum natura* as "greatly worthy of admiration".[5] Similarly, a few years later, the Italian philosopher Tommaso Campanella would seek in Telesio's *De rerum natura* a source to elaborate an alternative to Scholasticism.[6]

Framing Telesio's thought in the context of the (pre-?)modern polemics against Aristotelianism surely provides a more historically sound understanding of his works. Developing the image of Telesio as "first of the moderns", to use Bacon's phrase, scholars have sometimes popularized him as a forerunner

3 Ibid.

4 On this, see also Omodeo's introduction to this volume. Bondì provides a very good discussion of Telesio's influence in Bondi, *Introduzione a Telesio*.

5 See Fiorentino, *Bernardino Telesio*, 375–98.

6 See Badaloni, *Tommaso Campanella*.

of Newton, an advocate of sensory knowledge, a denier of the authority of Aristotle, and a critic of the esotericism of contemporary magical and hermetic knowledge (Cassirer is one example). At other times, they have instead deemed his reference to animism and universal sensibility a vestige of magical and pre-scientific thought (e.g. Gentile).[7] Against these narratives, projecting Telesio's work against the background of Aristotelianism might be a more insightful way to represent his work. At the same time, however, this makes the evaluation of Telesio's influence on early modern thought (and in particular on the development of seventeenth-century proto-inertial natural philosophy) more problematic. While scholars have shown that his work influenced major seventeenth-century intellectuals such as Bacon, Descartes, Gassendi, and Hobbes,[8] at the same time these authors (with the notable exception of Bacon) rarely referred directly to Telesio in their works, and almost never mentioned him as a source of their theories. Therefore, the problem of how to treat the relationship between Telesio and seventeenth-century natural philosophy seems to be an almost insurmountable methodological riddle.

This paper focuses on a notion closely connected to that of sentience, one which was indicated by Giglioni as the most original trait of Telesio's philosophy—self-preservation. In *De rerum natura iuxta propria principia*, Telesio argued for the existence of two antithetical active principles, heat and cold, which correspond to the sun and heaven and to the earth respectively, and yield motion, change, rest, and immutability. Heat and cold, rather than actively producing the matter on which they act, are the principles of its change, while the "mass or body" (*moles vero corpus*) on which the two active natures act remains, so to say, the stage of the battle between these two principles, although its nature and form changes (I, 5, 17).[9] Telesio attributed a crucial role to self-preservation, arguing that both principles are essentially antithetical and tend to counteract each other and seek their own preservation.

The aim of this paper is not to offer a genealogical reconstruction of Telesio's notion of self-preservation (as Martin Muslow has already remarkably done),[10] but rather to analyze this concept and its use within Telesio's natural

---

7 Cassirer, *Das Erkenntnisproblem*; Gentile, *Opere complete*, 133–206.

8 See for instance Bondì, *Introduzione a Telesio*; Schuhmann, "Hobbes and Renaissance Philosophy;" Mulsow, *Frühneuzeitliche Selbsterhaltung*.

9 Latin quotations are taken from Telesio, *La natura secondo i suoi principi* (2009).

10 In a book devoted to Telesio's notion of self-preservation, Martin Muslow (*Frühneuzeitliche Selbsterhaltung*) cautioned against viewing Telesio through the later narratives of the "Parmenidean sect" and of the revival of Stoic thought, which would put Telesio in direct connection with Vives and Spinoza. In contrast, he demonstrated a profound connection between Telesio's notion of self-preservation and the medical, astronomical and optical

philosophy with special regard to Telesio's doctrine of motion. While here I intend to provide a *working hypothesis* rather than a solid *Wirkungsgeschichte* (because, as I have mentioned, a study of Telesio's influence on early modern culture is made difficult by a patent scarcity of sources), I suggest that this doctrine might well have had a lasting influence on seventeenth-century developments of natural philosophy. I argue that Telesio's notion of self-preservation represents not only another element of critique of Aristotelianism but also a crucial turn in the way Scholastic physics understood activity in nature. Second, I show that Telesio's notion of self-preservation was likely motivated by the need to provide an alternative to the Aristotelian theory of motion, and in particular was rooted in some crucial inconsistencies in Aristotle's theory of falling objects. In the conclusion of my paper, I hint at some possible bearing of Telesio's doctrine of self-preservation on early modern proto-inertial natural philosophy, in particular on authors such as Descartes and Spinoza. While it is not possible to establish that Telesio's natural philosophy directly influenced these authors, and despite rejecting the narratives of Telesio as a "forerunner" or "anticipator" of modern mechanism, I argue that Telesio's critique of the Aristotelian doctrine of motion and activity, and its explanation in terms of self-preserving tendencies, likely contributed to the creation of the intellectual atmosphere from which early modern mechanism stemmed.

This paper is divided into four sections. First I offer a short reconstruction of the theme of self-preservation in Western philosophy, focusing in particular on its reception within Scholasticism. Second, I reconstruct Telesio's notion of self-preservation. Third, through a comparison with the Scholastic (and even pre-Scholastic) notion of self-preservation, I show that Telesio's detachment from the Scholastic tradition consisted of three traits: (1) the claim that this tendency to self-preservation is teleologically oriented but not intrinsically definite in time (i.e. does not have a *terminus ad quem*); (2) that while the Thomists grounded the drive to self-preservation within a structured theological framework (in which the instinct to self-preservation stems from the

discussions of the time, the diachronic transformations of Aristotelianism and Galenism, as well as the Avicennian and Averroistic influence in sixteenth-century Padua, where Telesio was educated. See Muslow, *Frühneuzeitliche Selbsterhaltung*, 397: "Die Phasen, die sich als Schwellenzeiten in der Entwicklung des Begriffs der conservado sui herausgestellt haben, mußten erst gegen die Blockade von Mythen der Rezeptionsgeschichte freigelegt werden. Die grands récits, die durch vorschnelle Vereindeutigungen der Positioniserung Telesios die faktischen Verhältnisse verdunkelt haben, sind im 17. und 18. Jahrhundert die Fabel von der Wiederbelebung der parmenideischen Sekte durch Telesio und im 19. Jahrhundert die Vorstellung einer frühneuzeitlichen natürlichen Geisteswissenschaft aufgrund einer durchgängigen Stoa-Rezeption von Vives bis Spinoza gewesen."

love of God for his creation), Telesio's project of a study of nature "within its own principles" excludes such a theological structure; (3) this solution brings about a decisive shift from a teleology conceived as the passage between different states and having goals extrinsic to the subject to one that I will define as "autotelic", in which the subject itself (its persistence in existence and increase in power) becomes the goal of activity. Fourth, I show how the notion of self-preservation helps to provide (and was likely motivated by the search for) an alternative explanation to Aristotle's theory of motion, which was rooted in the dualism between natural and violent motion. In Telesio's mind, self-preservation replaces the idea that the tendency to motion is brought about by the form of things. This conclusion is admittedly the most hypothetical part of my paper. Indeed, the autotelic drive to self-preservation characterizes many (proto-)inertial natural philosophies of the seventeenth century, such as those of Descartes and Spinoza. While it is uncertain to what extent Telesio's philosophy influenced these authors, I argue that Telesio's notion of self-preservation helped to create the intellectual atmosphere that led to further development of seventeenth-century mechanism.

1 The Concept of Self-preservation: an Ancient Legacy

The idea that living entities, and all things in general, tend to act in a self-preserving fashion is likely one of the most widespread ideas in the history of Western thought. One of the main points of contention between the Stoics and Epicureans consisted precisely in establishing the nature of what they called *hormé*, i.e. the first (or fundamental) action of living entities or their natural inclination to action. Stoics claimed that *hormé* was addressed to self-preservation (which in turn they believed was accomplished through the unfolding and realization of the animal essence, a process they termed *oikeiosis*) and resulted from the providential action of an immanent God, while the Epicureans (who denied divine providence by stating that the universe originated by chance) believed that it was addressed to the search for pleasure and the avoidance of pain. This debate was recorded by many ancient 'historians of philosophy', as for instance Diogenes Laërtius:

> [...] an animal's first impulse [πρώτην ὁρμήν], say the Stoics, is to self-preservation, because nature from the outset endears it to itself, as Chrysippus affirms in the first book of his work *On Ends*: his words are, 'The dearest thing to every animal is its own constitution and its consciousness thereof';[...] We are forced to conclude then that nature in

> constituting the animal made it near and dear to itself; for so it comes to repel all that is injurious and give free access to all that is serviceable or akin to it. As for the assertion made by some people that pleasure is the object to which the first impulse of animals is directed, it is shown by the Stoics to be false. For pleasure, if it is really felt, they declare to be a by-product, which never comes until nature by itself has sought and found the means suitable to the animal's existence or constitution[11] [...].

Animals, as much as men, appear to be endowed with a natural instinct to know what is harmful and what is helpful to them. Their actions in the world, despite not being determined by ratiocination, seem to be motivated by a fundamental instinct to survive and by an innate knowledge of what could be useful to this aim. The Stoics seem to have believed that such an instinct revealed the providential action of nature. For instance, Cicero noted that "Nature has provided with bounteous plenty for each species of animal that food which is suited to it [...]. But nature has also bestowed upon the beasts both sensation and desire, the one to arouse in them the impulse to appropriate their natural food [*conatum haberent ad naturales pastus capessendos*], the other to enable them to distinguish things harmful from things wholesome"(II, XLVIII).[12]

Wolfson, the great historian of ideas,[13] claimed that the Stoic idea of *hormé* was taken up by philosophers as diverse as Augustine,[14] Thomas Aquinas, and Dante Alighieri[15] in the Middle Ages. It is hard to establish whether this was a direct Stoic influence or a parallel development. Surely, the Christian conception of God as a loving and caring demiurge reinforced the connection between self-preservation and divine providence. Here, the Augustinian

11 Diogenes Laërtius, *Life of Eminent Philosophers*, 193.

12 Cicero, *De natura deorum*, 239. See also ibid., 178: "[...] as the other natural substances are generated, reared and sustained each by its own seeds, so the world-nature experiences all those motions of the will, those impulses of conation and desire ("conatus et adpetitiones"), that the Greeks call hormae, and follows these up with the appropriate action in the same way as do we ourselves, who experience emotions and sensations" (II, 58).

13 See Wolfson, *Philosophy of Spinoza*.

14 Augustine, *City of God*, 537: "What of animals in general, even irrational animals that have no power to reflect on these things? Do they not, from huge serpents down to tiny little worms, what that they want to go on being and, in order to do so, seek to escape death by every movement at their command? What of trees and shrubs of every kind that have no sensation to enable them to avoid destruction by perceptible movement, yet do they not ensure the growth of their topmost germinal shoots into the air by fixing another growth, of root, into the ground so as to draw nourishment from it and so, in their own fashion, preserve their existence?" (XI, 27).

15 Dante Alighieri, *Monarchy*, 20: "[...] everything that is desires its own being [...]" (I, 13).

notion of "natural love"—the love all creatures have for the means that are useful to their preservation, caused by and reflecting God's providential love for his creation—assumes crucial importance. Aquinas claimed that the greatest example of this was that each and every thing had a natural appetite for its own preservation.[16] In the *Summa Theologiae*, using the Aristotelian idea of "practical good", he described this natural appetite for self-preservation as the love "with which each and every particular thing loves its own good because of the common good of the whole universe, that is God" (I–II, q. 109 a. 3 co.).[17] In another passage of *De veritate*, he quotes Boethius to link self-preservation to divine providence: "Everything that already exists loves its particular being naturally, and preserves it with all its strength; from which Boethius says in the third book of *De consolatione*: divine providence gave to the things it created such greatest cause of persisting, so that, insofar as they can, they naturally desire to persist" (Q. 21 a. 2 co.).[18] However, while animals cannot avoid this natural drive toward their practical good, humans represent a more peculiar case. While each and every human being possesses a "natural appetite" by virtue of which (s)he wants to "exist and live",[19] "the disposition of human actions to the end is not according to nature, as it is with the irrational creatures, that act towards the end according to natural appetite only, but man acts towards an end through reason and will. So there is no natural law for human beings" (*Summa Theologiae* I–II, q. 91 a. 2 arg. 2).[20] While eventually the rational and voluntary determination of actions also depends on the love for good itself (God), Aquinas here is describing man as an "empire within an empire" (to use Spinoza's famous expression), whose actions do not strictly comply to any law.

16 Thomas Aquinas, *On the Power of God*, 79: "Everything has a natural appetence [appetitu naturali] for the preservation of its existence" (Q. 5, Art 1, 13).

17 Thomas Aquinas, *Summa Theologiae*, electronic edition, http://www.corpusthomisticum.org/: "Manifestum est autem quod bonum partis est propter bonum totius. Unde etiam naturali appetitu vel amore unaquaeque res particularis amat bonum suum proprium propter bonum commune totius universi, quod est Deus."

18 Thomas Aquinas, *De veritate*, electronic edition, http://www.corpusthomisticum.org/: "Omnia autem quae iam esse habent, illud esse suum naturaliter amant, et ipsum tota virtute conservant; unde Boetius dicit in III de consolatione: dedit divina providentia creatis a se rebus hanc vel maximam manendi causam, ut quoad possunt, naturaliter manere desiderent. [...]."

19 Thomas Aquinas, *Summa Theologiae*, electronic edition, http://www.corpusthomisticum.org/: "Uno quidem modo, appetitu naturali, sicut omnes homines volunt esse et vivere" (II–II, q. 30 a. 1 co.).

20 Ibid.: "... ordinatio humanorum actuum ad finem non est per naturam, sicut accidit in creaturis irrationabilibus, quae solo appetitu naturali agunt propter finem, sed agit homo propter finem per rationem et voluntatem. Ergo non est aliqua lex homini naturalis."

2 Telesio's Notion of Self-preservation

The passage from a teleology understood as tending to something beyond nature to one in which that something is reabsorbed within the domain of nature (described by Giglioni in the passage quoted in the introduction above) can be found, perhaps even more glaringly, in Telesio's notion of self-preservation. In Telesio's system, matter, per se neutral and amorphous, is dominated and continuously transformed by two principles: heat (having as its source the Sun and heaven and representing a principle of indefinite motion) and cold (having its place on the Earth and naturally deprived of motion). Heat and cold, rather than actively producing the matter on which they act, are the principles of its change, while the "mass or body" (*moles vero corpus*)—on which the two active natures act—remains, although its nature and form change (I, 5).[21] The earth, and in particular its surface, is the battlefield of these two opposite principles, from which all things, both organic and inorganic, originate.

In this system, the notion of self-preservation assumes a particular value. Telesio gives the tendency to self-preservation a fundamental role in his natural philosophy by placing two mutually antagonistic active principles, heat and cold, at the core of his view of nature and by making an informed and passive substance, matter, the battleground between these two principles. The two principles cannot fade away in time, leaving behind them an uninformed matter. In this context, the tendency to self-preservation becomes the key element to explain the mutual action of the two principles. In other words, self-preservation represents their essential, antagonistic way of acting upon each other to inform matter. As Telesio writes,

> Since it never forgets its disposition, each active nature [*natura agens*] never desists from acting [*nunquam agere cessat*] but fights and repels even similar natures to expand in their places, desiring to the greatest extent to be equal to itself and to preserve itself [*talis esse servarique*], and to expand to a greater extent [...] (I, 13)[22]

21 Telesio, *La natura secondo i suoi principi*, 16.

22 "[...] qualiscunque enim existit natura agens quaevis nunquam proprii ingenii oblita nunquam agere cessat, sed vel similes cognatasque oppugnat deturbatque, ut in earum se ipsam sedibus amplificet qualiscunque est talis esse servarique et diffundi amplius atque in subiectis produci omnibus summe appetens summeque contendens. Quod igitur dictum est agentia rerum calor esse frigus videntur." (I, 13, 38).

In this framework, justification for the presence of this sheer instinct to self-preservation is provided by the combination of several qualities. If natures are endowed with this drive to self-preservation, they must be endowed not only with an appetite for their own preservation (*conservationis appetitum*) and with hate of their own destruction (*propriae destructionis odium*) but also with a form of sensation, which Telesio describes as a "force" or "faculty" (*vim*) to recognize what is contrary and dissimilar to them and what can preserve or corrupt them. As he writes,

> [...] *the faculty of feeling* [sentiendi facultatem] *was given to both active natures, and only in this do the earth and the sun resemble each other* [...]. But if natures were to preserve themselves [*si servandae sint naturae*], not only was it necessary that they had a great desire to preserve themselves [*conservationis appetitum*] and a great hate of their own destruction [*propriae destructionis odium*] but also the faculty to recognize what is alike and similar and what is contrary and dissimilar [*cognata similiaque et contraria dissimiliaque dignoscendi vim*]. Indeed, they would not desire to preserve themselves [*conservari appetant*] nor they would fight corruption [*corrumpi aversentur*] if they could not recognize which things preserve and which corrupt them, and if they were not bound to the inclination to repel the former and to pursue the latter (I, 34).[23]

Since matter is inert and not endowed per se with any quality, all sensation (even that of animals or human beings) must derive from the natures that inform matter. As Telesio claims, as heat and cold were given the feeling of their own preservation and destruction [*propriae conservationis et propriae destructionis sensus*], it is necessary that all beings are also endowed with them, i.e. that the active natures present in each body feel; indeed, what beings are, act, and undergo is due to the natures that constitute them, because matter per se lacks any knowledge of itself [*conservationis et proprii boni sensus*] (I, 34).

23 "*Sentiendi facultatem naturae agenti utrique traditam esse, et in ea sola Caelo Terram coonvenire* [...]. Si servandae sunt naturae, non summum modo illis conservationis appetitum summumque propriae destructionis odium indi oportuit, sed cognata insuper similiaque et contraria dissimiliaque dignoscendi vim; frustra enim conservari appetant et corrumpi aversetur, nisi a quibus serventur et a quibus obleadantur disgnoscant, et fugiendi haec et sectandi illa studio teneantur" (Telesio, *La natura secondo i suoi principi*, 104).

Rooted in the very essences of the two natures, sensation embraces the whole of nature, from inanimate to animate bodies. What is more, the sensation of superior beings is not a characteristic they possess exclusively but it is due to the action of the two principles. Nerves and sensory organs are only paths that channel the action of the principles rather than representing the faculty of sensation per se. On this basis, Telesio's account of self-preservation embraces animate beings as well as inanimate ones.[24]

3 Telesian Self-preservation vs Scholastic Self-preservation

The use of an "anthropomorphized" lexicon in the natural discourse was surely not exclusive to the Telesian account of natural phenomena. Rather, the idea that natural things possessed "appetites", "impulses", or "tendencies" can be commonly found in Scholastic treatises on natural philosophy. In *Summa contra Gentiles*, Aquinas describes as follows the tendency of the stone to descend:

> [...] there is in every thing an appetite for good [*appetitus boni*]: for good is what every thing desires [*appetuntur*], as the philosophers teach. In this way, the appetite in things that lack thought is said to be a natural appetite, as for instance it is said that the stone desires [*appetit*] to be below. In things which have sensitive thought, it is said to be an animal appetite, which is divided into concupiscible and irascible. In things which have intelligence, it is said to be an intellectual or rational appetite, which is will (*Summa contra Gentiles*, II, 47, n. 2).[25]

Even in the seventeenth century, Goclenius, in his *Lexicon Philosophicum*, distinguished three proper usages ["*naturalis appetitus*", "*animalis appetitus*" and "*appetitus*" of the "*naturae intelligentis & voluntariae*"]. Goclenius himself underlined the "ambiguous nature of the term", showing that an "appetite" is said to be "of fire to occupy the superior places, of iron to conjoin with the magnet, of plants to absorb the lymph, of horses to Venus [to pleasure], of men to beatitude". "*Naturalis appetitus*", an acceptation that Goclenius describes however as somewhat inappropriate ["... *qui dicitur quodammodo improprie appetitus*", "that is called somewhat inappropriately appetite"], applies both to "plants which attract toward themselves and desire aliments *without*

24 See for instance I, 35. Here, Telesio claims that all beings, also those that do not have sensory organs, possess sensibility.

25 Alarcón, "Corpus Thomisticum", www.corpusthomisticum.org, 2000.

sensibility ["*... in stirpe, quae attrahit & appetit alimentum absque sensu* [...]" and to "inanimate things, such as the magnet ["*In inanimis, ut magnete*"]."[26]

The real novelty of the Telesian account of self-preservation consists in its lack of spatial and chronological determination, i.e. its lack of *telos* or *terminus ad quem*. In Scholastic philosophy, the preservative action of things was always identified with the achievement of the end that qualified the essence of the thing in question. As Aquinas epitomizes in his *Summa theologiae*, "... goodness is what all things desire [*appetunt*], that is, what they have as an end; it is clear that goodness is what is brought about in reason of the end."[27]

In contrast, in Telesio's picture of nature the search for "goodness", i.e. the search for the preservation and increase of the principles' essence, is an activity that is not restrained or confined in time and does not have a final conclusion—either temporal or spatial. While, say, the *appetitum boni* of the stone to reach its natural place is fulfilled (at least partially) once it touches ground, the appetite for preservation of Telesian principles characterizes their eternal, unrestrained activity. This marks a passage from a hetero-teleological picture of nature to an autotelic one, i.e. one in which the maintenance and increase of a thing's nature becomes the only way activity can be understood. As a consequence, the activity of a thing that does not suffer from the action of its contrary is potentially infinite:

> [...] in order for a thing to be for eternity and in infinite time, and therefore operate according to its own nature in infinite time, there is no need to be incorporeal or infinite itself, or that it needs to be supported by such a thing. On the contrary [it is only necessary that], as little as the thing can be, it does not suffer from its contrary and it is not changed (DN1, II, 55).

Another fundamental novelty of Telesio's notion of self-preservation is its disconnection from the theological framework that, as we have seen, traditionally associated it to the notion of divine providence. To provide a further

26 "Distinctio ambigae verbi naturae. Appetitus ὁμονύμω dicitur: In igne ad occupanda loca superiora:in ferro ad sui conjunctionem cum magnete: in planta ad humore sugendum: in equo ad Venerem: in homine ad beatitudinem", Goclenius, *Lexicon philosophicum*, 114. Similarly, Francesco Bonamici, in his *De Motu*, used the term '*appetitus*' and provided a general definition of it as the "inclination [*inclinatio*] which is necessary from the nature of every thing to agree to the good that convenes to itself according to nature." See Bonamici, *De Motu*, 28.

27 "[...] bonum sit quod omnia appetunt, hoc autem habet rationem finis; manifestum est quod bonum rationem finis importat" (Aquinas, *Summa Theologiae* I, q. 5 a. 4 co.).

example of this connection between divine providence and self-preservation, we can quote Aquinas once again. In *De potentia*, answering the first article of the *Quaestio 5*, Aquinas asks whether "things are preserved in being by God, or, any action of God excluded, they persist in being by themselves."[28] A possible argument in favor of the latter option, Aquinas states, is that "… the natural appetite cannot be vain and false. But each natural thing naturally desires the preservation of its own being. Therefore a thing can be preserved through itself, otherwise natural appetite is vain."[29] However (and this is Aquinas's position), things can well possess an appetite for their own preservation but have no power to preserve themselves in being: they are preserved in being by their cause—namely, by God. As he writes, "… it is admissible that every thing naturally desires its own preservation, though not that it is preserved by itself but by its cause."[30]

Through his notion of self-preservation and universal sensibility, Telesio completely overturns this perspective. As we have seen, in order to justify the fact that things act self-preservingly, "not only was it necessary that they had a great desire to preserve themselves [*conservationis appetitum*] and a great hate of their own destruction [*propriae destructionis odium*] but also the faculty to recognize what is alike and similar and what is contrary and dissimilar [*cognata similiaque et contraria dissimiliaque dignoscendi vim*]" (I, 34, Latin quoted above). In other words, the doctrine of universal sensibility, in addition to identifying the teleological tendency of things within nature and not without it, also provides a fundament to explain the self-preserving action of things outside a theological framework, or, to use Telesio's expression, according to the principles of nature alone. In this context, the reference to God (the only one in the whole second book of *De rerum natura*!) is just a reminder of the role of the author of nature, and it seems to be merely a justification ex-post of the order and harmony of nature. Although active natures fight for the affirmation of their being and reciprocally seek to destroy each other, there seems to be a sort of measure in nature, from which beautiful things stem in great number according to a certain order. Telesio explains that this is due to "God the perfect and greatest, not the oblique motion of the Sun; this motion

---

28 *De potentia*, Q.5 Art.1: "[…] utrum res conserventur in esse a Deo, an etiam circumscripta omni Dei actione, per se in esse remaneant."

29 Ibid., Q.5 Art.1 arg. 13 "Praeterea, appetitus naturalis non potest esse cassus et vanus. Sed quaelibet res naturaliter conservationem sui esse appetit. Potest ergo res per se ipsam conservari in esse; alias appetitus naturalis esset vanus."

30 Ibid., Q.5 Art.1 arg. 13 "Ad decimumtertium dicendum, quod licet quaelibet res naturaliter appetat sui conservationem, non tamen quod a se conservetur, sed a sua causa."

itself is indeed to be understood as a work of God and certainly as produced with admirable wisdom."[31] Here, though, God is just the author of the world: he does not act as the final end of a thing's action, as in Aquinas. Again, the teleological tendencies of active principles, as well as nature in general, are to be found entirely within nature and not without.

4 Self-preservation vs Natural and Violent Motion

How does this principle of self-preservation effectively come into play to explain natural phenomena? The most interesting application of the principle (and probably also the true motivation underlying Telesio's theory) is related to the explanation of the nature of motion, in particular as an objection to Aristotle's theory of natural motion. As is well known, the vulgate of Aristotle's physics proposed by the Scholastics divided motion into two kinds: natural, i.e. motion caused by the form of the object and aimed rectilinearly at the center of the corresponding sphere; and violent, i.e. motion caused by an external force and directed not in the direction otherwise followed by the thing in question. In this context, what we now call "acceleration of gravity" was seen as proportional to the weight of the object in question (in turn determined by its form) and the density of the medium across which the object travels.[32] The notion of natural motion, in particular, is at the center of Telesio's criticism. Things, according to Telesio, cannot move and then rest. If movement pertains to a certain thing, it would continue indefinitely, and vice versa if a thing is at rest there is nothing in the thing itself that would make it move in a second moment. As the incipit of I, 43 recites,

> The Earth is absolutely contrary to heaven and no motion corresponds to it, and the arguments of Aristotle according to which he attributes to it and to all other bodies a natural motion are absurd [...][33]

31 "Neque igitur sese mutuo perdant aut imminuant omnino quid, et longe plurima constituant longeque pulcherrima, et eadem semper ratione omnia. Hiuiusmodi Deus ipse Optimus Maximus, non obliqua Solis latio videatur, quae eiusdem Dei videri debet opus, et admirabili quidem constituta sapientia" (II, 37, 344).

32 Aristotle, *Physics* (2006), Book IV.

33 I, 43, 134: "Terram Coelo omnino contrariam esse, et nullum illi motum convenire, et absurdas esse Aristotelis rationes, quae motum illi naturalem et reliquis corporibus omnibus attribuunt, et cur deorsum sponte feratur."

As Telesio continues,

> One shall not think that things that appear by their own nature to lie down and to be immobile, though they have been seen to reach spontaneously that place, are carried there by their own natural motion; for immobility and motion cannot be of the same nature; instead, it is necessary that things that are by nature immobile are entirely devoid of any motion, and that, in contrast, things that move by nature are always in motion and never stop moving […][34]

This forces one to radically rethink the Aristotelian doctrine of motion, which is harshly criticized here. If a body, when dropped, falls down to the earth, this cannot be due to the fact that the body changes its state from 'motionless' to 'moving'. On the contrary, it is its desire to preserve its being that forces the body to fall down towards the centre of the earth. In Telesio's account, there, there are two main polarities: one between motion and immobility, [*motum and immobilitas*], understood as the nature charaterizing hot and cold things respectively, and therefore opposite one to another; and one between rest [*quies*] and labor [*labor*], characterizing a forced motion of things which are naturally at rest. As he writes,

> All motions […] appear to differ one from another negligibly, in contrast to motion and rest, which, as also Aristotle states, derives from a nature that is opposite to that of motion and indeed is really contrary and opposed to motion; indeed, rest [*quies*] is opposed to labor [*labor*] not to motion thanks to which mobile entities [*mobilia entia*] rest [*quiescunt*] and take much pleasure and gain strength.[35]

The notion of *labor* here is particularly relevant to explain the motion of cold things, which by nature would otherwise not move at all. Heavy things (things predominantly characterized by the principle of Earth) do not move

34 "At eque quae in proprio loco natura quiescere et immobilia esse videntur, sponte etiam suam ad illum deferri visa, proprio et naturali motu deferri videri debent; neque enim et immobilitas et motus naturae uni convenire, sed quae natura immobilia sunt motus omnis penitus expertia sint necesse est, quae contra natura moventur perpetuo moveantur oportet, et moveri nunquam cessant" (I, 44, 137–139).

35 I, 43, 138: "[…] motus omnes, differre etiam inter se visos quid quam motum atque immobilitatem, quae vel Aristotelis testimonio ab opposita motui manat natura, et quae vere motui contraria est oppositaque; nam quies labori opponitur, non motui in quo vel summe quiescunt summeque oblectantur et robur etiam sumut mobilia quae sunt entia."

downward because that is their natural motion, but because they seek their own preservation. In seeking their preservation, they force their natural disposition to rest (if they are cold) or to move with uniform motion (if they are hot), moving with increasing speed towards the Earth or moving non-uniformly. As he writes,

> The parts of the Earth, separated from their place and their totality, fall spontaneously and the more rapidly the bigger they are and the closer they are to their place, not because that motion is natural to the Earth itself or to its parts but because the desire, the care and love of all beings for their preservation and for similar beings, and the hate of their destruction and of contrary beings, and the sense of both things, are such that they seem to force their nature, and the immobile beings move and the mobile beings move with an unusual, albeit natural, motion.[36]

The change in motion of warm, moving things is provoked by the interaction of external forces that threaten the preservation of their nature. Similarly, the downward motion of cold, earthly things is not a "natural" form of motion; rather they are forcing their own nature (arguably through what Telesio has just defined as "labor") until they can be one with the principle that ensures their self-preservation and increases their being. As Telesio writes,

> [...] one shall not think that the parts of the Earth, while they fall towards their own place [*proprium locum*] and their own totality [*universalitatem*], move with natural motion, for once they have reached their totality they always become, like it, immobile; instead, one must think that they force their own nature and perform an action that is alien to their nature [*ipsae naturae inferre vim et alienam tantisper operari operationem*] until they reach their place and their totality.[37]

36 I, 43, 138: "Terra itaque partes, a proprio loco et a propria abstractae universitate, sponte delabuntur sua eoque velocius quo maiores et quo proprio loco propinquiores factae sint, non quod naturalis is vel Terrae ipsius vel eius partium motus sit, sed quod, ut expositum est, is entium omnium propriae conservationis et cognatorum appetitus studiumque est atque amor, propriaeque destructionis et contrariorum odium atque aversatio et utriusque sensus, ut propriae etiam naturae vim inferre videantur, et immobilia moveantur, et mobilia, si quidem naturali, insueto certe motu."

37 I, 43, 138–140: "Quaere nec Terrae partes ad proprium locum et ad propriam descententes universitatem naturali motu moveri videri debent, quae ad ipsam delatae perpetuo veluti et ipsa immobiles fiunt, sed et ipsae naturae inferre vim et alienam tantisper operari operationem, quo ad proprium locum et ad propriam ferantur universitatem."

This leads to a second, and more interesting, criticism of the dichotomy between violent and natural motion. Telesio claims that what Aristotle describes as the natural, downward motion of heavy things cannot be the case, for it gradually increases, as Aristotle claims when he observes the fall of a stone towards the center of its sphere. In contrast, the motion of things that move by nature is uniform. Analogously, violent motion cannot be characterized by a progressive decrease of its intensity. Instead, "natural motion" is constant by nature, since "the bodies that move by nature, since motion is the operation by which they preserve themselves to the highest degree, must always move with the same speed."[38] Here, we probably find an echo of the widespread Aristotelian and Ptolemaic idea according to which the motion of heavenly objects is characterized by constant speed. In this way, changes in motion are brought about by the presence of an external force (which Telesio identifies as a contrary nature) that threatens the preservation of the thing itself, or, conversely, by the thing's drive to self-preservation:

> [...] the bodies that increase continuously [their speed] move not by nature but either because they are pushed by some evil, and the more they hate it the more they increase their motion, or because they are attracted by some good, and the more they are enticed by its sweetness the more desirously they pursue it.[39]

On this basis, Telesio begins a very detailed confutation of the Aristotelian explanation of the increase of speed of a falling stone, which his doctrine of self-preservation is meant to replace. This confutation reveals, I believe, the motivations of Telesio's theories of sentience and self-preservation: his intention to provide an alternative to Aristotle's theory of motion. However, this motion cannot be deemed "natural", as natural motion, in Telesio's account, is characterized by constant speed and not by uniform acceleration. This overturns Aristotle's account of the difference between violent motion

---

38 "At neque qui motus assidue magis concitatur naturalis is propriusque, sed praeter naturam omnino videri debet, non siquidem in eo a naturali violentus differre, quod hic assidue magis languescat, ille vero assidue veluti robur sumat et magis concitetur, sed quod perpetuo idem est nihilque vel remittitur vel concitatur unquam. Quae scilicet natura moventur, quoniammotus propria ipsorum operatio est et qua summe in propria servantur natura, perpetuo eodem ferantur oportet" (I, 45, 142).

39 "Quae itaque assidue illum concitant, non natura moveri videri debent, sed vel a malo aliquo impulsa quod assiudue magis abhorrescentia motum assidue concitant magis, vel a bono quopiam tracta cuius dulcedine magis assidue illecta maiore id sectentur studio [...]" (145).

(characterized by progressive deceleration) and natural motion (characterized by constant acceleration).[40]

The increase of the speed of a falling object is at the center of Telesio's polemic. According to Aristotle, Telesio explains, the stone, while approaching the ground, progressively re-acquires its nature of heaviness, which it partially lost because of its distance from its natural place and the action of the contraries. But how can the air, which is very weak, deprive the stone of its nature, since once dropped the stone does not even stop for a second? And how can the Earth immediately re-introduce the nature of heaviness to the stone? What is more, in this way a stone that is lifted only a little should move more quickly than one that is lifted higher in the air, since its distance from the Earth is smaller; yet experience shows us the contrary.[41] In the second place, Telesio refutes the explanation based on the quantity of the medium. Accordingly, the stone would cut through the air below more easily the more it nears its natural place due to the decreasing quantity of air beneath it, and thus its motion would result in constant acceleration. Telesio deemed this explanation inconsistent. First, Telesio claims, the medium (air) is always the same substance. Second, the more the stone approaches its natural place, the more the medium must become dense due to the pressure that the falling object exerts. Third, the fact that the stone falls with less speed if dropped from a lower place represents a

40 "Propterea, quidam inquiunt, qui e sublimi dedicit lapis extremum et Terrae contiguum spatium velocius conficit, quod simili congataeque Terrae proximior factus, propriae naturae et propriae etiam gravitati quibus ab alieno loco et a contrariis veluti extus fuerat restituitur. Absurda sane ratio, et quae modis multis refelli possit" (145).

41 "Primum enim ne ipsi quidem qui ilam afferunt Peripatetici sibi ipsis (ut opinor) persuadeant, lapidem in sublime proiectum et nullo ibi temporis ammoratum momento a longe languidissimo aere propria natura propriaque exui gravitate, et leviorem omnino fieri, tum nullo itidem temporis momento a cognata Terra gravitatem illi indi; levitas enim gravitasque ne Aristoteli quidem agendi vi et se ipsas constituendi facultate praeditae videntur, sed illa quidem caloris, haec vero frigoris opus esse. Quoniam igitur paulo superior aer, non modo nihil inferiore calidior videri potest, sed vere frigidior existit, ut ad quem reflexa lux et languida et minus unita ascendit, nequaquam in eo levior fieri lapis videri potest. At ut calidior sit superior aer et levitas agendi vi praedita, non certe adeo nullo temporis momento lapidem tenuitate ulla aut ulla donent levitate, nec levior factus nullo itidem temporis momento gravior a cognata Terra fiat. Quin, si ea etiam omnia fiant, non vera sit ratio tamen, non scilicet Terrae proximior factus lapis propterea motum concitat, quod quae procul a Terra acto descendendi vires imminutae fuerant a propinqua Terra in integrum restituuntur; id enim si sit, ubi exiguum quid atollitur lapis nihil itaque levior fit spatium quo a Terra abest nihilo pigrius, forte etiam et velocius conficiat quam si e sublimiore decidat; secus omnino atque evenire videtur, quo enim a Terra minore sublatus est spatio, eo languidiore id delabitur motu, et eo idem vehementiore quo ex editiore descendat loco, ut manifeste descendendi vires non a Terra vicinitate, nec a re omnino alia ulla, at a descensus diurnitate augeri videantur" (147).

glaring confutation of this thesis.[42] Lastly, Telesio confutes the antiperistatic explanation. According to this theory, the motion of objects (both natural and violent) is caused by the displacement of air that the moving object provokes, since a vacuum cannot exist in nature and the air displaced by the moving object would fill the space left by the object, in turn propelling it. When applied to the fall of the object, the tendency it has to move downward combined with increased pressure and push of the air above it would explain the acceleration of the falling stone, which therefore would be proportional to the increase of the mass of air above it. Once again, Telesio harshly criticizes this explanation. First, as in the case of violent motion, things do not increase their speed by virtue of the action of the air but because of the force exerted on them.[43] The air still moves to fill any empty space left by the moving object but its impetus cannot be strong enough to cause the propulsion of the object itself. Moreover, how can the fact be explained that once the object leaves the hand it starts moving spontaneously, without any empty space having yet been created?[44] In addition, the increase of the quantity of air above the falling object cannot explain the increase in downward movement of the stone, since only the air close to the object would be set in motion and not the rest of the air above it, which would tend to stay still.[45] Furthermore, this would have paradoxical consequences in the case of violent motion. Even assuming that air, which is a thin and weak substance, possesses the force to move even a small object, and assuming that this force would increase proportionally to the quantity of the

42 147.: "[...] quoniam enim lapidis magni praesertim delapsus aeris concessionem, et quasi fugam, praevenit atque antevertit, in angustius agatur suppositus aer atque in se ipsum conspissetur oportet, qui itaque dividi magis repugnet motum omnino descendentis lapidis impediat quid retardetque. At ex editiore is delabens extremum et Terrae contiguum spatium celeritate longe maxima, ex humili vero languidissimo illud idem conficere videtur motu."

43 148: "[...] quae vi proiiciuntur, nequaquam ab aere ea impellente, ut Aristoteli placet, sed ab impressa vi deferri videntur."

44 148: "[...] satis enim sit aeri ne spatium ullum vacuum fiat providere, et nullo temporis momento eo accurrere unde necessit, quid, at non eo tamen impetu feratur oportet, ut non repleat modo relictum spatium, sed quod inde abiit impellat etiam stimuletque, siquidem propria natura in proprio loco immobilis Aristoteli aer, nequaquam prompte ac libens moveatur; at id modo quod moveri necesse est. Parum itaque a manu dimotus lapis, vel statim potius ac manu emissus est, coincidat; neque enim inter manum lapidemque spatium ullum inane factum est ad quod replendendum aeri sit convolandum."

45 "Tum si hac etiam ratione proiectorum motus fiat, non propterea aer universus ad Terram usque lapidem consectetur, sed proxima tantum assidue pars, quiescant reliquae et propria omnes gaudeant immobilitatem" (Ibid.).

medium (i.e. the higher the object is positioned), a stone thrust upward with violent motion would continue to move indefinitely, as the push imparted by the increasing quantity of air beneath would increase proportionally.[46]

On this basis, Telesio concludes that the Peripatetics could not find any convincing explanation for the non-uniform speed of falling objects. The explanation of the acceleration of the falling stone can thus only be based on the notion of sentience and on that of self-preservation: the stone hates contact with things that are contrary to its nature and increases its speed to enjoy, as soon as it can, its immobility among akin bodies.[47] As Telesio writes,

> The cause appears evident on the basis of what I have explained: that is, that the Earth, removed from its place and its totality, being immobile by nature, and hating very much this alien place and the contact with alien things, forces its nature and falls towards its own place and towards akin bodies; and this with increasing speed because, hating and being irritated by those [alien] things as much as by motion, it increases more and more the motion itself to enjoy as soon as possible its immobility among akin bodies.[48]

5 Conclusion

Telesio's argument relies on a strict analogy between human psychology and the natural world. The reason why earthly falling objects increase their motion is explained elsewhere through a comparison with the sense of anxiety (and at the same time of urgency) to reach pleasure or rest that we experience when we are doing something unpleasant:

---

46 "[...] perpetuo itaque moveantur quae vi moventur omnia, si siquidem modica aeris pars satis Aristoteli est sursum lapidem atollere, multo ubi longe facta est maiori idem possit. At non si ab impressa vi et ab indito ferantur motu, sed eo assidue languidius moveantur, quo a propellente removeantur magis, quo nimirum vis motusque inditus magis debilitetur magisque languescat" (I, 47, 150).

47 I, 47, 150.

48 "A proprio videlicet Terrae loco atque a propria abstracta universitate, propria quidem natura immobilis, at summe alienum locum atque alienorum exosa contactum, naturae vim infert et ad proprium locum atque ad cognata delabitur corpora, assideque id velocius quod et aliena illa et ipsum itidem exosa peraesaque motum assidue illum magis concitat, ut quam celerrime inter cognata propria gaudeat immobilitate" (I, 47, 150).

> […] the more we deal with things that provoke annoyance to us, but that nevertheless we have to deal with, and that once they are finished give us pleasure and rest, the more we increase our motion [or activity] despite hating and being bothered by the effort, since we increasingly desire the pleasure and rest that we are about to achieve.[49]

During the seventeenth century, this "anthropomorphization" of nature (or naturalization of human psychology) would no longer be taken as a valid means to explain natural phenomena. For instance, in *The Principles of Philosophy*, Descartes used the famous example of a stone revolving in a sling in order to explain his proto-inertial understanding of rectilinear motion. As Descartes claims, if released by the sling the stone would move rectilinearly. Circular motion is therefore constrained rectilinear motion. Descartes then applied this idea to his explanation of light, which he believed to be a pressure exerted by the particles composing the stars (which are indeed vortices), against the surrounding particles of air. He argued that the former ones possessed a "striving" (in Latin, a *conatus*) to move away in a rectilinear fashion from the centre of the vortex around which they revolve, then exerting a pressure against the air particles. But then he clarifies that his use of the expression "striving" does not entail any animistic hint:

> […] when I say that the globules of the second element 'strive' to move away from the centers (*'recedere conantur ab istis centris'*) around which they revolve, it should not be thought that I am implying that they have some thought from which this striving proceeds. I mean merely that they are positioned and pushed into motion in such a way that they will in fact travel in that direction, unless they are prevented by some other causes (II, 56.)[50]

The main aim of this passage was likely that of avoiding any possible charge of endorsing the animistic interpretation of natural phenomena that had characterized some of the most important natural philosophies of the Italian Renaissance (such as Telesio's) or in general the widespread use of psychological jargon in the physical discourse that characterized Scholastic physics.

---

49 "[…] id quod nobis usu evenire interdum videntur; nam quae dum tractanctur molestiam inferunt, et peragenda omnino sunt, et quibus peractis voluptas proposita est et quies, quo magis ea tractamus eo motum magis concitamus, et opus exosi pertaesique et appropinquantis voluptatis quietisque magis appetentes" (I, 45, 144).

50 Descartes, "Principles of Philosophy," 259.

However, while here we find ourselves miles away from Telesio's explanation of motion, it is hard not to see some interesting similarities between Telesio's understanding of motion in terms of self-preservation and cognate seventeenth-century theories. Indeed, the example of the falling stone shows that Telesio's notion of self-preservation is complementary to that of universal sensibility, and it is very likely motivated by the search for an explanation of the origin of motion alternative to that offered by Aristotelian physics. Moreover, its autotelic, non-finalistic and anti-providential character represents a trait of great discontinuity with respect to contemporary Scholastic physics. Providing a dichotomy between motion and rest, and understanding violent motion as a change of state due to the presence of an external force, it represents one of the first attempts to overcome the traditional Aristotelian picture of motion. In particular, when it comes to gravitation, Telesio's intuition that the downward motion of natural objects is not natural, but rather depends on the striving ("labor") of heavy objects to reconnect to their substance, strikes the readers as a clear detachment from Aristotelian-Scholastic physics.

While the notion of universal sensibility underlying this doctrine would generally be discarded by seventeenth-century mechanist philosophers, the idea that the behavior of physical objects can be explained in terms of a tendency to the preservation of states would characterize seventeenth-century proto-inertial natural philosophies. Two remarkable examples of this tendency can be found in Descartes and Spinoza. Descartes' first *law of nature* in *Principles of Philosophy* would be explained in terms of a tendency to the preservation of states ("everything tends, so far as it can, to persist in the same state" II, 37),[51] in turn rooted in the ordinary concourse of God. Spinoza would go as far as considering this striving towards self-preservation as the essential characteristic of all modes.[52] As I have shown, the explanation of the tendency of motion of bodies in terms of seeking self-preservation was not a unique trait of Telesio's natural philosophy but rather part of a widespread tradition. However, certain traits of Telesio's theory of self-preservation, such as the fact that the process lacks any kind of *terminus ad quem* and hetero-teleological determination, would become a fundamental trait of seventeenth-century mechanism. While it is hard to assess Telesio's influence on these (and other) authors, it seems to me safe to speculate that his theory of motion helped to create the intellectual atmosphere from which early modern mechanism stemmed.

51 Ibid., 259.

52 See for instance *Ethics* III, 6: "Everything, every thing, considered in itself, strives to persevere in its own being", in Spinoza, *Collected Works*, 498.

Bibliography

Primary Sources

Agelli, Antonio. *Commentarii in Psalmos et in divini officii cantica.* Rome: Typographia Vaticana, 1606.

Alberti, Leandro. *Descrittione di tutta Italia.* Bologna: Anselmo Giaccarelli, 1550.

Albumasar (Abu Ma'shar). *Introductorium in astronomiam.* [Venice]: Giacomo Penzio, 1506.

Alexander of Aphrodisias. *In quatuor libros meteorologicorum Aristotelis commentatio lucidissima*, transl. Alessandro Piccolomini. Venice: Girolamo Scoto, 1561.

Alexander of Aphrodisias. *Praeter commentaria scripta minora*, ed. Ivo Bruns. Berlin: Reimer, 1887.

Alexander of Aphrodisias. *In Aristotelis metaphysica commentaria*, ed. Michael Hayduck. Berlin: Reimer, 1891.

Alexander of Aphrodisias. *In Aristotelis meteorologicorum libros commentaria*, ed. Michael Hayduck, in *Commentaria in Aristotelem Graeca*, vol. 3,2. Berlin: Reimer, 1899.

Alexander of Aphrodisias. *Supplement* to *On the Soul*, ed. Robert W. Sharples. Ithaca, NY: Cornell University Press, 2004.

Alexandrinus, Hero. *Spiritali ... ridotti in lingua volgare da Alessandro Giorgi da Urbino.* Urbino: Bartolomeo & Simone Ragusi, 1592.

Alighieri, Dante. *Monarchy*, transl. Donald Nicholl. London: Weidenfeld & Nicolson, 1954.

Alpetragius (al-Bitruji). *De motibus celorum*, ed. Francis J. Carmody. Berkeley—Los Angeles: University of California Press, 1952.

Ameyden, Teodoro. *Trattato della natura del vino, e del ber caldo e freddo ... all'Illustrissimo et Reverendissimo Signor Cardinale Bianchetti.* Rome: Giacomo Mascardi, 1608.

Ammonius. *In Porphyrii isagogen*, ed. Adolf Busse. Berlin: Reimer, 1891.

Aquinas, Thomas. *Corpus Thomisticum*, ed. Enrique Alarcón, www.corpusthomisticum.org (accessed 7 January 2019). Aristotle. *De sensu and De memoria*, transl. George R. T. Ross. Cambridge: Cambridge University Press, 1906.

Aquinas, Thomas. *On the Power of God*, transl. by the English Dominican Fathers. London: Burns Oates & Washbourne, 1933.

Aristotle. *Meteorologica*, in *The Complete Works of Aristotle*, ed. Jonathan Barnes. Princeton: Princeton University Press, 1984.

Aristotle. *Meteorologica*, transl. Henry D. P. Lee. Cambridge, MA: Harvard University Press, 1952.

Aristotle. *Physics*. Stilwell: Digireads Publishing, 2006.

Augustine. *The City of God against Pagans*, transl. George McCracken. Cambridge, MA: Harvard University Press, 1968.

Averroes. *Grand commentaire de la* Métaphysique *d'Aristote, livre lam-lambda*, ed. Aubert Martin. Paris: Les Belles Lettres, 1984.

Azorius, Johannes. *Institutionum moralium in quibus universae quaestiones ad conscientiam recte aut prave factorum pertinentes, breviter tractantur. Pars prima.* Rome: Luigi Zanetti, 1600.

Azorius, Johannes. *Institutiones morales*. Cologne: Anton Hierat, 1602.

Azorius, Johannes. *Institutionum moralium quibus universae quaestione ad conscientiam recte aut prave factorum pertinentes, breviter tractantur*. Lyon: Horace Cardon, 1603.

Bacon, Francis. *Opera ..., tomus primus, qui continet De dignitate et augmentis scientiarum libros IX*, London: Haviland [1623].

Bacon, Francis. *Works*. 14 vols, ed. James Spedding, Robert L. Ellis, Douglas D. Heath. London: Longman, 1857–1874.

Bacon, Francis. *De principiis atque originibus secundum fabulas Cupidinis et Coeli. Sive Parmenidis et Telesii et precipue Democriti philosophia tractata in fabula de Cupidine*, in Bacon, *The Works*, vol. 5, 289–346. Boston: Houghton, Mifflin and Company, 1900.

Bacon, Francis. *Philosophical Studies, c.1611–c.1619*, ed. Graham Rees. Oxford: Clarendon Press, 1996.

Bacon, Francis. *The Advancement of Learning*, ed. Michael Kiernan. Oxford: Clarendon Press, 2000.

Bacon, Francis. *The "Instauratio magna" Part II: "Novum organum" and Associated Texts*, ed. Graham Rees. Oxford: Clarendon Press, 2004.

Barbetti, Giovan Battista. *Excursio de triplici causarum genere, ex Aristotelis, Ciceronis et Quinctiliani doctrina. Adiectis in fine selectis sententiis ex Isocratis orationibus de Regno administrando*. Rome: Stefano Paolini, 1601.

Bonamici, Francesco. *De Motu*. Florence: Bartolomeo Sermartelli, 1591.

Bonaventura, Federico. *De causa ventorum motus: Peripatetica disceptatio, in qua nullam esse inter Aristotelem, & Theophrastum in hac quaestione dissensionem, adversus communem sententiam demonstratur.* Urbino: Bartolomeo Ragusi, 1592.

Bonaventura, Federico. *Anemologiae pars prior: Id est De affectionibus, signis, causisque ventorum ex Aristotle, Theophrasto, ac Ptolemeo Tractatus*. Urbino: Bartolomeo & Simone Ragusi, 1593.

Borri, Girolamo. *Del flusso e reflusso del mare*. Lucca: Vincenzo Busdraghi, 1561.

Boyle, Robert. *A Free Inquiry into the Vulgarly Received Notion of Nature* [1686], ed. Edward B. Davis and Michael Hunter. Cambridge: Cambridge University Press, 1996.

Bozzio, Francesco. *De temporali ecclesiae monarchia et iurisdictione.* Rome: Luigi Zanetti, 1601.

Bozzio, Francesco. *De temporali ecclesiae monarchia et iurisdictione libri quinque*. Cologne: Johann Gymnich, 1602.

Brucker, Johann J. *Historia critica philosophiae*. Leipzig: Bernhard Christoph Breitkopf, 1742–1744.

Bruno, Giordano. *La cena de le Ceneri*, in *Opere Italiane*, vol 1, ed. Nuccio Ordine, 427–589. Turin: UTET, 2004.

Bruno, Giordano. *The Ash Wednesday Supper*, ed. and transl. Edward A. Gosselin and Lawrence S. Lerner. Toronto: University of Toronto Press, 1995.

Bruno, Giordano. *The Ash Wednesday Supper*, transl. Stanley L. Jaki. The Hague—Paris: Mouton, 1975.

Buongiovanni, Quinzio. *Quaestio de Divina Providentia iuxta Aristotelis mentem examinata publice in gymnasio Neapolitano*. Naples: Mattia Cancer, 1567.

Buongiovanni, Quinzio. *Peripateticarum disputationum de principiis naturae sectiones tres*. Venice: Pietro Deuchino, 1571.

Burton, Robert. *The Anatomy of Melanchol*. London-Toronto: J.M. Dent & Sons, 1977.

Cagnati, Marsilio. *De urbana febres curandi ratione commentarius apologeticus*. Rome: Luigi Zanetti, 1601.

Cagnati, Marsilio. *Opuscula varia*. Rome: Luigi Zanetti, 1603.

Campanella, Tommaso. *Philosophia sensibus demonstrata*. Naples: Orazio Salviani, 1591.

Campanella, Tommaso. *Epilogo magno (Fisiologia italiana): Testo italiano inedito, con le varianti dei codici delle edizioni latine*, ed. Carmelo Ottaviano. Rome: Reale Accademia d'Italia, 1939.

Campanella, Tommaso. *Poetica*, ed. Luigi Firpo. Rome: Reale Accademia d'Italia, 1944.

Campanella, Tommaso. *Del senso delle cose e della magia*, ed. Germana Ernst. Rome-Bari: Laterza, 2007.

Campanella, Tommaso. *Lettere*. Florence: Olschki, 2010.

Cardano, Girolamo. *De subtilitate*, ed. Elio Nenci. Milan: Franco Angeli, 2004.

Cartari, Vincenzo. *Immagini delli dei de gl'antichi* [*1556*]. Venice: Nicolò Pezzana, 1647.

Della Casa, Giovanni. *Lettere di Monsignor Giovanni Della Casa, arcivescovo di Benevento, a Carlo Gualteruzzi da Fano*, ed. Luigi Maria Rezzi. Imola: Tipografia del Seminario, 1824.

Cassiani, Piero. *Risposta ... al discorso sopra il bever fresco nuovamente stampato in Roma*. Bologna: Vittorio Benacci, 1603.

Cassiani, Piero. *De calidi potus apud Veteres usu ad Illustriss. et Reverendiss. D. Io. Antonium Abbatem Fachenetum ... Epistola*. Bologna: Vittorio Benacci, 1606.

Castelli, Benedetto. *Risposta alle opposizioni del S. Lodovico delle Colombe, e del S. Vincenzio di Grazia, contro al Trattato del Sig. Galileo Galilei, delle cose che stanno su l'Acqua, o che in quella si muovono*. Florence: Cosimo Giunti, 1615, in Galilei, *Opere*, vol. IV, 466–467.

Castiglione, Giacomo. *Discorso sopra il bever fresco cavato da Autori Antichissimi & principalissimi*. Rome: Bartolomeo Bonfadino, 1602.

Castiglione, Giuseppe. *Observationum in criticos decas prima*. Lyon: Horace Cardon, 1606.

Castiglione, Giuseppe. *De frigido et calido potu Apologeticus in quo Senecae, Tranquilli, Plauti et Martialis loca aliter atque a Lipsio accepta sunt, explicantur. Item Horatii, Vergilii, Athenei, Platonis et Aristotelis adversus Pierium Cassianum*. Rome: Guglielmo Facciotti, 1607.

Castriota Carrafa, Giovanna. *Rime et versi in lode della illustrissima et eccellentissima Signora Donna Giovanna Castriota Carrafa*. Vico Equense: Giuseppe Cacchi, 1585.

Cesalpino, Andrea. *Peripateticarum quaestionum libri quinque*. Venice: Lucantonio Giunta, 1571.

Cesalpino, Andrea. *Ars Medica*. Rome: Luigi Zanetti, 1602.

Chiocco, Andrea. "Quomodo a solis corpore in his inferibus calorem gigni censuerit Aristoteles," in *Quaestionum philosophicarum et medicarum*, Quaestio XIIII. Verona: Girolamo Discepolo, 1593.

Chrysogonus, Federicus. *De modo collegiandi, prognosticandi et curandi febres necnon de humana felicitate, ac denique de fluxu et refluxu maris*. Venice: Giovanni Antonio Nicolini da Sabbio & fratelli, 1528.

Cicero, Marcus Tullius. *De Natura Deorum. Academica*, transl. Harry Rackham. Cambridge, MA: Harvard University Press, 1967.

Cicogna, Emanuele Antonio. *Delle iscrizioni veneziane raccolte ed illustrate*. Venice: Giuseppe Orlandelli, 1824.

Cornelio, Flaminio. *Ecclesiae Venetae antiquis monumentis nunc etiam primum editis illustratae ac in decades distributae ... Decas nona et decima*. Venice: Giovanni Battista Pasquali, 1749.

Cornelio, Flaminio. *Catharus Dalmatiae civitas in Ecclesiastico et civili statu historicis documentis illustrata*. Padua: Tipografia del Seminario, 1759.

Cudworth, Ralph. *A True Intellectual System of the Universe*. London: Royston, 1678.

d'Abano, Pietro. *Conciliator*. Venice: Lucantonio Giunta (eredi), 1565.

D'Aquino, Giovanni Paolo. *Oratione ... in morte di Bernardino Telesio filosofo eccelentissimo agli Academici Cosentini*. Consenza: Leonardo Angrisano, 1596.

De Rubeis, Bernardo Maia. *De rebus congregationis sub titulo beati Jacob, Salomonii ... commentarius historicus*. Venice: Giovanni Battista Pasquali, 1751.

De Ystella, Luis. *Commentaria in Exodum in Parengrapham expositionem et scholia compacta*. Rome: Stefano Paolini, 1601.

Della Porta, Giambattista. *Magia naturalis sive de miraculis rerum naturalium*. Naples: Mattia Cancer, 1558.

Della Porta, Giambattista. *Magiae naturalis libri XX*. Naples: Orazio Salviani, 1589.

Della Porta, Giambattista. *De aeris transmutationibus* [1610], ed. Alfonso Paolella. Naples: Edizioni scientifiche italiane, 2000.

Della Porta, Giambattista. *Pneumaticorum libri III* [1601], ed. Oreste Trabucco. Naples: Edizioni scientifiche italiane, 2008.

Della Porta, Giambattista. *I tre libri de' spiritali* [1606], ed. Oreste Trabucco. Naples: Edizioni scientifiche italiane, 2008.

Della Porta, Giambattista. *Taumatologia e criptologia*, ed. Raffaele Sirri. Naples: Edizioni scientifiche italiane, 2013.

Delle Colombe, Ludovico. *Contro il moto della Terra* (1610–1611), in Galilei, *Opere*, vol. III, 251–290.

Delle Colombe, Ludovico. *Discorso apologetico d'intorno al Discorso di Galileo Galilei, circa le cose, che stanno su l'Acqua, o che in quella si muovono*. Florence: Zanobi Pignoni, 1612, in Galilei, *Opere*, vol. IV, 311–369.

Delphinus, Federicus. *De fluxu et refluxu aquae maris*. Venice: Accademia veneta, 1559.

Descartes, René. *The Principles of Philosophy*, in *The Philosophical Writings of Descartes*, Vol. 1, ed. John Cottingham, Robert Stoothoff and Dugald Murdoch. Cambridge: Cambridge University Press, 1985.

Descartes, René. *Les Méteores/Die Meteore. Faksimile der Erstausgabe 1637*, ed. Claus Zittel. Frankfurt: Klostermann, 2006.

Doni, Agostino. *De natura hominis libri duo: In quibus, discussa tum medicorum, tum philosophorum antea probatißimorum caligine, tandem quid sit homo, naturali ratione ostenditur*. Basel: Johann Froben, 1581.

Drebbel, Cornelis. *Ein kurzer Tractat von der Natur der Elementen und wie sie den Wind, Regen, Blitz und Donner verursachen und war zu sie nutzen*. Leiden: Henrichen von Haestens, 1608.

Durante, Castore. *Herbario nuovo ... Con figure che rappresentano le vive piante, che nascono in tutta Europa, & nell'Indie Orientali & Occidentali*. Rome: Giacomo Bericchia & Giacomo Tornieri, 1585.

Ennius, Quintus. *Fragmenta bzw. Poesis reliquae*, ed. Rolf Engelsing. Berlin: typoscript, 1983.

Enriquez, Roderico. *Commentariorum in Isaiam Prophetam Iuxta sensum litteralem et morale ... Tomus secundus*. Rome: Guglielmo Facciotti, 1602.

Fanucci, Camillo. *Trattato di tutte l'opere pie dell'alma città di Roma*. Rome: Lepido Faci & Stefano Paolini, 1601.

Fernel, Jean. *De abditis rerum causis*. Paris: André Wechel, 1551.

Fernel, Jean. *On the Hidden Causes of Things: Forms, Souls and Occult Diseases in Renaissance Medicine*, ed. John M. Forrester. Leiden: Brill, 2005.

Ferrerio, Giovanni Stefano. *Sancti Eusebii Vercellensis Episcopi et martyris eiusque in Episcopatu successorum vita et res gestae a I. S. F. Episc. Vercellensi collectae*. Rome: Luigi Zanetti, 1602.

Ficino, Marsilio. *Théologie platonicienne de l'immortalité des âmes*, ed. and transl. Raymond Marcel, 3 vols. Paris: Les Belles Lettres, 1964.

Ficino, Marsilio. *Three Books on Life*, ed. and transl. Carol V. Kaske and John R. Clark. Binghamton: The Renaissance Society of America, 1989.

Ficino, Marsilio. *Commentaire sur Le Banquet de Platon, De l'amour: Commentarium in convivium Platonis, De amore*, ed. Pierre Laurens. Paris: Les Belles Lettres, 2002.

Ficino, Marsilio. *Platonic Theology*, ed. James Hankins and William Bowen, transl. Michael J. B. Allen. Cambridge, MA: Harvard University Press, 2004.

Fulke, William. *A Goodly Gallerye with a Most Pleasant Prospect, into the Garden of Naturall Causes of All Kind of Meteors (1563)*, ed. Theodore Hornberger. Philadelphia: American Philosophical Society, 1979.

Fuscone, Pietro Paolo. *Trattato del bere caldo e freddo.* Genova: Giuseppe Pavoni, 1605.

Gaeta, Iacopo. *Ragionamento chiamato l'Academico overo della Bellezza*, ed. Anna Cerbo. Naples: Edizioni scientifiche italiane, 1996.

Galen. *Opera omnia*, ed. Karl G. Kühn. Leipzig: Knobloch, 1821–1833.

Galen. *On the Elements according to Hippocrates*, ed. Phillip De Lacy. Berlin: Akademie Verlag, 1996.

Galilei, Galileo. *Opere*, ed. Antonio Favaro, 20 vols. Florence: Barbera, 1890–1909.

Galilei, Galileo. *Discoveries and Opinions of Galileo*, transl. Stillman Drake. New York: Doubleday, 1957.

Galilei, Galileo. *Dialogue Concerning the Two Chief World Systems*, ed. Stillman Drake. New York: Modern Library, 2001.

Gemma, Cornelius. *De arte cyclognomica.* Antwerp: Christoffel Plantijn, 1569.

Gimma, Giacinto. *Idea della storia dell'Italia letterata esposta coll'ordine cronologico dal suo principio fino all'ultimo secolo*. Naples: Felice Mosca, 1723.

Goclenius, Rudolph. *Lexicon Philosophicum*. Frankfurt: Matthäus Becker, 1613.

Grassi, Orazio (Lotharius Sarsius Sigesanus). *De tribus cometis anni MDCXVIII disputatio astronomica.* Rome: Giacomo Mascardi, 1619, in Galilei, *Opere*, vol. VI, 21–35.

Grassi, Orazio (Lotharius Sarsius Sigesanus). *Libra astronomica ac philosophica* [1619], in Galilei, *Le Opere*, vol. VI, 109–180.

Grassi, Orazio (Lotharius Sarsius Sigesanus). *Ratio ponderum librae et simbellae*. Paris: Sébastien Cramoisy, 1626, in Galilei, *Opere*, vol. VI, 373–500.

Gualdo, Paolo. *Vita Ioannis Vincentii Pinelli*. Augsburg: Pinus (Markus Welser), 1607.

Guiducci, Mario. *Discorso delle comete fatto da lui nell'Accademia fiorentina nel suo medesimo consolato*. Florence: Pietro Cecconcelli, 1619, in Galilei, *Opere*, vol. VI, 39–108.

Hippocrates. *Œuvres complètes d'Hippocrate*, ed. Émile Littré. Paris: Bailière, 1839–1861.

Imperiale, Giovanni. *Musaeum historicum et physicum*. Venice: Lucantonio Giunta (eredi), 1640.

Isidor of Sevilla. *Eetymologiarvm sive originvm libri XX*, ed. Wallace Martin Lindsay. Oxford: Oxford University Press, 1911.

Jacob of Edessa. *Book of Treasures: Encyclopaedia of Philosophical and Natural Sciences as taught in Baghdad A.D. 817*, ed. and transl. Alphonse Mingana. Cambridge: Heffer, 1935.

Laërtius, Diogenes. *Life of Eminent Philosophers*, transl. Robert Drew Hicks. London-New York: Loeb Classical Library, 1925.

Lampe, Friedrich A. *Historia Ecclesiae Reformatae in Hungaria et Transylvania*. Utrecht: Jacob van Poolsum, 1728.

Leibniz, Gottfried Wilhelm. *Philosophical Papers and Letters*, ed. Leroy E. Loemker. Dordrecht: Reidel, 1969.

Liceti, Fortunio. *De novis astris et cometis libb. Sex*. Venice: Giovanni Guerligli, 1623.

Lipsius, Justus. *Physiologia stoicorum*. Antwerp: Christoffel Plantijn, 1604.

Longo, Giovanni Bernardino. *De cometis disputatio*. Naples: Orazio Salviani, 1578.

Longo, Giovanni Bernardino. *Lectiones in VIII Physicorum*. Vatican City, Biblioteca Apostolica Vaticana, MS. Reg. lat. 1968, item 1.

Lotter, Johann G. *De vita et philosophia Bernardini Telesii*. Leipzig: Breitkopf, 1733.

Lubieniecki, Stanisław. *Historia reformationis Polonicae*. Amsterdam: Johannes Aconius, 1685, Engl. transl., *History of the Polish Reformation and Nine Related Documents*, ed. George Huntston Williams, Minneapolis: Fortress Press, 1995.

Malvenda, Tomás. *De Antichristo libri undecim*. Rome: Carlo Vullietti, 1604.

Marini, Gaetano, *Degli archiatri pontifici*, 2 vols. Rome: Marco Pagliarini, 1784.

Marta, Giacomo Antonio. *Opuscula eccelentiss. Simonis Portii Neapol. Cum Iacobi Antonii Martae Philosophi Neapolitani Apologia immortalitate animae adversus opusculum de mente humana*. Naples: Orazio Salviani, 1578.

Marta, Giacomo Antonio. *Pugnaculum Aristotelis adversus principia Bernardini Telesii*. Rome: Bartolomeo Bonfadino, 1587.

Mascaretti, Giovan Battista. *Memorie istoriche di Grottammare* [1841]. In *Grottammare: Percorsi della Memoria*, ed. Vittorio Rivosecchi. Grottammare: Amministrazione comunale di Grottammare, 1994.

Masini, Nicolò. *De gelidi potus abusu libri tres*. Cesena: Bartolomeo Raverio, 1593.

Mattei, Saverio. *Quaesita per epistolam*. Naples: Simoniis fratribus, 1771.

Mizauld, Antoine. *Meteorologia*. Paris: Regnault & Claude Chaudière, 1547.

Mizauld, Antoine. *Miroueer du temps*. Paris: Regnault & Claude Chaudière, 1547.

Olympiodorus. *In meteora Aristotelis commentarii*, transl. Giovanni Battista Camozzi. Venice: Aldo Manuzio (eredi), 1551.

Olympiodorus. *In Aristotelis meteora commentaria*, ed. Wilhelm Stuve. Berlin: Georg Reimer, 1900.

Paracelsus (Theophrast von Hohenheim). *De meteoris*, in *Sämtliche Werke*, vol. I,13, 125–206. Hildesheim: Georg Olms, 1996.

Parrasio, Aulo Giano. *Liber De rebus per epistolam quaesitis*. Paris: Henri Estienne, 1567.

Pasio, Francesco. *Copia d'una breve relatione della christianità di Giappone, del mese di marzo del MDXCVIII insino ad Ottob. del medesimo anno*. Rome: Luigi Zanetti, 1601.

Patrizi, Francesco. *Nova de universis philosophia*. Ferrara: Benedetto Mammarello, 1590.

Patrizi, Francesco. *Obiectiones*, in Bernardino Telesio, *Varii de naturalibus rebus libelli*, ed. Luigi De Franco, 463–474. Florence: La Nuova Italia, 1981.

Patrizi, Francesco. *Obiectiones*, Facsimile repr. in Bernardino Telesio, *Delle cose naturali libri due; Opuscoli; Polemiche telesiane*, ed. Anna Laura Puliafito. 479–90. Rome: Carocci, 2013.

Peccana, Alessandro. *Del bever freddo libro uno. Con problemi intorno alla stessa materia …* Verona: Angelo Tamo, 1627.

Persio, Antonio. *Liber novarum positionum, in rhetoricis dialecticis ethicis iure civili iure pontificio physicis*. Venice: Giacomo Simbeni, 1575.

Persio, Antonio. *Trattato dell'ingegno dell'huomo*, Venice: Aldo Manuzio, 1576.

Persio, Antonio. *Disputationes libri novarum positionum Antonii Persii, triduo habitae Venetiis anno MDLXXV, mense maio*, ed. Andreas Alethinus, Florence: Giorgio Marescotti, 1576.

Persio, Antonio. *Del bever caldo costumato da gli antichi Romani Trattato … Nel quale si prova con l'historia, & essempio de gli antichi, & con la ragione, che il bere fatto caldo al fuoco è di maggior giovamento & forse anche gusto, che non è il freddo hoggidì usato …* Venice: Giovanni Battista Ciotti, 1593.

Persio, Antonio. *Apologia pro Bernardino Telesio adversus Franciscum Patritium*, in Bernardino Telesio, *Varii de naturalibus rebus libelli*, ed. Luigi De Franco, 474–495. Florence: La Nuova Italia, 1981.

Persio, Antonio. *Trattato dell'ingegno dell'huomo: In appendice, Del bever caldo*, ed. Luciano Artese. Pisa: Serra, 1999.

Persio, Antonio. *De natura ignis et caloris*. Rome, Biblioteca Corsiniana, MS. Linceo VI and VII.

Persio, Antonio. *Draft letter to Andrea Chiocco*. Rome, Biblioteca Corsiniana, MS. Linceo I.

Piccolomini, Alessandro. "Tractatus de iride," in *Alexander of Aphrodisias, In quatuor libros meteorologicorum Aristotelis commentatio lucidissima*, 117–129. Venice: Girolamo Scoto, 1561.

Piccolomini, Francesco. *In libros Aristotelis de Coelo lucidissima expositio*. Venice: Giovanni Antonio & Giacomo De Franceschi, 1607.

Pico della Mirandola, Giovanni. *Disputationes adversus astrologiam divinatricem*, in *Opera omnia*, vol. 1, ed. Cesare Vasoli. Hildesheim – Zürich – New York: Olms, 2005.

Philoponus, Johannes. *In I. meteorum Aristotelis expositionum in tres libros liber I. Ioanne Baptista Camotio interprete*. Venice: Aldo Manuzio (eredi), 1551.

Philoponus, Johannes. *In Aristotelis libros de generatione et corruption commentaria*, ed. Hieronymus Vitteli. Berlin: Reimer, 1897.

Philoponus, Johannes. *On Aristotle's On Coming-to-Be and Perishing*, ed. Christopher John Fardo Williams. Ithaca, NY: Cornell University Press, 1999.

Philoponus, Johannes. *Against Proclus: On the Eternity of the World 1–5*, transl. Michael Share. London: Duckworth, 2004.

Pomponazzi, Pietro. *Opera. De naturalium effectuum admiradorum causis, seu de incantationibus liber. Item de fato: libero arbitrio: praedestinatione: providentia Dei, libri V. In quibus difficillima capita et quaestiones theologicae et philosophicae ex sana orthodoxe fidei explicantur*. Basel: Sebastian Henricpetri, 1567.

Pomponazzi, Pietro. *Libri quinque de fato, de libero arbitrio et de praedestinatione*, ed. Richard Lemay. Lucani: Thesaurus Mundi, 1957.

Pomponazzi, Pietro. *Il fato, il libero arbitrio e la predestinatione*, transl. Vittoria Perrone Compagni. Turin: Aragno, 2004.

Porzio, Simone. *De rerum naturalium principiis*. Naples: Mattia Cancer, 1553.

Porzio, Simone. *An homo bonus vel malus volens fiat* [1551], ed. Eva Del Soldato. Rome: Edizioni di storia e letteratura, 2005.

Porzio, Simone. *Liber de fato*. Milan, Biblioteca Ambrosiana, MS. Q.122 sup.

Porzio, Simone. *Quaestio de materia caeli*. Milan, Biblioteca Ambrosiana, P. 197 sup., 33r–40r (item IV).

Quadrio, Francesco Saverio. *Della storia e della ragione d'ogni poesia*. Bologna: Pisarri, 1739–1744.

Quattromani, Sertorio. *La philosophia di Berardino Telesio ristretta in brevità et scritta in lingua toscana*. Naples: Giuseppe Cacchi, 1589.

Quattromani, Sertorio. *Lettere di Sertorio Quattromani gentil'huomo e academico cosentino*. Naples: Lazzaro Scoriggio, 1624.

Quattromani, Sertorio. *Scritti*, ed. F. Walter Lupi. Arcavacata di Rende: University of Calabria, 1999.

Raimondo, Annibale. *Trattato utilissimo e particolarissimo del flusso e riflusso del mare*. Venice: Domenico dei Niccolini, 1589.

Rao, Cesare. *I meteori*. Venice: Giovanni Varisco & Compagni, 1582.

Ringhieri, Innocentio. *Il Sole di M. Innocentio Ringhieri Gentil'huomo Bolognese*. Rome: Antonio Blado, 1550.

Rocca, Angelo. *De canonizatione Sanctorum commentarius, hoc est De diffinitione, auctoritate et antiquitate, deque causis et ordine iudiciario canonizndi Sanctos* ... Rome: Guglielmo Facciotti, 1601.

Sagri, Niccolò. *Ragionamenti sopra le varietà de i flussi et riflussi del mare oceano occidentale, fatti da Andrea di Noblisia, Pedotto Biscaino, et Vicenzo Sabici, nocchiero, &*

Ambrosio di Goze, ragusei; raccolti da Nicolo Sagri, et in un dialogo dall'istesso ridotti, diuiso in due parti, ad utilità di ciascuno navigante. Venice: Domenico & Giovanni Battista Guerra, 1574.

Santorio, Paolo Emilio. *Historia monasterii Carbonensis ordninis sancti Basilii*. Rome: Guglielmo Facciotti, 1601.

Scacchi, Francesco. *De salubri potu dissertatio*. Rome: Alessandro Zanetti, 1622.

Scipioni-Crostarosa, Noemi. "Lettere inedite di Bernardino Telesio e Giacomo Pelusio nel carteggio del Cardinale Guglielmo Sirleto." *Archivio storico per la Calabria e la Lucania* 7 (1937): 105–120.

Seneca, Lucius Annaeus. *Naturales quaestiones/Natural Questions*, transl. Thomas H. Corcoran. Cambridge, MA: Harvard University Press, 1971–1972.

Servetus, Michael. *Christianismi restitutio*. s.l. [Vienna], 1553.

Sfondrati, Pandolfo. *Causa aestus maris*. Ferrara: Benedetto Mammarello, 1590.

Simplicius. *Commentarius in libro de caelo Aristotelis*, transl. Guillelmus de Moerbeka. Venice: Girolamo Scoto, 1555.

Spinoza, Baruch. *The Collected Works*, ed. Edwin Curley. Princeton: Princeton University Press, 1985.

Steuco, Agostino. *Cosmopoeia*. Lyon: Sébastien Gryphius, 1535.

Tansillo, Luigi. *Capitoli giocosi e satirici*, ed. Scipione Volpicella. Naples: Libreria di Dura, 1870.

Tasso, Torquato. *Opere di Torquato Tasso colle controversie sulla Gerusalemme, poste in migliore ordine, ricorrette sull' edizione fiorentina*, ed. Giovanni Battista Manso and Giovanni Rosini, Vol. 23. Pisa: Capurro, 1832.

Tassoni, Alessandro. *Lettere*, vol. 1 and 2, ed. Pietro Puliatti. Rome – Bari: Laterza, 1978.

Tassoni, Alessandro. *Pensieri e scritti preparatori*, ed. Pietro Puliatti. Modena: Panini, 1986.

Telesio, Bernardino. *De natura iuxta propria principia liber primus et secundus*. Rome: Antonio Blado, 1565.

Telesio, Bernardino. *De rerum natura iuxta propria principia, liber primus, et secundus, denuo editi*. Naples: Giuseppe Cacchi, 1570.

Telesio, Bernardino. *De colorum generatione opusculum*. Naples: Giuseppe Cacchi, 1570.

Telesio, Bernardino. *De iis quae in aere fiunt et de terraemoribus*. Naples: Giuseppe Cacchi, 1570.

Telesio, Bernardino. *De mari*. Naples: Giuseppe Cacchi, 1570.

Telesio, Bernardino. *De rerum natura iuxta propria principia. Libri IX*. Naples: Orazio Salviani, 1586.

Telesio, Bernardino. *Varii de naturalibus rebus libelli ab Antonio Persio editi*. Venice: Felice Valgrisi, 1590.

Telesio, Bernardino. *De rerum natura: libri I–II–III* [1586]. Cosenza: Casa del Libro, 1965.

Telesio, Bernardino. *De fulmine*, in Carlo Delcorno, "Il commentario 'De fulmine' di Bernardino Telesio." *Aevum* 41 (1967): 480–502.

Telesio, Bernardino. *De rerum natura: libri IV–V–VI* [1586], ed. Luigi De Franco. Cosenza: Casa del Libro, 1974.

Telesio, Bernardino. *De rerum natura libri VII–VIII–IX* [1586], ed. Luigi De Franco. Florence: La Nuova Italia, 1976.

Telesio, Bernardino. *De rerum natura iuxta propria principia libri IX* [Naples, 1586], ed. Cesare Vasoli. Hildesheim – New York: Olms, 1971.

Telesio, Bernardino. *De rerum natura iuxta propria principia* [1586], ed. Luigi De Franco, Florence: La Nuova Italia, 1976.

Telesio, Bernardino. *Varii de naturalibus rebus libelli* [1590], ed. Luigi De Franco. Florence: La Nuova Italia, 1981.

Telesio, Bernardino. "De mari," in *De iis que in aere fiunt et de terremotibus; De mari*, transl. Francesco Martelli, ed. Luigi De Franco. Cosenza: Bios, 1990.

Telesio, Bernardino. *La natura secondo i suoi principi* [1570], ed. Roberto Bondì. Florence: La Nuova Italia, 1999.

Telesio, Bernardino. *De natura iuxta propria principia liber primus et secundus (Rome 1565)*, ed. Alessandro Ottaviani. Turin: Aragno, 2006.

Telesio, Bernardino. *La natura secondo i suoi principi* [1570], ed. Roberto Bondì. Milan: Bompiani, 2009.

Telesio, Bernardino. *De rerum natura iuxta propria principia, liber primus, et secundus, denuo editi (Napoli 1570)*, ed. Alessandro Ottaviani. Turin: Aragno, 2010.

Telesio, Bernardino. *De natura iuxta propria principia* [1565]; *Ad Felicem Moimonam iris* [1566], ed. Roberto Bondì. Rome: Carocci, 2011.

Telesio, Bernardino. *Varii de naturalibus rebus libelli ab Antonio Persio editi* [1590], ed. Miguel Ángel Granada. Rome: Carocci, 2012.

Telesio, Bernardino. *Sobre los cometas y la Vía Láctea; De cometis et lacteo circulo*, ed. and transl. Miguel Ángel Granada. Madrid: Tecnos Editorial, 2012.

Telesio, Bernardino. *De rerum natura iuxta propria principia, liber primus, et secundus, denuo editi* [1570]; *Opuscula* [1570], ed. Roberto Bondì. Rome: Carocci, 2013.

Telesio, Bernardino. *De rerum natura iuxta propria principia, libri IX* [1586], ed. Guido Giglioni. Rome: Carocci, 2013.

Telesio, Bernardino. *Delle cose naturali libri due; Opuscoli; Polemiche telesiane*, transl. Francesco Martelli, ed. Anna Laura Puliafito. Rome: Carocci, 2013.

Themistius. *Themistii in libros Aristotelis De anima paraphrasis*, ed. Richard Heinze. Berlin: Reimer, 1899.

Themistius. *On Aristotle's* On the Soul, ed. Robert B. Todd. Ithaca, NY: Cornell University Press, 1996.

Theodoric of Freiberg. *De iride et radialibus impressionibus: Über den Regenbogen und die durch Strahlen erzeugten Eindrücke*, ed. Joseph Würschmidt. Münster: Aschendorff, 1914.

Tiraboschi, Girolamo. *Storia della letteratura italiana*. 13 vols. Modena: Società tipografica, 1772–1782.

Vianelli, Girolamo. *Nuova serie de' vescovi di Malamocco e di Chioggia … parte seconda*. Venice: Nella Stamperia Baglioni, 1790.

Vimercati, Francesco. *In quatuor libros Aristotelis Meteorologicorum commentarii*. Paris: Michel de Vascosan, 1556.

Zabarella, Giacomo. *De calore coelesti*, in *De rebus naturalibus libri XXX*. Venice: Paolo Meietti, 1590.

Secondary Literature

Abbagnano, Nicola. *Bernardino Telesio*. Milan: Bocca, 1941.

Accademia Cosentina. *Atti del Convegno Internazionale di Studi su Bernardino Telesio*. Cosenza: Accademia Consentina, 1990.

Accademia dei Lincei. *Federico Cesi e la fondazione dell'accademia dei Lincei*. Naples: Accademia nazionale dei Lincei, 1988.

Agrimi, Marco. "Telesio nel Seicento napoletano," in Galasso et al. (eds.), *Bernardino Telesio e la cultura napoletana*, 331–372.

Aquilecchia, Giovanni. "I *Massimi Sistemi* di Galileo e *La Cena* di Bruno (per una comparazione tematico-strutturale)," in *Nuncius: Journal of the Material and Visual History of Science* 10/2 (1995): 485–496.

Arkoun, Mohammed. "Deux épîtres de Miskawayh (mort en 421/1030): édition avec introduction et notes," in *Bulletin d'études orientales* 17 (1961): 7–74.

Artese, Luciano. "Antonio Persio e la diffusione del ramismo in Italia," in *Atti e memorie dell'Accademia toscana di scienze e lettere 'La Colombaria'* 46 (1981): 83–116.

Artese, Luciano. "Una lettera di Antonio Persio al Pinelli: Notizie intorno all'edizione del primo tomo delle *Discussiones* del Patrizi," in *Rinascimento: Rivista dell'Istituto Nazionale di Studi sul Rinascimento* 26 (1986): 339–348.

Artese, Luciano. "Filosofia telesiana e ramismo in un inedito di Antonio Persio," in *Giornale critico della filosofia italiana* 46 (1987): 433–458.

Artese, Luciano. "Il rapporto Parmenide-Telesio, dal Persio al Maranta," in *Giornale critico della filosofia italiana* 60 (1991): 15–34.

Badaloni, Nicola. *Tommaso Campanella*. Milan: Feltrinelli, 1965.

Badaloni, Nicola. "Sulla costruzione e la conservazione della vita in Bernardino Telesio (1509–1588)," in *Studi Storici* 30 (1989): 25–42.

Badaloni, Nicola. "Sulla costruzione e sulla conservazione della vita in Bernardino Telesio," in *Bernardino Telesio nel 4° centenario della morte (1588)*, ed. Mario Santoro, 9–49. Naples: Istituto Nazionale di Studi sul Rinascimento Meridionale, 1989.

Bainton, Roland H. *Michel Servet hérétique et martyr: 1553–1953*. Geneva: Droz, 1953.

Balázs, Mihály. *Early Transylvanian Antitrinitarianism (1566–71)*. Baden-Baden: Koerner, 1996.

Baldini, Ugo. "Il pubblico della scienza nei permessi di lettura di libri proibiti delle congregazioni del Sant'Ufficio e dell'Indice (secolo XVI): Verso una tipologia professionale e disciplinare," in *Censura ecclesiastica e censura politica in Italia tra Cinquecento e Seicento*, ed. Cristina Stango, 171–201. Florence: Olschki, 2001.

Baldini, Ugo and Leen Spruit (eds). *Catholic Church and Modern Science. Documents from the Archives of the Roman Congregations of the Holy Office and the Index*. 4 vols. Rome: Libreria Editrice Vaticana, 2009.

Barker, Peter. "Stoic Contributions to Early Modern Science," in *Atoms, "pneuma" and Tranquillity. Epicurean and Stoic Themes in European Thought*, ed. Margaret J. Osler, 135–154. Cambridge: Cambridge University Press, 1991.

Bartelli, Francesco. *Note biografiche Bernardino Telesio—Galeazzo di Tarsia*. Consenza: Trippa, 1906.

Bartholmess, Chrétien. *De Bernardino Telesio*. Paris: Ducloux, 1849.

Bertoloni-Meli, Domenico. "Guidobaldo dal Monte and the Archimedean revival," in *Nuncius: Journal of the Material and Visual History of Science* 7/1 (1992): 3–34.

Biagioni, Mario and Lucia Felici. *La riforma radicale nell'Europa del Cinquecento*. Rome – Bari: Laterza, 2012.

Bianchi, Lorenzo. "'Des novateurs modernes en la philosophie': Telesio tra eruditi e libertini nella Francia del Seicento," in Galasso et al. (eds.), *Bernardino Telesio e la cultura napoletana*, 373–416.

Bianchi, Massimo and Marta Fattori (eds). *Spiritus: IV° colloquio internazionale*. Rome: Edizioni dell'Ateneo, 1984.

Bianchi, Massimo Luigi. "Scholastische Motive im ersten und zweiten Buch des *De subtilitate* Girolamo Cardanos," in *Girolamo Cardano: Philosoph, Naturforscher, Arzt*, ed. Eckhard Kessler, 115–130. Wiesbaden: Harrassowitz, 1994.

Biard, Joël. "La certitude des mathématiques et ses fondements selon Piccolomini," in *Alessandro Piccolomini (1508–1579): Un Siennois à la croisée des genres et des savoirs*, ed. Marie-Françoise Piéjus, Michel Plaisance and Matteo Residori, 247–257. Paris: Université Sorbonne nouvelle Paris 3, 2012.

Birnstein, Uwe. *Toleranz und Scheiterhaufen: Das Leben des Michael Servet*. Göttingen: Vandenhoeck & Ruprecht, 2013.

Black, Christopher F. *The Italian Inquisition*. New Haven: Yale University Press, 2009.

Blair, Ann. "The *Problemata* as a Natural Philosophical Genre," in *Natural Particulars: Nature and Disciplines in Renaissance Europe*, ed. Anthony Grafton and Nancy Siraisi, 171–204. Cambridge, MA: The MIT Press, 1999.

Boenke, Michaela. *Körper, Geist, Spiritus. Psychologie vor Descartes*. Munich: Fink, 2005.

Bondì, Roberto. "*Spiritus* e *anima* in Bernardino Telesio," in *Giornale critico della filosofia italiana* 72/3 (1993): 405–417.

Bondì, Roberto. "'Expurgatio impossibilis:' Filosofia e religione in Telesio," in *Rivista di storia della filosofia* 51/4 (1996): 881–894.

Bondì, Roberto. *Introduzione a Telesio.* Bari: Laterza, 1997.

Bondì, Roberto. "Introduzione" to Telesio, *La natura secondo i suoi princìpi.* Milan: Bompiani, 2009, VII–XXII.

Bondì, Roberto. "Dios, naturaleza y alma en Telesio," in Bondì et al., *Bernardino Telesio y la nueva imagen de la naturaleza*, 21–42.

Bondì, Roberto, *Il primo dei moderni: filosofia e scienza in Bernardino Telesio*. Rome; Edizioni di storia e letteratura, 2018.

Bondì, Roberto, Karl Schuhmann, Michel-Pierre Lerner, Miguel A. Granada and Susana Gómez López (eds). *Bernardino Telesio y la nueva imagen de la naturaleza en el Renacimiento*. Madrid: Siruela, 2013.

Borrelli, Arianna. "Pneumatics and the Alchemy of Weather: What is Wind and Why Does it Blow?," in *Variantology 3. On Deep Time Relations of Arts, Sciences and Technologies in China and Elsewhere*, ed. Siegfried Zielinski and Eckhard Fürlus, 27–72. Cologne: König, 2008.

Borrelli, Arianna. "The Weatherglass and its Observers in the Early Seventeenth Century," in *Philosophies of Technology: Francis Bacon and His Contemporaries*, vol. 1, ed. Claus Zittel, Gisela Engel, Nicole C. Karafyllis and Romano Nanni, 67–130. Leiden: Brill, 2008.

Borrelli, Arianna. "Giovan Battista Della Porta's Neapolitan Magic and his Humanistic Meteorology," in *Variantology 5. On Deep Time Relations of Arts, Sciences and Technologies*, ed. Siegfried Zielinski and Eckhard Fürlus, 103–130. Cologne: König, 2011.

Borrelli, Arianna. "Thinking with Optical Objects: Glass Spheres, Lenses and Refraction in Giovan Battista Della Porta's Optical Writings," in *Journal of Early Modern Studies* 3 (2014): 38–60.

Borrelli, Arianna. "The Recipe as a Heuristic Tool in Giovan Battista Della Porta's Pneumatic Writings," in *"A High Kind of Natural Magic:" Francis Bacon and Giovan Battista della Porta on "Philosophical Instruments" and the Creative Powers of Experimentation*, ed. Dana Jalobeanu and Doina-Cristina Rusu. Special Issue of *Centaurus* (2019, in press).

Bos, Gerrit and Charles Burnett (eds). *Scientific Weather Forecasting in the Middle Ages: Studies, Editions, and Translations of the Arabic, Hebrew and Latins Texts*. London: Kegan Paul International, 2000.

Boyer, Carl B. *The Rainbow: From Myth to Mathematics*. New York: Tamar Yoseloff, 1959.

Brucker, Johann Jakob. *Historia critica philosophiae*. Vol. IV, Part I. Leipzig: Breitkopf, 1743.

Brunschwig, Jacques. "Aristote et le statut épistémologique de la théorie de l'arc-en-ciel," in *Graceful Reason: Essays in Ancient and Medieval Philosophy Presented to Joseph Owens*, ed. Lloyd P. Gerson, 115–134. Toronto: Pontifical Institute for Mediaeval Studies, 1983.

Bucciantini, Massimo. *Galileo e Keplero: Filosofia, cosmologia e teologia nell'Età della Controriforma*. Turin: Einaudi, 2003.

Burton, Robert. *The Anatomy of Melancholy*. London-Toronto: J.M. Dent & Sons, 1977.

Bylebyl, Jerome J. "The School of Padua: Humanistic Medicine in the Sixteenth Century," in *Health, Medicine and Mortality in the Sixteenth Century*, ed. Charles Webster, 335–370. Cambridge: Cambridge University Press, 1979.

Caccamo, Domenico. *Eretici italiani in Moravia, Polonia, Transilvania (1558–1611): Studi e documenti*. Florence: Sansoni, 1970.

Camerota, Michele. *Galileo Galilei e la cultura scientifica nell'età della Controriforma*. Rome: Salerno Editrice, 2004.

Campanella, Tommaso. *Lettere*, ed. Germana Ernst et al. Florence: Olschki, 2010.

Canart, Paul. "Alvise Lollino et ses amis grecs," in *Studi veneziani* 12 (1970): 553–587.

Cantimori, Delio. *Eretici italiani del Cinquecento: Ricerche storiche*. Florence: Sansoni, 1939.

Casali, Elide. *Le spie del cielo: Oroscopi, lunari e almanacchi nell'Italia moderna*. Turin: Einaudi, 2003.

Cassirer, Ernst. *Das Erkenntnisproblem in der Philosophie und Wissenschaft der neueren Zeit*. Berlin: Bruno Cassirer, 1922.

Castelli, Daniela. "Tra ricerca empirica e osservazione scientifica: Gli studi ittiologici di Simone Porzio," in *Archives internationales d'histoire des sciences* 57 (2007): 105–123.

Castelli, Daniela. "Il *De' sensi* e il *Del sentire* di Simone Porzio: Due manoscritti ritrovati," in *Giornale critico della filosofia italiana* 87 (2008): 255–280.

Castelli, Daniela. "Tra aristotelismo, naturalismo e critica: Note in margine a Simone Porzio (1496–1554)," in *Critica e ragione/Critique et raison*, ed. Lorenzo Bianchi and Alberto Postigliola, 33–50. Naples: Liguori, 2011.

Castelli, Daniela. "Il *De conflagratione* di Simone Porzio: La collazione delle tre edizioni, un volgarizzamento e il ms. Phill. 12844 dell'HRC di Austin," in *Rinascimento Meridionale* 3 (2012): 81.

Cherchi, Paolo. "Il quotidiano, i *Problemata* e la meraviglia: Ministoria di un microgenere," in *Ministorie di microgeneri*, 11–40. Ravenna: Longo, 2003.

Chiara, Stanislao de. *Bricciche telesiane*. Cosenza: Aprea, 1879.

Cirino, Raffaele. "Bernardino Telesio e 'delle cose che in aria si fanno'," in *Bernardino Telesio: tra filosofia naturale e scienza moderna,* ed. Giuliana Mocchi, Sandra Plastina and Emilio Sergio, 65–70. Pisa: Serra, 2012.

Clutton-Brock, Martin and David Topper. "The Plausibility of Galileo's Tidal Theory," *Centaurus* 53/3 (2011): 221–235.

Cocco, Sean. "Locating the Natural Sciences in Early Modern Naples," in *Companion to Early Modern Naples*, ed. Tommaso Astarita, 453–76. Leiden: Brill, 2013.

Continisio, Chiara. "Scritture politiche urbinati nell'età di Francesco Maria II Della Rovere," in *I Della Rovere nell'Italia delle corti 1: Storia del ducato*, ed. Bonita Cleri, Sabine Eiche and John E. Law, 93–109. Urbino: Edizioni Quattro Venti, 2002.

Cornacchioli, Tobia. *Lineamenti di storia della cultura calabrese, Ipotesi su un frammento: l'Accademia Parrasiana*. Cosenza: Pellegrini, 1982.

Cornacchioli, Tobia. *Nobili, borghesi e intellettuali nella Cosenza del Quattrocento*. Cosenza: Edizioni Periferia, 1990.

Corsano, Antonio. "La psicologia di Telesio," in *Giornale critico della filosofia italiana* 21 (1940): 5–12.

Cortese, Nino. "L'Età spagnola," in *La storia della Università di Napoli*, ed. Francesco Torraca, 213–14. Naples: Ricciardi, 1924.

D'Amato, Alfonso. *I Domenicani a Bologna*, 2 vols. Bologna: Studio Domenicano, 1988.

De Anna, Gabriele. "Telesio e il naturalismo: le critiche alle tesi aristoteliche sull'immortalità dell'intelletto," in *Bernardino Telesio tra filosofia naturale e scienza moderna*, ed. Giuliana Mocchi, Sandra Plastina and Emilio Sergio, 85–98. Pisa – Rome: Fabrizio Serra Editore, 2012.

De Franco, Luigi. *L'eretico Agostino Doni, medico e filosofo cosentino del '500: In appendice A. Doni, "De natura hominis" con traduzione a fronte*. Cosenza: Pellegrini, 1973.

De Franco, Luigi. *Bernardino Telesio: La vita e l'opera*. Cosenza: Edizioni Pereferia, 1989.

De Franco, Luigi. "Telesio e la libertas philosophandi," in Luigi De Franco, *Bernardino Telesio: La vita e l'opera*, 109–121. Cosenza: Edizioni Periferia, 1989.

De Franco, Luigi. "La teoria della luce di B. Telesio," in *Telesio: La vita e l'opera*, 123–142. Cosenza: Edizioni Periferia, 1989.

De Franco, Luigi. "Nota introduttiva" to Bernardino Telesio, *De iis quae in aere fiunt et de terremotibus; De mari*. 9–16. Cosenza: Editoriale Bios, 1990.

De Franco, Luigi. "La teoria della luce di Bernardino Telesio," in Galasso et al. (eds.), *Bernardino Telesio e la cultura napoletana*, 53–77.

De Franco, Luigi. *Introduzione a Bernardino Telesio*. Soveria Mannelli: Rubbettino, 1995.

De Mas, Enrico. *Francesco Bacone da Verulamio: La filosofia dell'uomo*. Turin: Edizioni di Filosofia, 1964.

De Mas, Enrico. "Introduzione" to F. Bacone, *Dai naturalisti greci a Telesio,* vii–xxxviii. Cosenza: Laboratorio Edizioni, 1988.

De Mas, Enrico. "Bernardino Telesio e la falsità di Aristotele: il giudizio di Bacone e di Tobia Adami," in *Atti del Convegno Internazionale di Studi su Bernardino Telesio,* 167–179. Cosenza: Accademia Cosentina, 1990.

De Miranda, Girolamo. "Una lettera inedita di Telesio al cardinale Flavio Orsini," in *Giornale critico della filosofia italiana* 72 (1993): 361–375.

Debus, Allen G. "The Paracelsian Aerial Niter," in *Isis* 55 (1964): 43–61.

Debus, Allen G. "Chemistry and the Quest for a Material Spirit of Life in the Seventeenth Century," in *Spiritus. IV° colloquio internazionale,* ed. Massimo Bianchi and Marta Fattori, 245–263. Rome: Edizioni dell'Ateneo, 1984.

Del Col, Andrea. *L'Inquisizione in Italia dal XII al XXI secolo*. Milan: Mondadori, 2006.

Del Nero, Valerio. "Valla, Vives e Nizolio: Filosofia e linguaggio," in *Rinascimento* 34 (1994): 293–304.

Delcorno, Carlo. "Il commentario 'De fulmine' di Bernardino Telesio," in *Aevum,* 41 (1967): 474–506.

Des Chene, Dennis. *Physiologia: Natural Philosophy in Late Aristotelian and Cartesian Thought*. Ithaca, NY: Cornell University Press, 2000.

Di Bono, Mario. *Le sfere omocentriche di Giovan Battista Amico nell'astronomia del Cinquecento*. Genoa: Consiglio Nazionale delle Ricerche-Centro di Studio sulla Storia della Tecnica, 1990.

Di Napoli, Giovanni. *L'immortalità dell'anima nel Rinascimento*. Turin: Società Editrice Internazionale, 1963.

Domeyne, Pierre. *Michael Servet (1511–1553): Au risque de se perdre*. Paris: L'Harmattan, 2008.

Drake, Stillman. *Galileo at Work: His Scientific Biography*. Mineola, NY: Dover, 2003.

Duhem, Pierre. *Le Système du Monde: Histoire des Doctrines Cosmologiques de Platon à Copernic,* vol. 2. Paris: Hermann, 1914.

Dupré, Sven. "Ausonio's Mirrors and Galileo's Lenses: The Telescope and Sixteenth-Century Practical Optical Knowledge," in *Galilaeana* 2 (2005): 145–180.

Eamon, William. *Science and the Secrets of Nature: Books of Secrets in Medieval and Early Modern Culture*. Princeton: Princeton University Press, 1994.

Eamon, William. *The Professor of Secrets: Mystery, Medicine and Alchemy in Renaissance Italy*. Washington: National Geographic, 2010.

Eamon, William. "A Theater of Experiments: Giambattista Della Porta and the Scientific Culture of Renaissance Naples," in *The Optics of Giambattista Della Porta: A Reassessment,* ed. Arianna Borrelli, Giora Hon and Yaakov Zik, 11–38. Berlin: Springer, 2017.

Egidi, Vincenzo Maria and Mario Boretti. *I Telesio: Regesto dei documenti del sec.* XVI, ed. Raffaele Borretti. Cosenza: Calabria nobilissima, 1988.

Ernst, Germana. *Tommaso Campanella: Il libro e il corpo della natura*. Bari – Rome: Laterza, 2002.

Facca, Danilo. "Humana mens corruptibilis: L'antiaverroismo di Simone Porzio," in *Filosofia, filologia, biologia: itinerari dell'aristotelismo cinquecentesco*, ed. Danilo Facca and Giancarlo Zanier, 5–104. Rome: Edizioni dell'Ateneo, 1992.

Fauth, Wolfgang. *Helios Megistos: Zur synkretistischen Theologie der Spätantike*. Leiden: Brill, 1995.

Fédération Internationale des Sociétés et Instituts pour l'Étude de la Renaissance, ed. *Le Soleil à la Renaissance: Sciences et mythes*. Brussels: Editions de l'Université de Bruxelles, 1965.

Ferraro, Giovanni. "Dimostrazioni matematiche e conoscenza scientifica in Alessandro Piccolomini," in *Saggi di letteratura architettonica: Da Vitruvio a Winckelmann III*, ed. Howard Burns, Francesco Paolo Di Teodoro and Giorgio Bacci, 215–234. Florence: Olschki, 2010.

Firpo, Massimo. *Antitrinitari nell'Europa orientale del 1500: Nuovi testi di Szymon Budny, Niccolò Paruta e Iacopo Paleologo*. Florence: La Nuova italia, 1977.

Fiorentino, Francesco. *Bernardino Telesio, ossia studi storici sull'idea della natura nel Risorgimento italiano*. 2 vols. Florence: Le Monnier, 1872–1874, repr. Florence: Le Monnier, 1974.

Fiorentino, Francesco. *L'Accademia cosentina e la filosofia di Bernardino Telesio* [1872]. Grisolia: Marina di Belvedere, 1989.

Firpo, Luigi. *Bibliografia degli scritti di Tommaso Campanella*. Turin: Bona, 1940.

Firpo, Luigi. "Appunti campanelliani III: La perduta *Apologia pro Telesio*," in *Giornale critico della filosofia italiana* 21 (1940): 435–438.

Firpo, Luigi. "Filosofia italiana e controriforma IV: La proibizione di Telesio," in *Rivista di filosofia* 42 (1951): 30–47.

Firpo, Luigi. "La proibizione di Telesio," in *Rivista di filosofia* 42 (1951): 30–47.

Firpo, Luigi. "Bonaventura, Federico," in *Dizionario biografico degli italiani* 11 (1969): 644–646.

Firpo, Massimo. *Juan de Valdés and the Italian Reformation*. Aldershot: Ashgate, 2015.

Forrester, John M. and John Henry (eds). *Jean Fernel's* On the Hidden Causes of Things*: Forms, Souls and Occult Diseases in Renaissance Medicine*. Leiden: Brill, 2005.

Fragnito, Gigliola. "La censura libraria tra Congregazione dell'Indice, Congregazione dell'Inquisizione e Maestro del Sacro Palazzo (1571–1596)," in *La censura libraria nell'Europa del secolo XVI*, ed. Ugo Rozzo, 163–175. Udine: Forum Editrice Universitaria, 1997.

Fragnito, Gigliola. "Un archivio conteso: Le carte dell'Indice tra Congregazione e Maestro del Sacro Palazzo," in *Rivista Storica Italiana* 119/3 (2007): 1276–1318.

Frank, Martin and Pier Daniele Napolitani. "Il giovane Galileo e Guidobaldo dal Monte: discepolo e maestro?," in *Scienze e rappresentazioni: Saggi in onore di Pierre Souffrin*, ed. Pierre Caye, Romano Nanni and Pier Daniele Napolitani, 171–197. Florence: Olschki, 2015.

Freedberg, David. *The Eye of the Lynx: Galileo, His Friends, and the Beginnings of Modern Natural History*. Chicago: University of Chicago Press, 2004.

Freudenthal, Gad. *Aristotle's Theory of Material Substance: Heat and Pneuma, Form and Soul*. Oxford: Clarendon, 1995.

Freudenthal, Gad. "The Medieval Astrologization of Aristotle's Biology: Averroes on the Role of the Celestial Bodies in the Generation of Animate Beings," in *Arabic Sciences and Philosophy* 12/1 (2002), 111–137.

Fumaroli, Marc. "Les abeilles et les araignées," in *La Querelle des Anciens et des Modernes*, ed. Marc Fumaroli and Anne-Marie Le Coq, 7–218. Paris: Gallimard, 2001.

Funkenstein, Amos. *Theology and the Scientific Imagination from the Middle Ages to the Seventeenth Century*. Princeton: Princeton University Press, 1986.

Gabrieli, Giuseppe. "Notizia della vita e degli scritti di Antonio Persio Linceo," in *Rendiconti dell'Accademia dei Lincei* 6, IX (1933): 471–499.

Galasso, Giuseppe, Raffaele Sirri and Maurizio Torrini (eds). *Bernardino Telesio e la cultura napoletana*. Naples: Guida, 1992.

Galluzzi, Paolo. *Libertà di filosofare in Naturalibus: I mondi paralleli di Cesi e Galileo*. Rome: Scienze e Lettere, 2014.

Garin, Eugenio. "Nota telesiana: Antonio Persio," in *Giornale Critico della Filosofia Italiana* 28 (1949): 414–421.

Garin, Eugenio. "Noterella telesiana," in *Giornale critico della filosofia italiana* 36 (1957): 56–62.

Garin, Eugenio. "Fra '500 e '600: scienze nuove, metodi nuovi, nuove accademie," in *Nuncius: Journal of the Material and Visual History of Science* 1/1 (1986): 3–23.

Garin, Eugenio. *Umanisti, artisti, scienziati: Studi sul rinascimento italiano*. Rome: Editori Riuniti, 1989.

Garin, Eugenio. "Il termine 'spiritus' in alcune discussioni fra Quattrocento e Cinquecento," in idem, *Umanisti, artisti, scienziati: Studi sul Rinascimento italiano*, 295–303.

Garin, Eugenio. *Renaissance Characters*. Chicago-London: University of Chicago Press, 1991.

Garin, Eugenio. "Postilla telesiana," in *La cultura filosofica del Rinascimento italiano: Ricerche e documenti*, 442–450. Milan: Bompiani, 1994.

Garin, Eugenio. *History of Italian Philosophy*. Amsterdam: Rodopi, 2008.

Garin, Eugenio. *The Emergence of a Scientific Culture: Science and the Shaping of Modernity 1210–1685*. Oxford: Clarendon, 2010.

Gaukroger, Stephen. *Francis Bacon and the Transformation of Early-Modern Philosophy*. Cambridge: Cambridge University Press, 2001.

Gemelli, Benedino. *Aspetti dell'atomismo classico nella filosofia di Francis Bacon e nel Seicento*. Florence: Olschki, 1996.

Gentile, Giovanni. *Bernardino Telesio*. Bari: Laterza, 1911.

Gentile, Giovanni. "Bernardino Telesio," in *Opere complete di Giovanni Gentile: I problemi della Scolastica e il pensiero Italiano*. 133–206. Florence: Sansoni, 1963.

Giachetti Assenza, Valeria. "Bernardino Telesio: il migliore dei moderni. I riferimenti a Telesio negli scritti di Francesco Bacone." *Rivista critica di storia della filosofia* 35 (1980): 41–78.

Giacobbe, Giulio Cesare. "Il Commentarium de certitude mathematicarum disciplinarum di Alessandro Piccolomini," in *Physis: Rivista Internazionale di Storia della Scienza* 14/2 (1972): 162–193.

Giglioni, Guido. "The First of the Moderns or the Last of the Ancients? Bernardino Telesio on Nature and Sentience," in *Bruniana & Campanelliana* 1 (2010): 68–89.

Giglioni, Guido "Introduzione" to Telesio, *De rerum natura iuxta propria principia* [1586], xi–xxxii. Rome: Carocci, 2013.

Gilbert, Neal W. "Francesco Vimercato of Milan: A Bio-Bibliography," in *Studies in the Renaissance* 12 (1965): 188–217.

Gómez López, Susana. "Telesio y el debate sobre la naturaleza de la luz en el Renacimiento italiano," in Bondì et al., *Bernardino Telesio y la nueva imagen de la naturaleza*, 194–236.

Granada, Miguel Ángel (ed. and transl.). *Bernardino Telesio: Sobre los cometas y la Vía Láctea.* Madrid: Tecnos, 2012.

Granada, Miguel Ángel. "Introduzione" to Bernardino Telesio, *Varii de naturalibus rebus libelli*. xi–xxvi. Rome: Carocci editore, 2012.

Grendler, Marcella. "A Greek Collection in Padua: The Library of Gian Vincenzo Pinelli (1535–1601)," in *Renaissance Quarterly* 33/3 (1980): 386–416.

Gualdo, Ricardo. "Marco Fabio Calvo," in *Dizionario biografico degli italiani* 43 (1993): 723–727.

Guerrini, Luigi. *Galileo e gli aristotelici. Storia di una disputa*. Rome: Carocci editore, 2010.

Hanson, Norwood Russell. *Patterns of Discovery: An Inquiry into the Conceptual Foundations of Science*. Cambridge: Cambridge University Press, 1958.

Heilbron, John L. *Galileo*. Oxford: Oxford University Press, 2010.

Hellmann, Gustav. "Entwicklungsgeschichte des meteorologischen Lehrbuches," in *Beiträge zur Geschichte der Meteorologie*, ed. Gustav Hellmann, Vol. 2, 1–134. Berlin: Behrend, 1917.

Hellmann, Gustav. "Die Bauern-Praktik (1508). Facsimiliendruck mit einer Einleitung," in *Neudrucke von Schriften und Karten über Meteorologie und Erdmagnetismus*, no. 13. Nendeln/Liechtenstein: Kraus Reprint, 1969.

Hellmann, Gustav. "Meteorologische Beobachtungen vom XIV. bis XVII. Jahrhundert," in *Neudrucke von Schriften und Karten über Meteorologie und Erdmagnetismus*, no. 13. Nendeln/Liechtenstein: Kraus Reprint, 1969.

Hellmann, Gustav. "Wetterprognosen und Wetterberichte des XV. und XVI. Jahrhunderts. Facsimiliendruck mit einer Einleitung," in *Neudrucke von Schriften und Karten über Meteorologie und Erdmagnetismus*, no. 12, 7–26. Nendeln/Liechtenstein: Kraus Reprint, 1969.

Heninger, Simeon Kahn. *A Handbook of Renaissance Meteorology: With Particular Reference to Elizabethan and Jacobean Literature*. Durham, NC: Duke University Press, 1960.

Hirai, Hiro. "Concepts of Seeds and Nature in the Work of Marsilio Ficino," in *Marsilio Ficino: His Theology, His Philosophy, His Legacy*, ed. Michael J. B. Allen and Valery Rees, 257–284. Leiden-Boston: Brill, 2002.

Hirai, Hiro. *Le concept de semence dans les théories de la matière à la Renaissance: de Marsile Ficin à Pierre Gassendi*. Turnhout: Brepols, 2005.

Hirai, Hiro. *Medical Humanism and Natural Philosophy: Renaissance Debates on Matter, Life and the Soul*. Leiden: Brill, 2011.

Hirai, Hiro. "Il calore cosmico di Telesio fra il *De generatione animalium* di Aristotele e il *De carnibus* di Ippocrate," in *Bernardino Telesio tra filosofia naturale e scienza moderna*, ed. Giuliana Mocchi, Sandra Plastina and Emilio Sergio, 71–83. Pisa-Rome: Serra, 2012.

Ingegno, Alfonso. *Saggio sulla filosofia di Cardano*. Florence: La Nuova Italia, 1980.

Ingegno, Alfonso. "Corpo, spiritus, anima: il problema della libertà in Telesio," in *Bernardino Telesio nel 4º centenario della morte (1588)*, ed. Mario Santoro, 51–70. Naples: Istituto Nazionale di Studi sul Rinascimento Meridionale, 1989.

Iovine, Maria Fiammetta. "Henry Savile lettore di Bernardino Telesio: L'esemplare 537.C.6 del De rerum natura 1570," in *Nouvelles de la République des Lettres* 18/2 (1998): 51–84.

Jacquart, Danielle. "Médecine et astrologie à Paris dans la première moitié du XIVe siècle," in *Filosofia, scienza e astrologia nel Trecento europeo*, ed. Graziella Federici Vescovini and Francesco Barocelli, 121–134. Padua: Il Poligrafo, 1992.

Keller, Alex G. "Mathematics, Mechanics and Experimental Machines in Northern Italy in the Sixteenth Century," in *The Emergence of Science in Western Europe*, ed. Maurice P. Crosland, 15–34. London: Macmillan, 1975.

Keßler, Eckhard. "Von der Psychologie zur Methodenlehre: Die Entwicklung des methodischen Wahrheitsbegriffs in der Renaissancepsychologie," in *Zeitschrift für Philosophische Forschung* 41 (1987): 548–570.

Keßler, Eckhard. "Selbstorganisation in der Naturphilosophie der Renaissance," in *Jahrbuch für Komplexität in den Natur-, Sozial- und Geisteswissenschaften* 3 (1992): 15–29.

Keßler, Eckhard. "Metaphysics or Empirical Science? The two Faces of Aristotelian Natural Philosophy in the Sixteenth Century," in *Renaissance Readings of the Corpus Aristotelicum*, ed. Marianne Pade, 79–101. Copenhagen: Museum Tusculanum Press, 2001.

Kingsley, Peter. *Ancient Philosophy, Mystery, and Magic: Empedocles and Pythagorean* Tradition. Oxford: Clarendon, 1996.

Klier, Gerhardt. *Die drei Geister des Menschen. Zur sogenannten Spirituslehre in der frühen Neuzeit*. Stuttgart: Steiner, 2002.

Kristeller, Paul Oskar. *Eight Philosophers of the Italian Renaissance*. London: Chatto & Windus, 1965, Stanford: Stanford University Press, 1964.

Kristeller, Paul Oskar. *Iter Italicum* II. Leiden: Brill, 1967.

Kuhn, Heinrich C. *Venetischer Aristotelismus im Ende der aristotelischen Welt: Aspekte der Welt und des Denkens des Cesare Cremonini*. Frankfurt a.M.: Lang, 1996.

Kusukawa, Sachiko. *The Transformation of Natural Philosophy: The Case of Philip Melanchthon*. Cambridge: Cambridge University Press, 2011.

Leggatt, Stuart. *Aristotle: On the Heavens I–II*. Warminster: Ars & Phillips, 1995.

Leijenhorst, Cees, Christoph Lüthy and Johannes M. M. H. Thijssen (eds). *The Dynamics of Aristotelian Natural Philosophy from Antiquity to the Seventeenth Century*. Leiden: Brill, 2002.

Leijenhorst, Cees. "Bernardino Telesio. New fundamental principles of nature," in *Philosophers of the Renaissance*, ed. Paul R. Blum, 168–180. Washington D.C.: The Catholic University of America Press, 2010.

Lennox, James G. *Aristotle's Philosophy of Biology: Studies in the Origins of Life Sciences*. Cambridge: Cambridge University Press, 2001.

Lepore, Ugo. "Per la biografia di Aulo Giano Parrasio (1470–1521)," in *Biblion* 1/1 (1959): 27–44.

Lerner, Michel-Pierre. "Aristote 'oblieux de lui même' selon B. Telesio," in *Les études philosophiques* 3 (1986): 371–389.

Lerner, Michel-Pierre. "Le problème de la matière céleste après 1550: aspects de la bataille des cieux fluide," in *Revue d'histoire de science* 42/3 (1989): 255–280.

Lerner, Michel-Pierre. "La physique céleste de Telesio: problèmes d'interprétation," in *Atti del Convegno Internazionale di Studi su Bernardino Telesio*. 83–114. Cosenza: Accademia Cosentina, 1990.

Lerner, Michel-Pierre. "Le parménidisme de Telesio: origine et limites d'une hypothèse," in Galasso et al. (eds.), *Bernardino Telesio e la cultura napoletana*, 79–105.

Lettinck, Paul. *Aristotle's "Meteorology" and Its Reception in the Arab World*. Leiden: Brill, 1999.

Liebeschütz, Wolf. "The Significance of the Speech of Praetextatus," in *Pagan Monotheism in Late Antiquity*, ed. Polymnia Athanassiadi and Michael Frede, 185–205. Oxford: Clarendon, 1999.

Lohr, Charles H. *Renaissance Latin Aristotle Commentaries II: Renaissance Authors.* Florence: Olschki, 1988.

Lohr, Charles H. "Renaissance Latin Translations of the Greek Commentaries on Aristotle," in *Humanism and Early Modern Philosophy,* ed. Jill Kraye and Martin W. F. Stone, 24–40. London: Routledge, 2000.

Long, Pamela O. *Artisan/Practitioners and the Rise of the New Sciences, 1400–1600.* Corvallis, OR: Oregon State University Press, 2011.

Lotter, Johann Georg. *De vita et philosophia Bernardini Telesii commentarius ad inlustrandas historiam philosophicam universim et litterariam saeculi XVI christiani sigillatim comparatus.* Leipzig: Breitkopf, 1733.

Lupi, F. Walter. "Il sublime secondo Telesio," in *Atti del convegno internazionale di studi su Bernardino Telesio.* 47–68. Cosenza: Accademia Cosentina, 1990.

Lupi, F. Walter. *Alle origini della Accademia Telesiana.* Cosenza: Brenner, 2011.

Maccagni, Carlo and Giovanna Derenzini. "Libri Apollonii qui ... desiderantur," in *Scienza e filosofia: Saggi in onore di Ludovico Geymonat,* ed. Corrado Mangione, 668–696. Milan: Garzanti, 1985.

Maierù, Luigi. "Alcune riflessioni sul contesto in cui leggere il *De rerum natura* di Bernardino Telesio," in *Bernardino Telesio: tra filosofia naturale e scienza moderna,* ed. Giuliana Mocchi, Sandra Plastina, and Emilio Sergio, 51–64. Pisa: Serra, 2012.

Manzoni, Claudio. *Umanesimo ed eresia: Michele Serveto.* Naples: Guida, 1974.

Marchetti, Valerio. "Figure di esuli italiani del Cinquecento: Fabio Nifo," in *Critica Storica* 8/6 (1969): 691–705.

Margolin, Jean-Claude. "Bacon, lecteur critique d'Aristote et de Telesio," in *Atti del Convegno Internazionale di Studi su Bernardino Telesio.* 135–166. Cosenza: Accademia Cosentina, 1990.

Márquez Villanueva, Francisco. "Bernardino Telesio y el 'antiguo sacerdote' de *La Galatea,*" in *Cervantes en letra viva: estudios sobre la vida y la obra.* Barcelona: Reverso, 2005.

Martin, Craig. *Renaissance Meteorology: Pomponazzi to Descartes.* Baltimore: The Johns Hopkins University Press, 2011.

Martínez de Bujanda, Jesús, Claude Sutto, René Davignon, Ela Stanek, Marcella Richter (eds). *Index de Rome 1590, 1593, 1596: Avec étude des index de Parme, 1580 et Munich, 1582.* Geneva: Droz, 1994.

Martorelli Vico, Romana. *Medicina e filosofia: Per una storia dell'embriologia medievale nel XIII e XIV secolo.* Milan: Guerini, 2002.

Maylender, Michele. *Storia delle accademie d'Italia.* 5 vols. Bologna: Cappelli, 1926–1930.

Mercati, S. G. "Appunti telesiani." *Archivio storico per la Calabria e la Lucania* 7 (1937): 215–241.

Micheli, Gianni. *Le origini del concetto di macchina*. Florence: Olschki, 1995.

Minieri Riccio, Camillo. *Notizie biografiche e bibliografiche degli scrittori napoletani*. Vol. 2. Naples: Rinaldi e Sellitto, 1877.

Mondrain, Brigitte. "Éditer et traduire les médecins grecs au XVIe siècle: l'exemple de Janus Cornarius," in *Les voies de la science grecque: Études sur la transmission des textes de l'Antiquité au dix-neuvième siècle*, ed. Danielle Jacquart, 391–417. Geneva: Librairie Droz, 1997.

Moraux, Paul. "À propos du νοῦς θύραθεν chez Aristote," in *Autour d'Aristote: Recueil d'études de philosophie ancienne et médiévale offert à Monseigneur A. Mansion*, ed. Augustin Mansion, 255–295. Louvain: Publications Universitaires, 1955.

Muccillo, Maria. "La storia della filosofia presocratica nelle *Discussiones peripateticae* di Francesco Patrizi da Cherso," in *La Cultura* 13 (1975): 48–105.

Mulsow, Martin. *Frühneuzeitliche Selbsterhaltung: Telesio und die Naturphilosophie der Renaissance*. Tübingen: Niemeyer, 1998.

Mulsow, Martin. "Philosophia italica als reduzierte Prisca-Sapientia-Ideologie: Antonio Persios und Francesco Patrizis Rekonstruktionen der Elementenlehre," in *Das Ende des Hermetismus*, 253–280. Tübingen: Mohr Siebeck, 2002.

Mulsow, Martin. "Reaktionärer Hermetismus vor 1600? Zum Kontext der venezianischen Debatten über die Datierung von Hermes Trismegistos," in *Das Ende des Hermetismus*, 161–185. Tübingen: Mohr Siebeck, 2002.

Mulsow, Martin. "The *Historia* of Religions in the Seventeenth Century," in *Historia: Empiricism and Erudition in Early Modern Europe*, ed. Gianna Pomata and Nancy G. Siraisi, 181–209. Cambridge, MA: MIT Press, 2005.

Mulsow, Martin. Ein unbekanntes Gespräch Telesios: Aristoteleskritik, Sensualismus und Theorie des Lichts in der Spätrenaissance" In: *Sol et Homo. Mensch und Natur in der Renaissance. Festschrift für Eckhard Keßler*. Sabrina Ebbersmeyer, Helga Pirner-Pareschi und Thomas Ricklin (Hrsg.), 511–536. München: Wilhelm Fink, 2008.

Mulsow, Martin. *Die unanständige Gelehrtenrepublik: Wissen, Libertinage und Kommunikation*. Stuttgart: J.B. Metzler, 2010.

Mulsow, Martin and Marcelo Stamm (eds). *Konstellationsforschung*. Frankfurt a.M.: Suhrkamp, 2005.

Nardi, Bruno. *Saggi sull'aristotelismo Padovano dal secolo XIV al XVI*. Florence: Sansoni, 1958.

Nardi, Bruno. *Studi di filosofia medievale*. Rome: Edizioni di Storia e Letteratura, 1960.

Nardi, Michele Giuseppe. *Problemi d'embriologia umana antica e medioevale*. Florence: Sansoni, 1938.

Nauta, Lodi. "Anti-Essentialism and the Rhetoricization of Knowledge: Mario Nizolio's Humanist Attack on Universals," in *Renaissance Quarterly* 65/1 (2012): 31–66.

Nauta, Lodi. "De-essentializing the World: Valla, Agricola, Vives, and Nizolio on Universals and Topics," in *Essays in Renaissance Thoughts and Letters: In Honor of*

John Monfasani, ed. Alison Frazier and Patrick Nold, 196–215. Leiden-Boston: Brill, 2015.

Nuovo, Angela. "The Creation and Dispersal of the Library of Gian Vincenzo Pinelli," in *Books on the Move: Tracking Copies through Collections and the Book Trade*, ed. Robin Myers, Michael Harris and Giles Mandelbrote, 39–68. New Castle – London: Oak Knoll Press/The British Library, 2007.

Ogilvie, Brian W. *The Science of Describing: Natural History in Renaissance Europe*. Chicago: University of Chicago Press, 2006.

Omodeo, Pietro Daniel. "Giordano Bruno and Nicolaus Copernicus: The Motion of the Earth in *The Ash Wednesday Supper*," in *Nuncius: Journal of the Material and Visual History of Science* 24/1 (2009): 35–59.

Omodeo, Pietro Daniel. Review of Granada, *Bernardino Telesio: Sobre los cometas y la Vía Láctea*. *Journal for the History of Astronomy* 44/2 (2013): 216.

Omodeo, Pietro Daniel. *Copernicus in the Cultural Debates of the Renaissance: Reception, Legacy, Transformation*. Leiden: Brill, 2014.

Omodeo, Pietro Daniel. "Riflessioni sul moto terrestre nel Rinascimiento: tra filosofia natural, meccanica e cosmologia," in *Scienza e rappresentazione. Saggi in memoria di Pierre Souffrin*, ed. Pierre Caye and Pier Daniele Napolitani, 285–300. Florence: Olschki, 2016.

Ongaro, Giuseppe. "La scoperta della circolazione polmonare e la diffusione della Christianismi restitutio di Michele Serveto nel XVI secolo in Italia e nel Veneto," in *Episteme* 5 (1971): 5–44.

Panichi, Nicola. "Premesse teoriche della filosofia politica di Federico Bonaventura," in *Federico Bonaventura tra politica e scienza*, ed. Flavio Vetrano, Nicola Panichi and Viviana Bonazzoli, 7–58. Urbino: Accademia Raffaello, 2006.

Paolella, Alfonso. "Prefazione," in Giambattista Della Porta, *De aeris transmutationibus*, Alfonso Paolella, xi–xlii. Naples: Edizioni scientifiche italiane, 2000.

Paschetto, Eugenia. *Pietro d'Abano, medico e filosofo*. Florence: Vallecchi, 1984.

Pissavino, Paolo. "L'altro sole di Francesco de' Vieri," in *Atti del convegno internazionale di studi su Bernardino Telesio*. 207–220. Cosenza: Accademia Cosentina, 1990.

Plastina, Sandra. "Un moderno eretico in filosofia: Agostino Doni". *Bruniana & Campanelliana* 16/1 (2010): 149–160.

Plastina, Sandra. "Bernardino Telesio nell'Inghilterra del Seicento," in *Bernardino Telesio: Tra filosofia naturale e scienza moderna*, 133–143. Pisa-Rome: Serra, 2012.

Poppi, Antonino. *Introduzione all'aristotelismo padovano*. Padua: Antenore, 1970.

Poppi, Antonino. *La dottrina della scienza in Giacomo Zabarella*. Padua: Antenore, 1972.

Poppi, Antonino. *Cremonini e Galileo inquisiti a Padova nel 1604: Nuovi documenti d'archivio*. Padova: Antenore, 1992.

Pousseur, Jean-Marie. "Bacon, a Critic of Telesio," in *Francis Bacon's Legacy of Texts: "the art of discovery grows with discovery,"* ed. W.A. Sessions, 105–17. New York: AMS Press, 1990.

Pousseur, Jean-Marie. "La notion baconienne de principe dans le 'De principiis'," in *Nouvelles de la République des Lettres* 1 (2001): 105–120.

Preus, Anthony. "Science and Philosophy in Aristotle's *Generation of Animals*," in *Journal of the History of Biology* 3/1 (1970): 1–52.

Puliafito, Anna Laura. "Introduzione" to Bernardino Telesio, Francesco Patrizi and Antonio Persio (eds.), *Delle cose naturali libri due, volgarizzamento di Francesco Martelli; Opuscoli, volgarizzamento di Francesco Martelli, Polemiche telesiane*, xiii–xliii. Rome: Carocci, 2013.

Pupo, Spartaco. *L'anima immortale di Telesio: Per una storia delle interpretazioni*. Cosenza: Pellegrini, 1999.

Putscher, Marielene. *Pneuma, spiritus, Geist. Vorstellungen vom Lebensantrieb in ihren geschichtlichen Wandlungen*. Wiesbaden: Steiner, 1973.

Randall, John Herman. *The School of Padua and the Emergence of Modern Science*. Padua: Antenore, 1961.

Rebellato, Elisa. *La fabbrica dei divieti: Gli Indici dei libri proibiti da Clemente VIII a Benedetto XIV*. Milan: Bonnard, 2008.

Rebellato, Elisa. "Il miraggio dell'espurgazione: L'indice di Guanzelli del 1607." *Società e Storia* 31/122 (2008): 715–742.

Redondi, Pietro. *Galileo eretico*. Turin: Einaudi, 1983.

Redondi, Pietro. *Galileo Heretic*. Princeton: Princeton University Press, 1987.

Rees, Graham. "Bacon's speculative philosophy," in *The Cambridge Companion to Bacon*, ed. M. Peltonen, 121–145. Cambridge: Cambridge University Press, 1996.

Reeves, Eileen A. *Evening News: Optics, Astronomy, and Journalism in Early Modern Europe*. Philadelphia: Penn Press, 2014.

Renn, Jürgen (ed.). *Galileo in Context*. Cambridge: Cambridge University Press, 2001.

Renn, Jürgen and Peter Damerow. *The Equilibrium Controversy: Guidobaldo del Monte's Critical Notes on the Mechanics of Jordanus and Benedetti and their Historical and Conceptual Background*. Berlin: Edition Open Access, 2012.

Riondato, Ezio and Antonino Poppi. *Cesare Cremonini: Aspetti del pensiero e scritti. Atti del Convegno di studio (Padova, 26–27 febbraio 1999)*. Padua: Accademia Galileiana di Scienze, Lettere ed Arti in Padova, 2000.

Rossi, Paolo. "Il *De principiis* di Mario Nizolio," in *Archivio di filosofia* 3 (1953): 57–92.

Rossi, Paolo. *Il tempo dei maghi: Rinascimento e modernità*. Milan: Raffaello Cortina Editore, 2006.

Rossi, Paolo. *Francis Bacon: From Magic to Science*. London: Routledge & Kegan Paul, [1968] 2009.

Rotondò, Antonio. "Cultura umanistica e difficoltà di censori. Censura ecclesiastica e discussioni cinquecentesche sul platonismo," in *Le pouvoir et la plume: Incitation, contrôle et répression dans l'Italie du 16. siècle*, 15–50. Paris: Université de la Sorbonne Nouvelle, 1982.

Rotondò, Antonio. *Studi e ricerche di storia ereticale italiana del Cinquecento*. Turin: Giappichelli, 1974.

Rotondò, Antonio. "Doni, Agostino," in *Dizionario Biografico degli Italiani*, 41 (1992), *sub voce*.

Saitta, Giuseppe B. *Il pensiero italiano nell' Umanesimo e nel Rinascimento*. Vol. 3. Florence: Sansoni, 1961.

Sambursky, Shmuel. "Philoponus' Interpretation of Aristotle's Theory of Light," in *Osiris* 7 (1958): 114–126.

Santoro, Mario (ed.). *Bernardino Telesio nel 4° centenario della morte (1588)*. Naples: Istituto Nazionale di Studi sul Rinascimento Meridionale, 1989.

Savelli, Rodolfo. "Allo scrittoio del censore: Fonti a stampa per la storia dell'espurgazione dei libri di diritto in Italia tra Cinque e Seicento," in *Società e storia* 26/100–101 (2003): 293–230.

Savelli, Rodolfo. "La biblioteca disciplinata: Una 'libraria' cinque-seicentesca tra censura e dissimulazione," in *Tra diritto e storia: Studi in onore di Luigi Berlinguer promossi dalle Università di Siena e di Sassari* II, 865–944. Soveria Mannelli (Catanzaro): Rubbettino, 2008.

Savorelli, Alessandro. "Letture telesiane da Fiorentino a Gentile," in Galasso et al. (eds.), *Bernardino Telesio e la cultura napoletana*, 445–473.

Schmitt, Charles B. "Experience and Experiment: A Comparison of Zabarella's View with Galileo's in *De motu*," in *Studies in the Renaissance* 16 (1969): 80–138.

Schmitt, Charles B. "Theophrastus," in *Catalogus Translationum et Commentariorum: Mediaeval and Renaissance Latin Translations and Commentaries*, vol. II, ed. Paul Oskar Kristeller and F. Edward Cranz: 239–322. Washington D.C.: The Catholic University of America Press, 1971.

Schmitt, Charles B. *Aristotle and the Renaissance*. Cambridge, MA: Harvard University Press, 1983.

Schmitt, Charles B. "Philoponus' Commentary on Aristotle's Physics in the Sixteenth Century," in *Philoponus and the rejection of Aristotelian Science*, ed. Richard Sorabji, 210–227. London: Duckworth, 1987.

Schneider-Carius, Karl. *Wetterkunde—Wetterforschung. Geschichte ihrer Probleme und Erkenntnisse in Dokumenten aus drei Jahrhunderten*. Freiburg: K. Alber, 1955.

Schuhmann, Karl. "Hobbes and Renaissance Philosophy," in *Hobbes Oggi*, ed. Andrea Napoli. Milan: Angeli, 1990.

Schuhmann, Karl. "Telesio's Concept of Matter," in *Atti del Convegno Internazionale di Studi su Bernardino Telesio*, 115–134. Cosenza: Accademia Consentina, 1990.

Schuhmann, Karl. "Le concept de l'espace chez Telesio," in Galasso et al. (eds.), *Bernardino Telesio e la cultura napoletana*, 141–167.

Schuhmann, Karl. "Le concept de l'espace chez Telesio," in *Selected Papers on Renaissance philosophy and on Thomas Hobbes*, ed. Piet Steenbakkers and Cees Leijenhorst, 117–133. Dordrecht: Kluwer, 2004.

Scribano, Maria Emanuela. "Il problema del libero arbitrio del *De fato* di Pietro Pomponazzi," in *Annali dell'Instituto di Filosofia* 3 (1981): 23–69.

Selmi, Elisabetta. "'Formazione' e 'ricezione' del pensiero telesiano nel dialogo con i filosofi e i letterati dello studio patavino," in: *Bernardino Telesio: Tra filosofia naturale e scienza moderna*, 37–50. Pisa–Rome: Serra, 2012.

Sergio, Emilio. "Parrasio in Calabria (1511–1515) e la fondazione dell'Accademia Cosentina," in *Bollettino Filosofico*, Dipartimento di Filosofia, Università della Calabria, 23 (2007): 419–436.

Sergio, Emilio. "Parrasio, Antonio Telesio e l'Accademia Cosentina," in *Bernardino Telesio tra filosofia naturale e scienza moderna*, ed. Giuliana Mocchi, Sandra Plastina, Emilio Sergio, 15–22. Pisa–Rome: Serra, 2012.

Sergio, Emilio. "L'Accademia Cosentina e l'eredità del naturalismo telesiano," in *Storia del pensiero filosofico in Calabria da Pitagora ai giorni nostri*, ed. Mario Alcaro, 259–272. Soveria Mannelli (Catanzaro): Rubbettino, 2012.

Sezgin, Fuat. *Geschichte des arabischen Schrifttums*. Vol. 7, *Astrologie—Meteorologie und verwandtes bis ca. 430 H.* Leiden: Brill, 1979.

Shea, William R. *Galileo's Intellectual Revolution*. London: Macmillan, 1972.

Simonutti, Luisa. "Après Michel Servet: Hérésie et antitrinitarisme," in *Michel Servet (1511–1553): Hérésie et pluralisme du XVIe au XXIe siècle*, ed. Valentine Zuber, 185–212. Paris: Champion, 2007.

Siraisi, Nancy G. *The Clock and the Mirror: Girolamo Cardano and Renaissance Medicine*. Princeton: Princeton Unviersity Press, 1997.

Smith, A. Mark. *From Sight to Light: The Passage from Ancient to Modern Optics*. Chicago: University of Chicago Press, 2015.

Soleri, Giacomo. *Telesio*. Brescia: La Scuola, 1944.

Solmsen, Friedrich. "The Vital Heat, the Inborn Pneuma, and the Aether," in *Journal of Hellenic Studies* 77/1 (1957): 119–123.

Sorabji, Richard. "Neoplatonists and Christians: Place and Bodies in the Same Place," in *Matter, space and motion: theories in antiquity and their sequel*, 106–122. Ithaca, NY: Cornell University Press, 1988.

Spoerri, Walter. "L'anthropogonie du *Peri sarkon* (et Diodore, i 7, 3 s.)," in *Formes de pensée dans la collection Hippocratique*, ed. François Lasserre and Philippe Mudry, 57–70. Geneva: Droz, 1983.

Spruit, Leen. "Elementi aristotelici e polemica anti-peripatetica nella dottrina dell'anima divina di Telesio," in *Verifiche: Rivista di scienze umane* 21/3 (1992): 351–370.

Spruit, Leen. "Telesio's Psychology and the Northumberland Circle," occasional paper in *The Durham Thomas Harriot Seminar Papers* 25 (1998).

Stella, Aldo. *Dall'anabattismo alsocinianesimo nel Cinquecento veneto*. Padua: Liviana, 1967.

Suitner, Riccarda. "Radical Reformation and Medicine in the Late Renaissance: The Case of the University of Padua," in *Nuncius: Journal of the Material and Visual History of Science* 31/1 (2016): 11–31.

Suter, Rufus. "The Scientific Work of Alessandro Piccolomini," in *Isis* 60/2 (1969): 210–222.

Taub, Liba. *Ancient Meteorology*. London: Routledge, 2003.

Thomas, Craig. "Experience of the New World and Aristotelian Revisions of the Earth's Climates during the Renaissance," in *History of Meteorology* 3 (2006): 1–16.

Trabucco, Oreste. *"L'opere stupende dell'arti più ingegnose": La recezione degli Pneumatiká di Erone Alessandrino nella cultura italiana del Cinquecento*. Florence: Olschki, 2010.

Trabucco, Oreste. "Nel cantiere della *Magia*," in *La "mirabile" natura. Magia e scienza in Giovan Battista Della Porta (1615–2015)*, ed. Marco Santoro, 219–232. Pisa: Serra, 2016.

Tristano, Caterina. *La biblioteca di un umanista calabrese: Aulo Giano Parrasio*. Manziana: Vecchiarelli, 1995.

Troilo, Erminio. *Bernardino Telesio*. Modena: Formiggini, 1910.

Valente, Michaela. "Libertas philosophandi: Agostino Doni da Cosenza a Cracovia," in *Archivio Storico per la Calabria e la Lucania* 69 (2002): 117–131.

Valleriani, Matteo. "From *Condensation* to *Compression*: How Renaissance Italian Engineers Approached Hero's *Pneumatics*," in *Übersetzung und Transformation*, ed. Hartmut Böhme, Christoph Rapp and Wolfgang Rösler, 333–353. Berlin – New York: De Gruyter, 2011.

Van Deusen, Neil Cleveland. "Telesio: The First of the Moderns." PhD diss., Columbia University, 1932.

Van den Broecke, Steven. *The Limits of Influence. Pico, Louvain, and the Crisis of Renaissance Astrology*. Leiden: Brill, 2004.

Vasoli, Cesare. "Un episodio della disputa cinquecentesca su Cicerone e il ciceronianesimo: Mario Nizolio e Marcantonio Maioragio," in *Civitas mundi: Studi sulla cultura del Cinquecento*, 235–260. Rome: Edizioni di Storia e Letteratura, 1996.

Vasoli, Cesare. "Tra Aristotele, Alessandro di Afrodisia e Juan de Valdés: Note su Simone Porzio," in *Revista di Storia della Filosofia* 56/4 (2001): 561–607.

Ventrice, Pasquale. *La discussione sulle maree tra astronomia, meccanica e filosofia nella cultura veneto-padovana del Cinquecento.* Venice: Istituto Veneto di Scienze, Lettere ed Arti, 1989.

Verbeke, Gérard. *L'évolution de la doctrine du pneuma du stoicisme á S. Augustin.* Paris: Desclée de Brouwer, 1945.

Verdigi, Mariano. *Simone Simoni: Filosofo e medico nel '500.* Lucca: Maria Pacini Fazzi, 1997.

Vermij, Rienk. "A Science of Signs: Aristotelian Meteorology in Reformation Germany," in *Early Science and Medicine* 15 (2010): 648–674.

Verrycken, Koenraad. "The Development of Philoponus' Thought and Its Chronology," in *Aristotle Transformed: The ancient commentators and their influence*, ed. Richard Sorabji. Ithaca, NY: Cornell University Press, 1990.

Walker, Daniel P. *Spiritual and Demonic Magic from Ficino to Campanella.* London: Warburg Institute, 1958.

Walker, Daniel P. "The Astral Body in Renaissance Medicine," in *Journal of the Warburg and Courtauld Institutes* 21 (1958): 119–133.

Whitaker, Virgil K. "Francesco Patrizi and Francis Bacon," in *Studies in the Literary Imagination* 4/1 (1971): 107–120.

Wilbur, Earl Morse. *A History of Unitarianism: Socinianism and its Antecedents.* Cambridge, MA: Harvard University Press, 1946.

Wildberg, Christian. *John Philoponus' Criticism of Aristotle's Theory of Aether.* Berlin: De Gruyter, 1988.

Williams, Arnold. *The Common Expositor: An Account of the Commentaries on Genesis 1527–1633.* Chapel Hill: University of North Carolina Press, 1948.

Williams, George Huntston. *The Radical Reformation.* Philadelphia: Westminster Press, 1962.

Wolfson, Harry Austryn. *The Philosophy of Spinoza.* Cambridge, MA: Harvard University Press, 1934.

Zambelli, Paola. "Fine del mondo o inizio della propaganda? Astrologia, filosofia della storia e propaganda politico-religiosa nel dibattito sulla congiunzione del 1524," in *Scienze, credenze occulte, livelli di cultura. Convegno internazionale di studi*, ed. Giancarlo Garfagnigni, 291–368. Florence: Olschki, 1982.

Zik, Yaakov and Giora Hon. "Giambattista Della Porta: a Magician or an Optician?," in *The Optics of Giambattista Della Porta: A Reassessment*, ed. Arianna Borrelli, Giora Hon and Yaakov Zik, 39–55. Berlin: Springer.

Zilsel, Edgar. "The Origins of William Gilbert's Scientific Method," in *Journal for the History of Ideas* 2/1 (1941): 1–32.

Zilsel, Edgar. "The Sociological Roots of Science [1942]," in *Social Studies of Science* 30/6 (2000): 935–939.

Zuber, Valentine, ed. *Michel Servet (1511–1553): Hérésie et pluralisme du XVIe au XXIe siècle.* Paris: Honoré Champion, 2007.

Index

www.ingramcontent.com/pod-product-compliance
Lightning Source LLC
LaVergne TN
LVHW010641300926
849161LV00005B/1007